Representative Procedures in Quantitative Chemical Analysis

Representative Procedures in Quantitative Chemical Analysis

Contributors

Xiao-Ting Liu, Xu-Guang Wang et al.

AURIS
Reference

www.aurisreference.com

Representative Procedures in Quantitative Chemical Analysis

Contributors: Xiao-Ting Liu, Xu-Guang Wang et al.

Published by Auris Reference Limited
www.aurisreference.com

United Kingdom

Representative Procedures in Quantitative Chemical Analysis

ISBN: 978-1-78154-897-4

British Library Cataloguing in Publication Data
A CIP record for this book is available from the British Library

Printed in the United Kingdom

Exclusively distributed by CBS Publishers & Distributors Pvt. Ltd.

Sales & Distribution Rights only for India, Pakistan, Bangladesh, Sri Lanka, Nepal and Bhutan.This book is not to be sold outside these territories.

Contents

List of Abbreviations

AD	Applicability Domain
ADME	Absorption, Distribution, Metabolism, Excretion
AFM	Atomic Force Microscopy
BBB	Blood-Brain Barrier
CDR	Circular Dorsal Ruffles
CE	Collision Energy
CML	Chronic Myelogenous Leukemia
CMM	Chinese Medicinal Materials
CNKI	Chinese National Knowledge Infrastructure
CNS	Central Nervous System
EGF	Epidermal Growth Factor
EGFR	Epidermal Growth Factor Receptor
ESI	Electrospray Ionization
FACS	Fluorescence-Activated Cell Sorting
GS	Glutamine Synthetase
LMD	Laser Microdissection
LOD	Limit of Detection
LOQ	Limit of Quantification
LSC	Leukemia Stem Cells
MDA	Monoester Diterpene Alkaloids
MRM	Multiple Reaction Monitoring
NEA	Nonester Alkaloids
ORAC	Oxygen Radical Absorbance Capacity
PDGF	Platelet-Derived Growth Factor
PHC	Petroleum Hydrocarbons
PID	Photo Ionization Detector
QSPR	Quantitative Structure-Property Relationships
RSD	Relative Standard Deviation
SAR	Structure-Activity Relationships
SFDA	State Food and Drug Administration
SPR	Surface Plasmon Resonance
SQC	SuoQuan Capsules
SQP	SuoQuan Pills
TCM	traditional Chinese Medicines
TIC	Total Ion Chromatogram
TPH	Total Petroleum Hydrocarbons

List of Contributors

Xiao-Ting Liu
Beijing Institute of Pharmacology and Toxicology, Beijing 100850, China
School of Traditional Chinese Materia Medica, Shenyang Pharmaceutical University, Shenyang 110016, China

Xu-Guang Wang
Beijing Institute of Pharmacology and Toxicology, Beijing 100850, China

Yu Yang
Beijing Institute of Pharmacology and Toxicology, Beijing 100850, China

Rui Xu
Beijing Institute of Pharmacology and Toxicology, Beijing 100850, China

Fan-Hua Meng
Beijing Institute of Pharmacology and Toxicology, Beijing 100850, China

Neng-Jiang Yu
Beijing Institute of Pharmacology and Toxicology, Beijing 100850, China

Yi-Min Zhao
Beijing Institute of Pharmacology and Toxicology, Beijing 100850, China

Patricio Sobrero
Laboratorio de Bioquı´mica, Microbiologı´a e Interacciones Biolo´gicas en el Suelo, Departamento de Ciencia y Tecnologı´a, Universidad Nacional de Quilmes, Buenos Aires, Argentina

Jan-Philip Schlu"ter
Institute of Biology III, Faculty of Biology, University of Freiburg, Freiburg, Germany
LOEWE Center for Synthetic Microbiology, Marburg, Germany

Ulrike Lanner3,
3 Core Facility Proteomics, Center for Biological Systems Analysis (ZBSA), Freiburg, Germany

Andreas Schlosser
Core Facility Proteomics, Center for Biological Systems Analysis (ZBSA), Freiburg, Germany
Rudolf Virchow Center, University of Wuerzburg, Wuerzburg, Germany

Anke Becker
Institute of Biology III, Faculty of Biology, University of Freiburg, Freiburg, Germany
LOEWE Center for Synthetic Microbiology, Marburg, Germany

Claudio Valverde
Laboratorio de Bioquı́mica, Microbiologı́a e Interacciones Biolo´gicas en el Suelo, Departamento de Ciencia y Tecnologı́a, Universidad Nacional de Quilmes, Buenos Aires, Argentina

Erik Bernitt
Institut für Biophysik, Universität Bremen, Bremen, Germany

Cheng Gee Koh
School of Biological Sciences, Nanyang Technological University, Singapore, Singapore

Nir Gov
Department of Chemical Physics, Weizmann Institute of Science, Rehovot, Israel

Hans-Günther Döbereiner
Institut für Biophysik, Universität Bremen, Bremen, Germany

Wei-Ting Kuo
Institute of Biomedical Engineering, National Taiwan University, Taipei, Taiwan

Wen-Chun Lin
Institute of Biomedical Engineering, National Taiwan University, Taipei, Taiwan

Kai-Chun Chang
Graduate Institute of Clinical Dentistry, National Taiwan University, Taipei, Taiwan

Jian-Yuan Huang
Institute of Biomedical Engineering, National Taiwan University, Taipei, Taiwan

Ko-Chung Yen
Institute of Biomedical Engineering, National Taiwan University, Taipei, Taiwan

InChi Young
Institute of Biomedical Engineering, National Taiwan University, Taipei, Taiwan

Yu-Jun Sun
Institute of Biomedical Engineering, National Taiwan University, Taipei, Taiwan

Feng-Huei Lin
Institute of Biomedical Engineering, National Taiwan University, Taipei, Taiwan
Institute of Biomedical Engineering and Nanomedicine, National Health Research Institutes, Miaoli, Taiwan

Beibei Zhang
Department of Molecular and Cellular Pharmacology, Pharmacogenomics and Pharmacoinformatics, Mie University Graduate School of Medicine, Edobashi, Tsu, Mie, Japan

Yasuhito Shimada
Department of Molecular and Cellular Pharmacology, Pharmacogenomics and Pharmacoinformatics, Mie University Graduate School of Medicine, Edobashi, Tsu, Mie, Japan
Mie University Medical Zebrafish Research Center, Edobashi, Tsu, Mie, Japan
Department of Bioinformatics, Mie University Life Science Research Center, Edobashi, Tsu, Mie, Japan
Department of Omics Medicine, Mie University Industrial Technology Innovation, Edobashi, Tsu, Mie, Japan
Department of Systems Pharmacology, Mie University Graduate School of Medicine, Edobashi, Tsu, Mie, Japan

Junya Kuroyanagi
Department of Molecular and Cellular Pharmacology, Pharmacogenomics and Pharmacoinformatics, Mie University Graduate School of Medicine, Edobashi, Tsu, Mie, Japan

Noriko Umemoto
Department of Molecular and Cellular Pharmacology, Pharmacogenomics and Pharmacoinformatics, Mie University Graduate School of Medicine, Edobashi, Tsu, Mie, Japan
Department of Systems Pharmacology, Mie University Graduate School of Medicine, Edobashi, Tsu, Mie, Japan

Yuhei Nishimura
Department of Molecular and Cellular Pharmacology, Pharmacogenomics and Pharmacoinformatics, Mie University Graduate School of Medicine, Edobashi, Tsu, Mie, Japan
Mie University Medical Zebrafish Research Center, Edobashi, Tsu, Mie, Japan
Department of Bioinformatics, Mie University Life Science Research Center, Edobashi, Tsu, Mie, Japan
Department of Omics Medicine, Mie University Industrial Technology Innovation, Edobashi, Tsu, Mie, Japan
Department of Systems Pharmacology, Mie University Graduate School of Medicine, Edobashi, Tsu, Mie, Japan

Toshio Tanaka
Department of Molecular and Cellular Pharmacology, Pharmacogenomics and Pharmacoinformatics, Mie University Graduate School of Medicine, Edobashi, Tsu, Mie, Japan
Mie University Medical Zebrafish Research Center, Edobashi, Tsu, Mie, Japan3Department of Bioinformatics, Mie University Life Science Research Center, Edobashi, Tsu, Mie, Japan
Department of Omics Medicine, Mie University Industrial Technology Innovation, Edobashi, Tsu, Mie, Japan
Department of Systems Pharmacology, Mie University Graduate School of Medicine, Edobashi, Tsu, Mie, Japan

Cho X. J. Chan
Biology Department, Brooklyn College City University of New York, New York, New York, United States of America
The Graduate Center, City University of New York, New York, New York, United States of America
Haskins Laboratories and the Department of Chemistry and Physical Sciences, Pace University, New York, New York, United States of America

Ivor G. Joseph
Biology Department, Brooklyn College City University of New York, New York, New York, United States of America

Andy Huang
Biology Department, Brooklyn College City University of New York, New York, New York, United States of America

Desmond N. Jackson
Biology Department, Brooklyn College City University of New York, New York, New York, United States of America

Peter N. Lipke
Biology Department, Brooklyn College City University of New York, New York, New York, United States of America
The Graduate Center, City University of New York, New York, New York, United States of America

Yujie Chen
School of Chinese Medicine, Hong Kong Baptist University, Kowloon, Hong Kong Special Administrative Region, People's Republic of China
Department of Resources Science of Traditional Chinese Medicines, State Key Laboratory of Modern Chinese Medicines, College of Traditional Chinese Medicines, China Pharmaceutical University, Tongjiaxiang-24, Gulou District, Nanjing 210009, People's Republic of China

Liang Xu
School of Pharmacy, Liaoning University of Traditional Chinese Medicine, Dalian, China.

Yuancen Zhao
School of Chinese Medicine, Hong Kong Baptist University, Kowloon, Hong Kong Special Administrative Region, People's Republic of China

Zhongzhen Zhao
School of Chinese Medicine, Hong Kong Baptist University, Kowloon, Hong Kong Special Administrative Region, People's Republic of China

Hubiao Chen
School of Chinese Medicine, Hong Kong Baptist University, Kowloon, Hong Kong Special Administrative Region, People's Republic of China

Tao Yi
School of Chinese Medicine, Hong Kong Baptist University, Kowloon, Hong Kong Special Administrative Region, People's Republic of China

Minjian Qin
Department of Resources Science of Traditional Chinese Medicines, State Key Laboratory of Modern Chinese Medicines, College of Traditional Chinese Medicines, China Pharmaceutical University, Tongjiaxiang-24, Gulou District, Nanjing 210009, People's Republic of China

Zhitao Liang
School of Chinese Medicine, Hong Kong Baptist University, Kowloon, Hong Kong Special Administrative Region, People's Republic of China

Feng Chen
School of Pharmacy, Hainan Medical University, Hainan Provincial Key Laboratory of R&D of Tropical Herbs, Haikou 571101, China

Hai-long Li
School of Pharmacy, Hainan Medical University, Hainan Provincial Key Laboratory of R&D of Tropical Herbs, Haikou 571101, China

Yong-Hui Li
School of Pharmacy, Hainan Medical University, Hainan Provincial Key Laboratory of R&D of Tropical Herbs, Haikou 571101, China

Yin-Feng Tan
School of Pharmacy, Hainan Medical University, Hainan Provincial Key Laboratory of R&D of Tropical Herbs, Haikou 571101, China

Jun-Qing Zhang
School of Pharmacy, Hainan Medical University, Hainan Provincial Key Laboratory of R&D of Tropical Herbs, Haikou 571101, China

Guy Schwartz
Porter School of Environmental Studies, Tel-Aviv University, Tel-Aviv 69978, Israel
Remote Sensing Laboratory, Tel-Aviv University, Tel-Aviv 69978, Israel
Geography and Human Environment Department, Tel-Aviv University, P.O. Box 39040, Tel-Aviv 69978, Israel

Eyal Ben-Dor
Remote Sensing Laboratory, Tel-Aviv University, Tel-Aviv 69978, Israel

Gil Eshel
The Soil Erosion Research Station, Ruppin Institute, Emeck Hefer 40250, Israel

Na Guo
Experimental Research Center, China Academy of Chinese Medical Sciences, Beijing 100700, China

Mingtao Liu
SRI International, Menlo Park, CA 94025, USA

Dawei Yang
Key Laboratory of Biofuels, Qingdao Institute of Bioenergy and Bioprocess Technology, Chinese Academy of Sciences, Songling road 189, Qingdao 266101, China

Ying Huang
Experimental Research Center, China Academy of Chinese Medical Sciences, Beijing 100700, China

Xiaohong Niu
Experimental Research Center, China Academy of Chinese Medical Sciences, Beijing 100700, China

Ruifan Wu
College of Pharmacy, Xinjiang Medical University, Urumqi 830011, China

Ying Liu
Key Laboratory of Bioactive Substances and Resource Utilization of Chinese Herbal Medicine, Ministry of Education, Institute of Materia Medica, Chinese Academy of Medical Sciences and Peking Union Medical College, Beijing 100050, China

Guizhi Ma
College of Pharmacy, Xinjiang Medical University, Urumqi 830011, China

Deqiang Dou
Department of Chinese Medicine Chemistry, Liaoning University of Traditional Chinese Medicine, Dalian 116600, China.

Gloria Castellano
Departamento de Ciencias Experimentales y Matemáticas, Facultad de Veterinaria y Ciencias Experimentales, Universidad Católica de Valencia San Vicente Mártir, Guillem de Castro-94, E-46001 València, Spain

Francisco Torrens
Institut Universitari de Ciència Molecular, Universitat de València, Edifici d'Instituts de Paterna, E-46071 València, Spain

Preface

Representative Procedures in Quantitative Chemical Analysis discusses procedures in relation to their essential features, underlying principles, and varied applications. In first chapter, a combinative method using HPLC-QTOF-MS and HPLC-UV was first established for the identification and simultaneous quantification of the major constituents whose chemical structures are presented. Second chapter focuses on quantitative proteomic analysis of the HFQ-regulon in sinorhizobium meliloti. In third chapter, we present a detailed study on CDR phenomenology and quantitative analysis of wavefront dynamics that allows us to characterize several properties of the mechanism underlying CDRs. The main aim of fourth chapter is to estimate the binding affinity and adhesion force of two targeting molecules, anti-EGFR monoclonal antibody (mAb LA1) and the peptide GE11 (YHWYGYTPQNVI), with respect to EGFR and to compare these values with those obtained for the ligand, EGF. Fifth chapter aims to quantitative phenotyping-based in vivo chemical screening in a zebrafish model of leukemia stem cell xenotransplantation. Sixth chapter explores on quantitative analyses of force-induced amyloid formation in candida albicans ALS5P. In seventh chapter, tissue-specific chemicals of P. quinquefolium were analyzed by laser microdissection and ultra-high performance liquid chromatography- quadrupole/time-of-flight-mass spectrometry (UHPLC-Q/TOF–MS) to elucidate the distribution pattern of ginsenosides in tissues. Eighth chapter aims to analyze the main constituents using ultra-fast performance liquid chromatography coupled to tandem mass spectrometry (UFLC-MS/MS). Ninth chapter main goal is to evaluate reflectance spectroscopy as a tool for TPH assessment, as compared with three commercial certified laboratories using traditional methods. The primary aim of tenth chapter is to develop a direct and rapid RRLC-MS/MS method for simultaneously quantifying the ten constituents in Shen-Fu decoction, namely, ginsenosides-Rb1, Rb2, Rc, Rd, Rg1, Re and Rf and Aconitum alkaloids including AC, MA and HA. Eleventh chapter presents a theoretical study on quantitative structure-antioxidant activity models of isoflavonoids.

Chapter 1

QUALITATIVE AND QUANTITATIVE ANALYSIS OF LIGNAN CONSTITUENTS IN CAULIS TRACHELOSPERMI BY HPLC-QTOF-MS AND HPLC-UV

Xiao-Ting Liu [1,2], Xu-Guang Wang [1], Yu Yang [1], Rui Xu [1], Fan-Hua Meng [1], Neng-Jiang Yu [1] and Yi-Min Zhao [1]

[1]Beijing Institute of Pharmacology and Toxicology, Beijing 100850, China

[2]School of Traditional Chinese Materia Medica, Shenyang Pharmaceutical University, Shenyang 110016, China

ABSTRACT

A high-performance liquid chromatography coupled with quadrupole tandem time-of-flight mass (HPLC-QTOF-MS) and ultraviolet spectrometry (HPLC-UV) was established for simultaneous qualitative and quantitative analysis of the major chemical constituents in Caulis Trachelospermi, respectively. The analysis was performed on an Agilent Zorbax Eclipse Plus C18 column (4.6 mm × 150 mm, 5 µm) using a binary gradient system of water and methanol, with ultraviolet absorption at 230 nm. Based on high-resolution ESI-MS/MS fragmentation behaviors of the reference standards, the characteristic cleavage patterns of lignano-9, 9›-lactones and lignano-8›-hydroxy-9, 9›-lactones were obtained. The results demonstrated that the characteristic fragmentation patterns are valuable for identifying and differentiating lignano-9,9›-lactones and lignano-8›-hydroxy-9,9›-lactones. As such, a total of 25 compounds in Caulis Trachelospermi were unambiguously or tentatively identified via comparisons with reference standards or literature. In addition, 14 dibenzylbutyrolatone lignans were simultaneously quantified in Caulis Trachelospermi by HPLC-UV method. The method is suitable for the qualitative and quantitative analyses of dibenzylbutyrolatone lignans in Caulis Trachelospermi.

INTRODUCTION

Caulis Trachelospermi, the stems and leaves of *Trachelospermum jasminoides* (Lindl.) Lem, is mainly distributed in Henan, Anhui, Hubei, Shandong and Guangxi provinces in China. It has been used in traditional Chinese medicine for the treatment of rheumatic arthralgia, aching of the loins and knees, traumatic injuries [1], and its medicinal properties such as anticancer and anti-inflammation have been reported [2,3]. Chemical investigations indicated that it mainly contains lignans, flavonoids and triterpenoids [4,5,6]. In our previous study, the extract of Caulis Trachelospermi and its main dibenzylbutyrolactone lignan constituents exhibited marked anti-inflammatory activity in animal model [7], moderate inhibiting activity on NF-κB signaling pathway induced by TNF-α [8] as well as strong inhibiting activity on JAK/STAT pathway [9]. As mentioned above, the major bioactive constituents of Caulis Trachelospermi are disclosed to be dibenzylbutyrolactone lignans.

Up to now, high-performance liquid chromatography (HPLC) [10,11,12,13,14,15,16] and ultraviolet spectrophotometry (UV) [17,18] have been developed and focused on the quantitative analysis of total flavonoid, total lignans and a few active compounds such as trachelogenin and tracheloside in Caulis Trachelospermi. However, the fingerprint analysis in our previous research [19] has led to the discovery of more than 15 characteristic peaks, the content of which is still unequivocal. To the best of our knowledge, there have been no reports for the simultaneous determination of 14 main dibenzylbutyrolactone lignans by HPLC so far. Therefore, it is necessary to develop a sensitive and selective method to quantify the dibenzylbutyrolactone lignan constituents in Caulis Trachelospermi.

Currently, HPLC coupled with quadrupole time-of-flight mass spectrometry (QTOF-MS) is used in composition analysis and quantification of a wide variety of natural product compounds [20,21,22]. On the other hand, QTOF allows the generation of mass information with greater accuracy and precision, and it provides both elemental compositions and fragmentation patterns in a highly sensitive and convenient way [23]. However, no attempts have been made to identify the constituents in Caulis Trachelospermi based on accurate mass measurements using HPLC-QTOF-MS.

In the present study, a combinative method using HPLC-QTOF-MS and HPLC-UV was first established for the identification and simultaneous quantification of the major constituents whose chemical structures are presented in Figure 1in 14 batch samples of Caulis Trachelospermi, which could provide an alternative, feasible approach for the quality assessment of Caulis Trachelospermi.

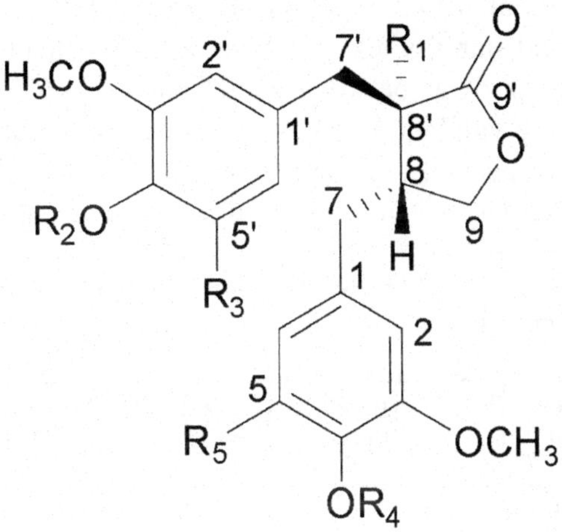

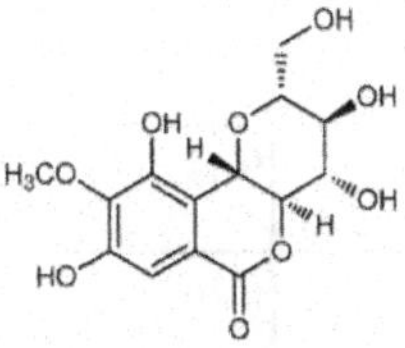

bergenin (15)

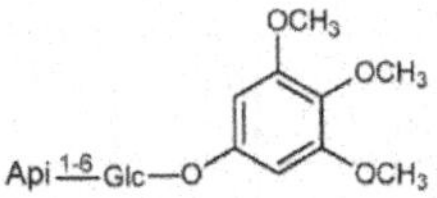

kelampayoside A (16)

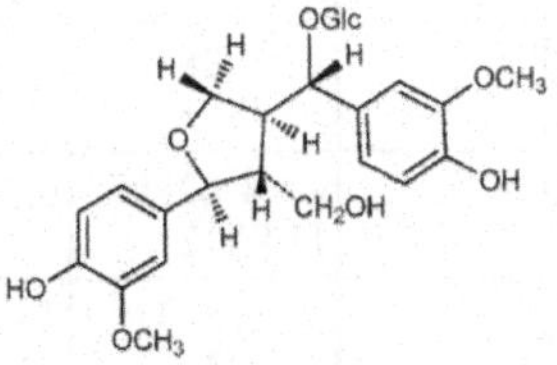

tanegoside A (17)

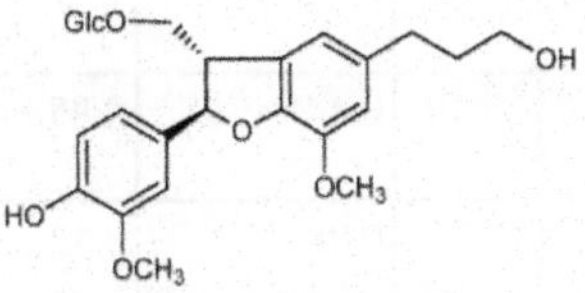

(7R, 8S)-dihydrodehydrodiconiferyl alcohol
9-O-β-D-glucopyranoside (23)

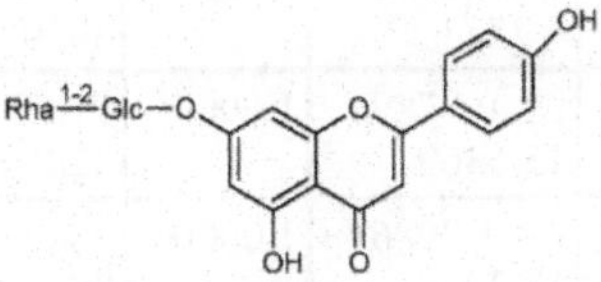

apigenin 7-O-β-neospheroside (25)

Figure 1. Chemical structures of compounds identified in Caulis Trachelospermi. Glc: glucose, Api: apiose, Rha: rhamnose.

RESULTS AND DISCUSSION

Quantitative Analysis of Dibenzylbutyrolatone Lignans

Validation of the Developed Method

The quantification method was validated in terms of linearity, LODs and LOQs, precision, repeatability, stability, and accuracy. The results are listed in Table 1 and Table 2. All calibration curves showed good linearity ($r^2 > 0.9997$) within the test ranges, and the LODs and LOQs were 1.24–9.00 and 3.71–31.71 ng, respectively. The intra and inter-day precision of the standard solutions were found in the range of 0.17%–0.75% and 0.15%–2.87%, respectively. Both for repeatability and stability test, the RSD were less than 2.94% and 2.78%, respectively. Recovery was between 96.68% and 103.63% with RSD values below 1.87%. These validation results indicated that the present method was sensitive, precise, repeatable, stable and accurate for the quantitative analysis of 14 dibenzylbutyrolatone lignans in Caulis Trachelospermi.

Table 1. Linear regression, LOD and LOQ, intra-day and inter-day precisions of the 14 dibenzylbutyrolatone lignans

Com-pound	Regression Equation [b]	R^2	Range (ng)	LOD (ng)	LOQ (ng)	Intra-day (RSD%, $n = 6$)	Inter-day (RSD%, $n = 6$)
1 [a]	$y = 0.9845x - 3.0467$	1.0000	36.18–1266.30	4.27	8.54	0.33	1.70
2	$y = 1.2843x - 12.4895$	1.0000	91.80–3213.00	9.00	18.01	0.44	0.15
3	$y = 1.0255x - 11.8232$	0.9999	39.03–1366.05	8.37	18.30	0.42	0.41
4	$y = 0.8371x - 4.3034$	1.0000	46.29–1620.15	4.99	14.97	0.33	0.26
5	$y = 1.1938x - 4.7773$	1.0000	86.31–3020.85	1.39	6.95	0.17	0.47
6	$y = 0.9789x - 15.6369$	1.0000	91.60–3206.00	6.25	27.49	0.44	0.85
7	$y = 1.8986x + 12.2728$	1.0000	280.56–9819.60	2.94	12.73	0.28	0.42
8	$y = 1.0888x - 3.4518$	1.0000	107.38–3758.30	7.93	31.74	0.24	0.46
9	$y = 1.7141x - 29.0498$	1.0000	61.05–2136.75	8.06	24.93	0.22	0.41
10	$y = 1.2220x - 1.4093$	1.0000	86.40–3024.00	7.72	24.95	0.31	0.85
11	$y = 1.0257x - 24.0442$	0.9997	49.50–1732.50	1.24	3.71	0.19	0.48
12	$y = 2.1915x - 41.1740$	1.0000	125.20–4382.00	3.13	10.42	0.22	0.43
13	$y = 1.5909x - 0.0134$	1.0000	12.35–432.25	4.06	13.95	0.17	2.87
14	$y = 1.7761x - 4.4637$	0.9999	22.76–796.43	3.35	11.18	0.75	2.22

[a] The compounds are the same as in Figure 1; [b] y is the peak area, x is the concentration (ng) of compound.

Table 2. Repeatability, stability and recovery of 14 dibenzylbutyrolatone lignans in Caulis Trachelospermi

Com-pound	RSD (%, $n = 6$)		Recovery (%, $n = 6$)				
	Repeatability	Stability	Original (µg)	Spiked (µg)	Observed (µg)	Mean	RSD (%)
1	0.71	1.51	163.01	164.02	329.03	101.22	1.02
2	0.46	0.31	343.80	362.30	706.23	100.04	0.24
3	2.94	0.85	127.43	143.63	273.94	102.01	1.87
4	0.89	2.08	170.29	175.29	344.42	99.33	0.45
5	0.41	1.23	375.40	335.38	718.15	102.20	0.75
6	2.07	2.78	483.13	474.85	946.76	97.64	1.09
7	0.52	0.28	1325.26	1242.48	2542.17	97.94	0.34
8	0.70	1.19	224.28	257.71	478.29	98.56	0.38
9	0.90	1.03	198.16	234.43	429.08	98.51	0.17
10	0.76	2.21	421.27	421.63	858.20	103.63	0.39
11	1.47	0.80	292.02	277.20	575.80	102.37	0.30
12	1.12	0.26	439.82	420.67	859.56	99.78	0.27
13	2.70	1.14	15.88	20.75	36.93	101.43	0.71
14	1.98	1.00	89.00	87.38	174.48	96.68	0.48

Sample Analysis

The developed method was successfully applied to the simultaneous determination of 14 dibenzylbutyrolatone lignans in 14 batches of Caulis Trachelospermi. Representative chromatograms are shown in Figure 2C. Quantification of each compound in the samples was calculated with the external standard using the calibration curves. Information regarding the content is summarized in Table 3. According to Table 3, all the 14 compounds were detected from the 14 batches of Caulis Trachelospermi samples. Their contents varied dramatically with RSD (%) ranging from 30.54%–61.54%, but the variation of total content of all compounds was not that large. The average total content of these 14 compounds was 28.531 mg/g. The results also showed that in all samples, tracheloside was the maximal constituent, with a mean content of 7.237 mg/g. Besides, as shown in Table 3, most compounds in Y2-4 revealed relatively lower contents than others, and the contents of Y2-12 differed greatly from other batches. This was probably caused by the poor native quality of the analyzed samples. In general, the full scale multiple compounds quantification method developed in this paper has not been provided so far which will shed some new light on the quality control of Caulis Trachelospermi.

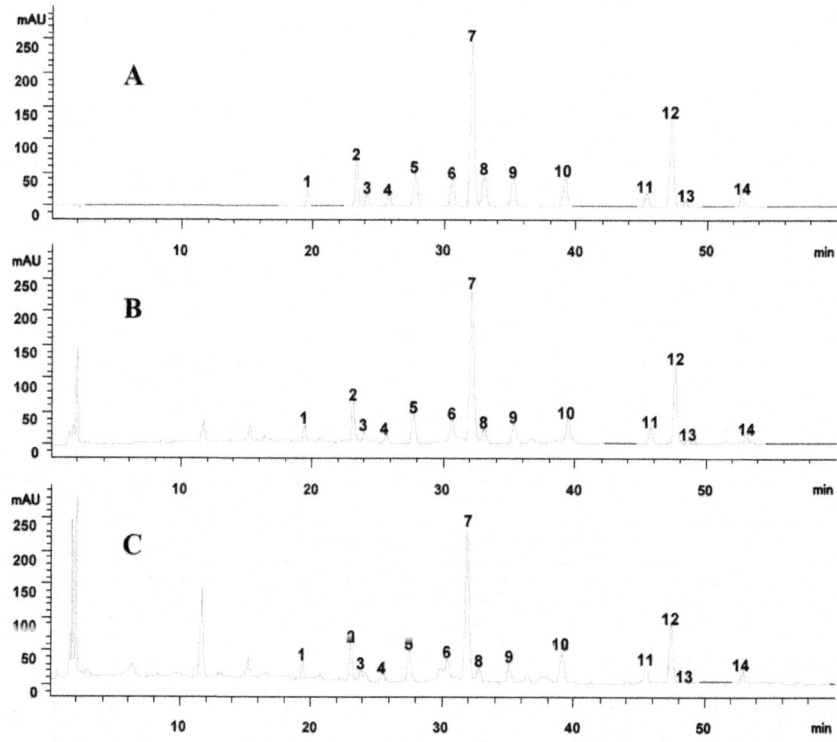

Figure 2. (**A**) HPLC-UV chromatogram of mixed standards. (**B**) HPLC-UV chromatogram of the sample extracted from Caulis Trachelospermi. (**C**) HPLC-UV chromatogram of Caulis Trachelospermi. The number of peaks marked in Figure 2 is corresponding to Figure 1.

Table 3. Content ($n = 3$, mg/g) of 14 dibenzylbutyrolatone lignans in the tested samples

Sample	1	2	3	4	5	6	7	8	9	10	11	12	13	14	Sum
Y2-1	0.785	1.706	0.667	0.758	1.949	2.121	6.602	1.130	0.932	2.184	1.388	2.041	0.092	0.405	22.761
Y2–2	1.355	4.712	1.397	1.639	2.643	5.294	11.301	2.136	0.966	2.678	1.440	1.833	0.116	0.423	37.934
Y2–3	1.068	3.394	2.013	1.757	1.533	5.627	7.218	1.808	0.313	3.166	0.584	0.917	0.048	0.348	29.795
Y2–4	0.706	1.032	0.735	0.306	0.333	1.172	2.777	0.314	1.043	1.155	1.171	2.222	0.147	0.303	13.415
Y2–5	0.989	2.491	1.113	0.540	1.327	2.408	8.054	1.040	1.517	2.579	1.498	3.505	0.159	0.484	27.703
Y2–6	1.000	3.984	0.622	0.563	1.378	2.113	8.742	0.814	1.932	0.570	1.409	3.422	0.303	0.259	27.112

Y2–7	1.402	3.601	1.427	1.585	2.748	4.234	10.339	1.769	1.116	2.047	1.554	2.255	0.161	0.413	34.650
Y2–8	1.080	4.098	1.108	1.742	2.698	5.124	10.344	2.377	0.955	2.256	1.383	2.003	0.131	0.453	35.751
Y2–9	0.676	1.029	0.862	0.616	1.314	1.829	4.505	0.805	1.381	1.839	2.251	4.168	0.210	0.664	22.148
Y2–10	0.927	4.273	1.931	2.174	1.625	6.832	8.613	2.021	0.471	3.789	0.865	1.314	0.056	0.573	35.464
Y2–11	0.805	3.544	1.234	1.105	1.414	3.796	7.718	1.280	1.262	1.878	0.847	1.534	0.099	0.442	26.958
Y2–12	0.397	0.984	0.779	0.608	0.606	1.364	3.054	0.677	1.267	0.701	2.360	3.906	0.247	1.644	18.594
Y2–13	0.827	2.164	1.239	1.070	1.498	2.774	7.275	1.045	2.546	1.484	3.102	5.335	0.348	0.900	31.607
Y2–14	0.616	0.998	0.895	0.660	1.398	1.744	4.775	0.943	1.341	1.736	2.279	4.352	0.206	0.810	22.752
Average	0.902	2.715	1.145	1.080	1.605	3.317	7.237	1.297	1.217	2.004	1.581	2.772	0.166	0.580	28.531
RSD (%)	30.54	50.94	38.17	54.69	44.48	54.86	36.79	47.74	45.80	44.37	43.69	47.92	53.41	61.54	26.20

Qualitative Analysis of Caulis Trachelospermi

Fragmentation Characteristics of Dibenzylbutyrolatone Lignans

The mass spectra of reference compounds indicated that the accurate molecule weight of the quasi-molecule ions was highly consistent with that of the calculated ones (see Table S1). Thus, the molecule formula of the compounds in the sample can be uniquely deduced with the accurate molecule weight. In the MS spectra, all dibenzylbutyrolatone lignan standards showed strong $[M+Na]^+$ signals in the positive ion mode. The selected precursor ions were dissociated using MS/MS to generate a series of abundant fragment ions. According to the MS data, the fragmentation patterns of lignano-9, 9'-lactones and lignano-8'-hydroxy-9, 9'-lactones exhibited diagnostic distinction between each other. The fragmentation pathways are summarized in Figure 3 and characteristic fragment ions of reference standards in MS/MS spectra are shown in Table 4 and Table 5.

The fragmentation pathways (Figure 3) of lignano-9, 9›-lactones such as matairesinol were in agreement with Schmidt's research that the $[A]^+$ ion, the analogous product ion $[A']^+$ and the ion $[B]^+$ is observed with significant abundance [24]. Furthermore, the high-resolution MS experiment of our research provides sufficient confirmation for Schmidt's conclusions. The formation of two characteristic fragments ($[A]^+$ and $[B]^+$) provides information allowing the distinction between isomers with exchanged substitution of the two benzyl moieties [25]. Notably, the fragment ion at m/z 223.0968 (corresponding to $[M+H-A]^+$ of matairesinol) which was first observed in our research can also determine the substitution of the two benzyl moietites of lignano-9,9›-lactones. Thus, the ion $[M+H-A]^+$ can be interpreted as a complementary of $[B]^+$ for the characteristic product ions.

The MS/MS spectrum of trachelogenin which is representative of lignano-8›-hydroxy-9,9›-lactones generated an abundant fragment at m/z 371.1496 $[M+H−H_2O]^+$ to yield a lign-7-eno-9,9'-lactones intermediate. The subsequent fragmentation pathways are identical with that of lign-7-eno-9,9›-lactones proposed by Schmidt *et al.* [25] and very abundant ions $[C+H]^+$ and $[A']^+$ were observed. However, fragmentary ion at m/z 137.0613 (termed $[A]^+$ in Figure 3) which has not been reported by Schmidt *et al.* in lign-7-eno-9,9'-lactones can also be detected. Consequently, the fragment described above is of diagnostic value in the straightforward assignment of aromatic substitution in lignano-8'-hydroxy-9,9›-lactones. In the MS/MS spectra of lignano-9,9›-lactone and lignano-8›-hydroxy-9,9›-lactone *O*-glycosides, ion $[M+H−162]^+$ or $[M+H−162−162]^+$ was obtained after eliminating the glucose residue $(−162$ Da) and subsequent MS behavior is in line with the typical fragmentation pathways of the aglycone moieties except matairesinol 4›-*O*-*β*-gentiobioside (**4**). On the other hand, nortrachelogenin 5'-*C*-*β*-d-glucoside (**1**) showed the characteristic fragment ions at 441.1555 $[M+H−2H_2O−60]^+$ and 423.1459 $[M+H−3H_2O−60]^+$ which was result from a cross-link cleavage of the C-glycoside moiety. Unfortunately, the diagnostic ion cannot be detected in 5-methoxytrachelogenin (**13**) and arctigenin (**14**). It was likely due to the inadequate MS/MS conditions or some unknown factors which requires further investigation. Overall, the characteristic $[B]^+$ and $[C+H]^+$ ions in combination with the $[A]^+$ and $[A']^+$ ions allows the unambiguous identification and distinction between lignano-9,9'-lactones and lignano-8'-hydroxy-9,9'-lactones.

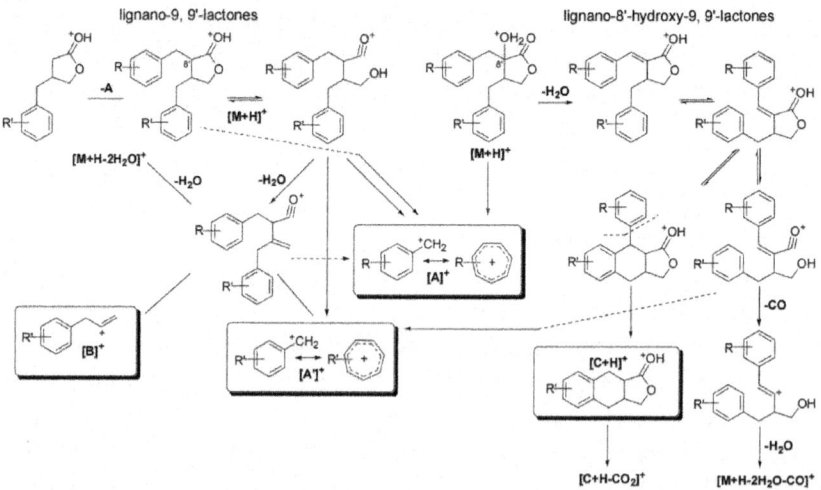

Figure 3. MS/MS fragmentation pathways of lignano-9, 9'-lactones and lignano-8'-hydroxy-9, 9'-lactones.

Table 4. Key MS/MS fragmentation data of reference lignano-9,9›-lactones compounds

Peak No.	[A]⁺ (Measured)	[A]⁺ (Calculated)	Error (ppm)	[A']⁺ (Measured)	[A']⁺ (Calculated)	Error (ppm)	[B]⁺ (Measured)	[B]⁺ (Calculated)	Error (ppm)	[M+H−A]⁺ (Measured)	[M+H−A]⁺ (Calculated)	Error (ppm)
4	137.0603	137.0603	0	137.0603	137.0603	0						
6	137.0599	137.0603	−2.92	137.0599	137.0603	−2.92	163.0755	163.0754	0.61	223.0973	223.0965	3.55
8	137.0601	137.0603	−1.46							237.1117	237.1121	−1.69
10	137.0604	137.0603	0.73	151.0759	151.0754	3.31	177.0923	177.0910	7.34	237.1127	237.1121	2.53
11	137.0607	137.0603	2.92	137.0607	137.0603	2.92	163.0762	163.0762	4.91	223.0968	223.0965	1.35
14	137.0599	137.0603	−2.92	151.0753	151.0754	−0.66						

Table 5. Key MS/MS fragmentation data of reference lignano-8'-hydroxy-9,9›-lactones compounds

Peak No.	[A]⁺ (Measured)	[A]⁺ (Calculated)	Error (ppm)	[A']⁺ (Measured)	[A']⁺ (Calculated)	Error (ppm)	[C+H]⁺ (Measured)	[C+H]⁺ (Calculated)	Error (ppm)
1				137.0607	137.0603	2.92			
2	137.0610	137.0603	5.12	137.0610	137.0603	5.12	233.0816	233.0808	3.43
3	137.0602	137.0603	−0.73	137.0602	137.0603	−0.73	233.0812	233.0808	1.72
5	137.0619	137.0603	11.67	151.0775	151.0754	13.90	247.0982	247.0965	6.88
7	137.0605	137.0603	1.46	151.0763	151.0754	5.96	247.0973	247.0965	3.24
9	137.0614	137.0603	8.03	137.0614	137.0603	8.03	233.0820	233.0808	5.15
12	137.0613	137.0603	7.30	151.0770	151.0754	10.59	247.0977	247.0965	4.86
13	137.0613	137.0603	7.30	181.0861	181.0859	1.10			

Identification of Constituents in the Sample Extracted from Caulis Trachelospermi

Before qualitative analysis of the constituents in Caulis Trachelospermi by HPLC-QTOF-MS, purification from Caulis Trachelospermi was performed by HP-20 macroporous resin column chromatography to obtain an extract for analysis in order to reduce the matrix interference. As shown in the HPLC-UV chromatogram (Figure 2B), the sample after purification successfully remains the major constituents in Caulis Trachelospermi. Its total ion chromatogram (TIC) from HPLC-QTOF-MS analysis is shown in Figure 4. Under the present chromatographic and MS conditions, in total 25 compounds were detected (as shown in Table 6 and Figure 4). Among them, 15 compounds were unambiguously identified by comparing with the retention time and MS data of reference standards.

Compound **21** showed an accurate mass of $[M+Na]^+$ ion at m/z 559.1808 corresponding to the molecular formula $C_{26}H_{32}O_{12}$. In the MS/MS spectrum, signals for ion at m/z 397.1328 $[M+Na-162]^+$ via the loss of glucose and the characteristic ion at 159.0415 $[A-H+Na]^+$ implied that it was nortrachelogenin glucoside. According to the different LC retention behaviors from the known isomers nortracheloside (**2**) and nortrachelogenin 8'-O-β-d-glucoside (**3**), it was assigned as nortrachelogenin 4-O-β-d-glucoside. Similarly, compound **18**, **22** and **24** was deduced to be nortrachelogenin 4, 4)-di-O-β-d-glucoside, 4-demethyltraxillaside and traxillageside, respectively. Their structures were confirmed by NMR techniques after isolaton and purification in our previous research [9,26].

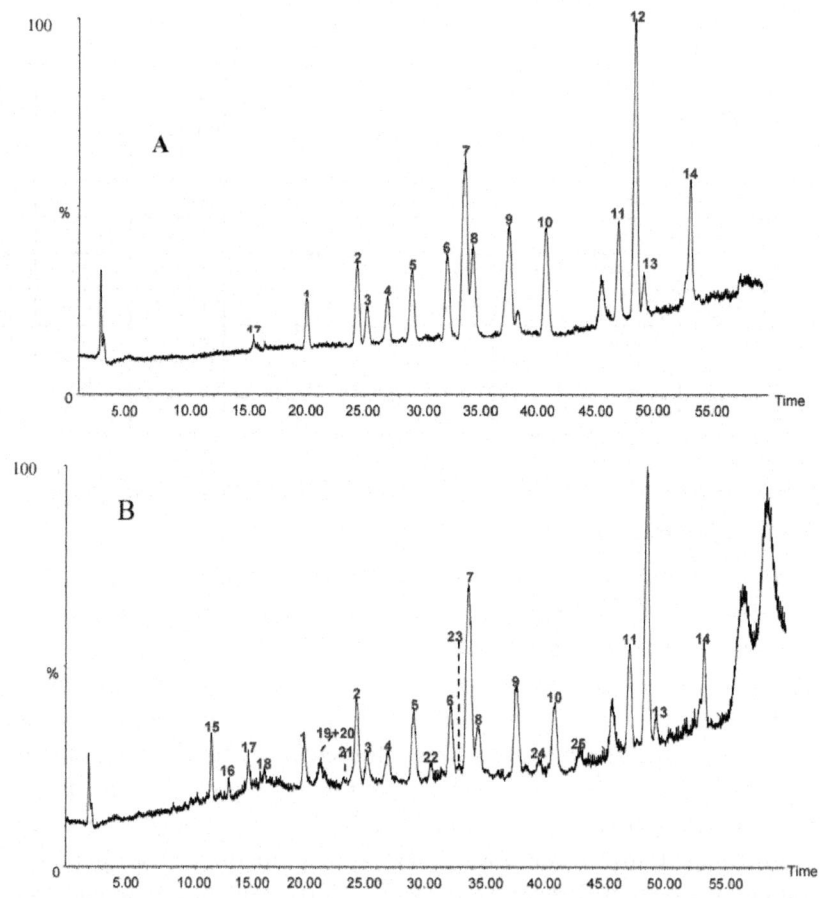

Figure 4. (**A**) Total ion current (TIC) of mixed standards. (**B**) Total ion current (TIC) of the sample extracted from Caulis Trachelospermi. The number of peaks marked in Figure 4 is corresponding to Figure 1.

Table 6. Key MS/MS fragmentation data of other constituents in the sample extraced from Caulis Trachelospermi

Peak No.	t_R(min)	Precursor ion (m/z)	Error (ppm)	Formula	Fragments (m/z)	Elem. comp.	Pathways	Identity
15	12.349	329.0865 [M+H]$^+$	-2.43	$C_{14}H_{16}O_9$	293.0663 275.0549 263.0555 247.0610 233.0442	$C_{14}H_{13}O_7{}^+$ $C_{14}H_{11}O_6{}^+$ $C_{13}H_{11}O_6{}^+$ $C_{13}H_{11}O_5{}^+$ $C_{12}H_9O_5{}^+$	[M+H−2H$_2$O]$^+$ [M+H−3H$_2$O]$^+$ [M+H−2H$_2$O− HCOH]$^+$ [M+H−3H$_2$O− CO]$^+$ [M+H−2H$_2$O− CH$_3$COOH]$^+$	bergenin
16	13.795	501.1551 [M+Na]$^+$	-5.79	$C_{20}H_{30}O_{13}$	411.1230 369.1147	$C_{17}H_{24}O_{10}Na^+$ $C_{15}H_{22}O_9Na^+$	[M+Na−3HCOH]$^+$ [M+Na−Api]$^+$	kelampayoside A
17	15.450	561.1963 [M+Na]$^+$	2.63	$C_{26}H_{34}O_{12}$	381.1310	$C_{20}H_{22}O_6Na^+$	[M+Na−Glc− H$_2$O]$^+$	tanegoside A
18	16.775	721.2315 [M+Na]$^+$	-0.69	$C_{32}H_{42}O_{17}$	559.1771 397.1249 159.0419	$C_{26}H_{32}O_{12}Na^+$ $C_{20}H_{22}O_7Na^+$ $C_8H_8O_2Na^+$	[M+Na−Glc]$^+$ [M+Na−2Glc]$^+$ [A−H+Na]$^+$/[A'− H+Na]$^+$	nortrachelogenin 4,4'- di- O-β-d- glucoside
19	21.416	705.2371 [M+Na]$^+$	0	$C_{32}H_{42}O_{16}$	543.1870 381.1335 159.0435	$C_{26}H_{32}O_{11}Na^+$ $C_{20}H_{22}O_6Na^+$ $C_8H_8O_2Na^+$	[M+Na−Glc]$^+$ [M+Na−2Glc]$^+$ [A−H+Na]$^+$/[A'− H+Na]$^+$	matairesinol 4,4'- di- O-β-d- glucoside
20	21.416	721.2309 [M+Na]$^+$	-1.53	$C_{32}H_{42}O_{17}$	559.1743 397.1295 159.0421	$C_{26}H_{32}O_{12}Na^+$ $C_{20}H_{22}O_7Na^+$ $C_8H_8O_2Na^+$	[M+Na−Glc]$^+$ [M+Na−2Glc]$^+$ [A−H+Na]$^+$/[A'− H+Na]$^+$	nortrachelogenin 4'- O-β- gentiobioside
21	22.976	559.1792 [M+Na]$^+$	0.18	$C_{26}H_{32}O_{12}$	397.1328 159.0415	$C_{20}H_{22}O_7Na^+$ $C_8H_8O_2Na^+$	[M+Na−Glc]$^+$ [A−H+Na]$^+$/[A'− H+Na]$^+$	nortrachelogenin 4- O-β-d- glucoside
22	30.531	573.1968 [M+Na]$^+$	3.49	$C_{27}H_{34}O_{12}$	411.1426	$C_{21}H_{24}O_7Na^+$	[M+Na−Glc]$^+$	4-demethyltraxillaside
23	32.928	545.1992 [M+Na]$^+$	-1.28	$C_{26}H_{34}O_{11}$	383.1596 159.0366	$C_{20}H_{24}O_6Na^+$ $C_8H_8O_2Na^+$	[M+Na−Glc]$^+$ [A−H+Na]$^+$	dihydrodehydrodicon iferyl alcohol-9- O-β- d-glucoside
24	39.434	587.2125 [M+Na]$^+$	3.58	$C_{28}H_{36}O_{12}$	425.1583 159.0412	$C_{22}H_{26}O_7Na^+$ $C_8H_8O_2Na^+$	[M+Na−Glc]$^+$ [A−H+Na]$^+$	traxillageside
25	42.339	579.1719 [M+H]$^+$	0.86	$C_{27}H_{30}O_{14}$	433.1125 271.0605	$C_{21}H_{21}O_{10}{}^+$ $C_{15}H_{11}O_5{}^+$	[M+H−Rha]$^+$ [M+H−Rha−Glc]$^+$	apigenin 7- O-β- neospheroside

Compound **20** yielded product ions at 559.1743 [M+Na−162]$^+$ and 397.1295 [M+Na−162−162]$^+$ which correspond to the aglycone form after loss of hexosyl moiety, together with the fragment ion at 159.0421 [A−H+Na]$^+$, sugguesting the aglycone moiety is nortrachelogenin. According to the biogenetic regularity that the sugar residue is mostly glucosyl or gentiobiosyl residue and sugar residue is preferentially connected to C-4', compound **20** was deduced as nortrachelogenin 4, 4›-di-*O*-*β*-d-glucoside or nortrachelogenin 4'-*O*-*β*-gentiobioside. Since its LC retention time is different from that of the known isomer nortrachelogenin 4, 4›-di-*O*-*β*-d-glucoside (**18**), its structure was assigned as nortrachelogenin 4'-*O*-*β*-gentiobioside. Similarly, Compound **19** was deduced as matairesinol 4, 4›-di-*O*-*β*-d-glucoside according to its MS data and different LC retention time from that of its known isomer matairesinol 4›-*O*-*β*-gentiobioside (**4**).

Other compounds (**15**, **16**, **23** and **25**) were tentatively identified by comparing MS data and LC retention behavior with the literature published by our group [9,26,27].

However, the dibenzylbutyrolatone lignans with low content in sample did not show key information such as the characteristic ions $[B]^+$ and $[C+H]^+$ having critical importance in structural elucidation of dibenzylbutyrolatone lignans. This was probably due to the trace content of these compounds which were easily interfered with by matrix.

EXPERIMENTAL SECTION

Chemicals, Reagents and Materials

Reference standards (Figure 1), nortrachelogenin 5'-*C*-*β*-d-glucoside (**1**), nortracheloside (**2**), nortrachelogenin 8'-*O*-*β*-d-glucoside (**3**), matairesinol 4'-*O*-*β*-gentiobioside (**4**), trachelogenin 4'-*O*-*β*-gentiobioside (**5**), matairesinoside (**6**), tracheloside (**7**), arctigenin 4'-*O*-*β*-gentiobioside (**8**), nortrachelogenin (**9**), arctiin (**10**), matairesinol (**11**), trachelogenin (**12**), 5-methoxytrachelogenin (**13**), arctigenin (**14**) and tanegoside A (**17**) were isolated from Caulis Trachelospermi and identified in our previous research [8,26,27,28]. The purities of the standards were determined to be above 95% by normalization of the peak areas detected by HPLC analyses.

HPLC grade methanol was obtained from Fisher Scientific (Fair Lawn, NJ, USA). Purified water was purchased from Wahaha Ltd. (Hangzhou, China). Other reagents, purchased from Sinopharm Chemical Reagent Co. Ltd. (Beijing, China) were of analytical grade. HP-20 macroporous resin was purchased from Mitsubishi Chemical Co. (Tokyo, Japan). 14 batches of Caulis Trachelospermi were acquired from different pharmaceutical companies in China (Table 7). All of the samples were authenticated by senior engineer Qi-yun Ma, Beijing Institute of Pharmacology and Toxicology, Beijing, China. Voucher specimens were deposited in the Department of Natural Products

Chemistry, Beijing Institute of Pharmacology and Toxicology, Beijing, China. Batch no. 080130 (purchased from Beijing Qijing Chinese Herbs Factory, Beijing, China in 2008) was selected as the sample for method validation.

Table 7. Sample information for 14 batches Caulis Trachelospermi

Sample no.	Batch no.	Source
Y2–1	080130	Zhejiang (QJ [a])
Y2–2	0511011	Zhejiang (CG)
Y2–3	060530	Anhui (SF)
Y2–4	061329	Anhui (SL)
Y2–5	05090101	Shandong (PSL)
Y2–6	05081205	Zhejiang (QJ)
Y2–7	20030928	Yunnan (BTS)
Y2–8	20060415	Yunnan (BTS)
Y2–9	5050028	Guangxi (TRT)
Y2–10	1040804	Henan (TRT)
Y2–11	060626	Jiangsu (LRT)
Y2–12	070301	Zhejiang (XD)
Y2–13	F060290	Sichuan (DRT)
Y2–14	20070102	Jiangsu (QX)

[a] Abbreviated for different pharmaceutical companies.

Standard Solutions Preparation

Individual stock solutions of reference standards were prepared by accurately weighing into a 10 mL volumetric flask and dissolving the reference compounds in methanol. A mixed solution used for quantitative analysis was prepared by placing a certain amount of each stock solution in a 10 mL volumetric flask and diluted to volume with 50% methanol aqueous solution at the concentration of 36.18 $\mu g \cdot mL^{-1}$ nortrachelogenin 5›-C-β-d-glucoside (**1**), 91.80 $\mu g \cdot mL^{-1}$ nortracheloside (**2**), 39.03 $\mu g \cdot mL^{-1}$ nortrachelogenin 8›-O-β-d-glucoside (**3**), 46.29 $\mu g \cdot mL^{-1}$ matairesinol 4›-O-β-gentiobioside (**4**), 86.31 $\mu g \cdot mL^{-1}$ trachelogenin 4'-O-β-gentiobioside (**5**), 91.60 $\mu g \cdot mL^{-1}$ matairesinoside (**6**), 280.56 $\mu g \cdot mL^{-1}$ tracheloside (**7**), 107.38 $\mu g \cdot mL^{-1}$ arctigenin 4'-O-β-gentiobioside (**8**), 61.56 $\mu g \cdot mL^{-1}$ nortrachelogenin (**9**), 86.40 $\mu g \cdot mL^{-1}$ arctiin (**10**), 49.50 $\mu g \cdot mL^{-1}$ matairesinol (**11**), 125.20 $\mu g \cdot mL^{-1}$ trachelogenin (**12**), 12.35 $\mu g \cdot mL^{-1}$ 5-methoxytrachelogenin (**13**), 22.76 $\mu g \cdot mL^{-1}$ arctigenin (**14**). Meanwhile, a mixed solution including above solution and 58.46

$\mu g \cdot mL^{-1}$ tanegoside A (17) was prepared for qualitative analysis. An aliquot of 10 μL was injected for HPLC-UV analysis and 1 μL for HPLC-QTOF-MS analysis. All the solutions were stored at 4 °C and brought to room temperature before use.

Preparation of Caulis Trachelospermi for HPLC-UV Analysis

Fourteen batches of Caulis Trachelospermi samples were pulverized and passed through a 100 mesh screen. Four hundred milligrams of the obtained fine powder was accurately weighed into a 50 mL capped conical flask, and 20 mL 50% aqueous methanol was accurately added. Sonication was performed at room temperature for 30 min, and then the same solvent was added to compensate for the lost weight during the extraction. The extracts were filtered with a 0.45 μm membrane filter prior to HPLC analysis, discarding the first part of the filtrate. An aliquot of 20 μL was injected for HPLC-UV analysis.

Preparation of the Sample of Caulis Trachelospermi for HPLC-QTOF-MS Analysis

Caulis Trachelospermi (Batch no. 080130, 160 g) was extracted two times with 80% alcohol at boiling temperature. The extract was concentrated and diluted in 1600 mL 5% alcohol. The solution was first centrifuged to remove the insoluble substance and then was passed through a HP-20 macroporous resin column (100 mL) and eluted by 500 mL water and 500 mL 70% alcohol successively. The 70% alcohol elution was concentrated and dried to produce 5.3 g product. 10.0 mg of the product weighed accurately was dissolved into a 10 mL volumetric flask and adjusted to volume with methanol-water (50:50, v/v) to obtain the sample solution. Prior to injection, the solution was passed through a 0.45 μm membrane filter. An aliquot of 1 μL was injected for HPLC-QTOF-MS analysis.

HPLC-UV Conditions for Quantitative Analysis

Quantitative analysis was performed on an Agilent 1200 series HPLC-UV system (Agilent Technologies, Santa Clara, CA, USA), compressing a quaternary pump, a vacuum degasser, an autosampler, a thermostatted column compartment and a UV-vis detector. Separation was done on an Agilent Zorbax Eclipse Plus C18 column (4.6 mm × 150 mm, 5 μm) and column temperature was maintained at 30 °C. The mobile phase was water (A) and methanol (B) with a linear gradient program as follows: 0–15 min, 10%–30% B; 15–40 min, 30%–40% B; and 40–60 min, 40%–60% B. Re-equilibration duration was 30 min between individual runs and the flow rate was kept at 0.8 $mL \cdot min^{-1}$. The detector wavelength was set at 230 nm [19].

HPLC-QTOF-MS Conditions for Qualitative Analysis

Chromatography was performed using a Waters ACQUITY UHPLC system (Waters Corporation, Milford, MA, USA), equipped with a binary solvent delivery system and an autosampler. HPLC conditions were the same as those for quantitative analysis.

The Waters ACQUITY XEVO G2 QTOF mass spectrometer (Waters Corporation, Manchester, UK) was interfaced to the UHPLC system via an electrospray ionization (ESI) source. The source was operated in positive ionization mode. The desolvation gas was set to 600 L·h^{-1} at temperature of 300 °C, the cone gas set to 50 L·h^{-1}, and the source temperature set to 100 °C. The capillary voltage and cone voltage were set to 3000 V and 20 V, respectively. The TOF data were collected between m/z 50 and 1200. The MS/MS experiments were performed using variable collision energy (20–30 eV). The accurate mass and composition for the precursor and fragment ions were calculated using Masslynx 4.1 software (Waters Corp., Milford, MA, USA) that was incorporated in the instrument.

Validation of the Quantitative Method

Calibration Curve, Limits of Detection and Quantification

For the calibration curves, a 1, 2, 3, 5, 10, 15, 20, 25, 35 µL volume of the mixed standard solution was injected respectively, and then the calibration curves were constructed by plotting the peak area *versus* the concentration (ng) of each analyte. The limit of detection (LOD) and the limit of quantification (LOQ) under the present chromatographic conditions were determined by injecting a series of diluted standard solutions when the signal-to-noise ratio (S/N) of analytes were about 3 and 10, respectively.

Precision and Accuracy

Precision of the developed method was evaluated in six replicates of the mixed standard solutions within one and three consecutive days to determine intra and inter-day precision, respectively. Variations of the peak area were taken as the measures of precision and expressed as relative standard deviation (RSD).

Recovery test was used to evaluate the accuracy of this method. The test was performed by adding accurate amounts of the mixed standard solutions into 200 mg of Caulis Trachelospermi (Batch no. 080130) in sextuplicate. The mixture were then extracted and analyzed as described in Section 3.3 and Section 3.5. The average recovery percentage was calculated by the formula: recovery (%) = (observed amount − original amount)/spiked amount × 100%.

Repeatability and Stability

To confirm the repeatability, six independent samples were prepared and analyzed from the same sample (Batch no. 080130). Stability was assessed through analyzing replicate injections of the same sample at 0 h, 2 h, 4 h, 6 h, 8 h and 24 h, which were stored at 25 °C. The relative standard deviation (RSD) was used to evaluate the results.

CONCLUSIONS

In the present study, a HPLC-UV method was first developed for the simultaneous determination of 14 dibenzylbutyrolatone lignans in Caulis Trachelospermi. The developed method was validated for all parameters and has been successfully applied to analyze 14 batches Caulis Trachelospermi samples, which could be helpful in quality assessment and standardization of Caulis Trachelospermi and its product. Meanwhile, a HPLC-QTOF-MS method was employed for the identification and structural characterization of major constituents in the sample extracted from Caulis Trachelospermi. The specific fragment ions obtained by MS/MS provide sufficient information for structure elucidation. Moreover, the fragmentation patterns of lignano-8'-hydroxy-9, 9'-lactones were investigated for the first time in this work. This qualitative identification method provide essential data for further chemical or pharmacological studies of Caulis Trachelospermi, and may be applied for the identification of bioactive dibenzylbutyrolatone lignans from other related plants.

SUPPLEMENTARY MATERIALS

Table S1. MS/MS data and proposed fragmentation pathways of reference compounds

Peak No.	Compounds	t_R (min)	Precursor Ion (m/z)	Error (ppm)	Formula	Fragments (m/z)	Elem. comp.	Pathways
1	nortrachelogenin 5'-C-β-D-glucoside	20.090	559.1801 [M+Na]$^+$	1.79	$C_{26}H_{32}O_{12}$	483.1670	$C_{26}H_{27}O_9^+$	[M+H−3H₂O]$^+$
						441.1555	$C_{24}H_{25}O_8^+$	[M+H−2H₂O−(CHO−CH₂OH)]$^+$
						465.1556	$C_{26}H_{25}O_8^+$	[M+H−4H₂O]$^+$
						423.1459	$C_{24}H_{23}O_7^+$	[M+H−3H₂O−(CHO−CH₂OH)]$^+$
						203.0711	$C_{12}H_{11}O_3^+$	[C+H−HCOH]$^+$
						137.0607	$C_8H_9O_2^+$	[A']$^+$
2	nortracheloside	24.370	559.1769 [M+Na]$^+$	−3.93	$C_{26}H_{32}O_{12}$	397.1255	$C_{20}H_{22}O_7Na^+$	[M+Na−Glc]$^+$
						375.1438	$C_{20}H_{23}O_7^+$	[M+H−Glc]$^+$
						357.1332	$C_{20}H_{21}O_6^+$	[M+H−Glc−H₂O]$^+$
						329.1388	$C_{19}H_{21}O_5^+$	[M+H−Glc−H₂O−CO]$^+$
						311.1281	$C_{19}H_{19}O_4^+$	[M+H−Glc−2H₂O−CO]$^+$
						233.0816	$C_{13}H_{13}O_4^+$	[C+H]$^+$
						189.0924	$C_{12}H_{13}O_2^+$	[C+H−CO₂]$^+$
						137.0610	$C_8H_9O_2^+$	[A]$^+$/[A']$^+$
3	nortrachelogenin 8'-O-β-D-glucoside	25.227	559.1805 [M+Na]$^+$	2.50	$C_{26}H_{32}O_{12}$	375.1446	$C_{20}H_{23}O_7^+$	[M+H−Glc]$^+$
						357.1342	$C_{20}H_{21}O_6^+$	[M+H−Glc−H₂O]$^+$
						329.1390	$C_{19}H_{21}O_5^+$	[M+H−Glc−H₂O−CO]$^+$
						311.1281	$C_{19}H_{19}O_4^+$	[M+H−Glc−2H₂O−CO]$^+$
						233.0812	$C_{13}H_{13}O_4^+$	[C+H]$^+$
						189.0915	$C_{12}H_{13}O_2^+$	[C+H−CO₂]$^+$
						137.0602	$C_8H_9O_2^+$	[A]$^+$/[A']$^+$
4	matairesinol 4'-O-β-gentiobioside	26.719	705.2364 [M+Na]$^+$	−0.99	$C_{32}H_{42}O_{16}$	543.1831	$C_{26}H_{32}O_{11}Na^+$	[M+Na−Glc]$^+$
						381.1292	$C_{20}H_{22}O_6Na^+$	[M+Na−2Glc]$^+$
						159.0422	$C_8H_8O_2Na^+$	[A+Na]$^+$/[A'+Na]$^+$
						137.0603	$C_8H_9O_2^+$	[A]$^+$/[A']$^+$

Peak No.	Compounds	t_R (min)	Precursor Ion (m/z)	Error (ppm)	Formula	Fragments (m/z)	Elem. comp.	Pathways
5	trachelogenin 4'-O-β-gentiobioside	29.039	730.2925 [M+NH₄]$^+$	0.41	$C_{33}H_{44}O_{17}$	551.2136	$C_{27}H_{35}O_{12}^+$	[M+H−Glc]$^+$
						389.1615	$C_{21}H_{25}O_7^+$	[M+H−2Glc]$^+$
						371.1508	$C_{21}H_{23}O_6^+$	[M+H−2Glc−H₂O]$^+$
						343.1562	$C_{20}H_{23}O_5^+$	[M+H−2Glc−H₂O−CO]$^+$
						325.1445	$C_{20}H_{21}O_4^+$	[M+H−2Glc−2H₂O−CO]$^+$
						247.0982	$C_{14}H_{15}O_4^+$	[C+H]$^+$
						203.1082	$C_{13}H_{15}O_2^+$	[C+H−CO₂]$^+$
						151.0775	$C_9H_{11}O_2^+$	[A']$^+$
						137.0619	$C_8H_9O_2^+$	[A]$^+$
6	matairesinoside	32.188	543.1857 [M+Na]$^+$	2.76	$C_{26}H_{32}O_{11}$	359.1500	$C_{20}H_{23}O_6^+$	[M+H−Glc]$^+$
						341.1393	$C_{20}H_{21}O_5^+$	[M+H−Glc−H₂O]$^+$
						323.1289	$C_{20}H_{19}O_4^+$	[M+H−Glc−2H₂O]$^+$
						223.0973	$C_{12}H_{16}O_3^+$	[M+H−Glc−A]$^+$
						163.0755	$C_{10}H_{11}O_2^+$	[B]$^+$
						137.0599	$C_8H_9O_2^+$	[A]$^+$/[A']$^+$
7	tracheloside	33.679	573.1951 [M+Na]$^+$	0.52	$C_{27}H_{34}O_{12}$	389.1603	$C_{21}H_{25}O_7^+$	[M+H−Glc]$^+$
						371.1501	$C_{21}H_{23}O_6^+$	[M+H−Glc−H₂O]$^+$
						343.1548	$C_{20}H_{23}O_5^+$	[M+H−Glc−H₂O−CO]$^+$
						325.1444	$C_{20}H_{21}O_4^+$	[M+H−Glc−2H₂O−CO]$^+$
						247.0973	$C_{14}H_{15}O_4^+$	[C+H]$^+$
						203.1075	$C_{13}H_{15}O_2^+$	[C+H−CO₂]$^+$
						151.0763	$C_9H_{11}O_2^+$	[A']$^+$
						137.0605	$C_8H_9O_2^+$	[A]$^+$

Peak No.	Compounds	t_R (min)	Precursor Ion (m/z)	Error (ppm)	Formula	Fragments (m/z)	Elem. comp.	Pathways
8	arctigenin 4'-O-β-gentiobioside	34.517	714.2961 [M+NH₄]⁺	−1.68	C₃₃H₄₄O₁₆	373.1645	C₂₁H₂₅O₆⁺	[M+H−2Glc]⁺
						355.1540	C₂₁H₂₃O₅⁺	[M+H−2Glc−H₂O]⁺
						237.1117	C₁₃H₁₇O₄⁺	[M+H−2Glc−A]⁺
						137.0601	C₈H₉O₂⁺	[A]⁺
9	nortrachelogenin	37.508	397.1262 [M+Na]⁺	−0.25	C₂₀H₂₂O₇	357.1342	C₂₀H₂₁O₆⁺	[M+H−H₂O]⁺
						329.1392	C₁₉H₂₁O₅⁺	[M+H−H₂O−CO]⁺
						311.1288	C₁₉H₁₉O₄⁺	[M+H−2H₂O−CO]⁺
						233.0822	C₁₄H₁₃O₄⁺	[C+H]⁺
						189.0925	C₁₂H₁₃O₂⁺	[C+H−CO₂]⁺
						175.0766	C₁₁H₁₁O₂⁺	[C+H−CO−HCOH]⁺
						137.0614	C₈H₉O₂⁺	[A]⁺/[A']⁺
10	arctiin	40.640	557.1993 [M+Na]⁺	−1.08	C₂₇H₃₄O₁₁	373.1649	C₂₁H₂₅O₆⁺	[M+H−Glc]⁺
						355.1544	C₂₁H₂₃O₅⁺	[M+H−Glc−H₂O]⁺
						337.1439	C₂₁H₂₁O₄⁺	[M+H−Glc−2H₂O]⁺
						295.1338	C₁₉H₁₉O₃⁺	[M+H−Glc−H₂O−2HCOH]⁺
						237.1127	C₁₃H₁₇O₄⁺	[M+H−Glc−A]⁺
						177.0923	C₁₁H₁₃O₂⁺	[B]⁺
						151.0759	C₉H₁₁O₂⁺	[A']⁺
						137.0604	C₈H₉O₂⁺	[A]⁺
11	matairesinol	46.937	359.1493 [M+H]⁺	−0.56	C₂₀H₂₂O₆	341.1392	C₂₀H₂₁O₅⁺	[M+H−H₂O]⁺
						323.1286	C₂₀H₁₉O₄⁺	[M+H−2H₂O]⁺
						305.1178	C₂₀H₁₇O₃⁺	[M+H−3H₂O]⁺
						291.1019	C₁₉H₁₅O₃⁺	[M+H−2H₂O−CH₃OH]⁺
						231.0810	C₁₇H₁₁O⁺	[M+H−2H₂O−2CH₃OH−CO]⁺
						223.0968	C₁₂H₁₅O₄⁺	[M+H−A]⁺
						163.0762	C₁₀H₁₁O₂⁺	[B]⁺
						137.0607	C₈H₉O₂⁺	[A]⁺/[A']⁺
						131.0503	C₉H₇O⁺	[B−CH₃OH]⁺

Peak No.	Compounds	t_R (min)	Precursor Ion (m/z)	Error (ppm)	Formula	Fragments (m/z)	Elem. comp.	Pathways
12	trachelogenin	48.366	411.1422 [M+Na]⁺	0.49	C₂₁H₂₄O₇	371.1496	C₂₁H₂₃O₆⁺	[M+H−H₂O]⁺
						343.1548	C₂₀H₂₃O₅⁺	[M+H−H₂O−CO]⁺
						325.1443	C₂₀H₂₁O₄⁺	[M+H−2H₂O−CO]⁺
						247.0977	C₁₄H₁₅O₄⁺	[C+H]⁺
						203.1082	C₁₃H₁₅O₂⁺	[C+H−CO₂]⁺
						189.0918	C₁₂H₁₃O₂⁺	[C'+H−CO₂]⁺
						151.0770	C₉H₁₁O₂⁺	[A']⁺
						137.0613	C₈H₉O₂⁺	[A]⁺
13	5-methoxytrachelogenin	49.403	441.1527 [M+Na]⁺	0.45	C₂₂H₂₆O₈	181.0861	C₁₀H₁₃O₃⁺	[A']⁺
						159.0433	C₈H₈O₂Na⁺	[A+Na]⁺
						137.0613	C₈H₉O₂⁺	[A]⁺
14	arctigenin	53.235	395.1488 [M+Na]⁺	4.30	C₂₁H₂₄O₆	137.0599	C₈H₉O₂⁺	[A]⁺
						159.0405	C₈H₈O₂Na⁺	[A+Na]⁺

Table S2. Constituents comparing with reference compounds detected in the sample extraced from Caulis Trachelospermi

Peak No.	t_R (min)	Precursor Ion (m/z)	Error (ppm)	Formula	Fragments (m/z)	Elem. comp.	Pathways	Identity
1	20.090	559.1797 [M+Na]⁺	1.07	C₂₆H₃₂O₁₂	501.1779	C₂₆H₂₉O₁₀⁺	[M+H−2H₂O]⁺	nortrachelogenin 5'-C-β-D-glucoside
					483.1664	C₂₆H₂₇O₉⁺	[M+H−3H₂O]⁺	
					465.1555	C₂₆H₂₇O₈⁺	[M+H−4H₂O]⁺	
					441.1555	C₂₄H₂₅O₈⁺	[M+H−2H₂O−(CHO−CH₂OH)]⁺	
					423.1448	C₂₄H₂₃O₇⁺	[M+H−3H₂O−(CHO−CH₂OH)]⁺	
					399.1450	C₂₃H₂₃O₇⁺	[M+H−H₂O−C₄H₈O₃−H₂O]⁺	
					203.0725	C₁₂H₁₁O₃⁺	[C+H−HCOH]⁺	
					137.0614	C₈H₉O₂⁺	[A']⁺	

Peak No.	t_R (min)	Precursor Ion (m/z)	Error (ppm)	Formula	Fragments (m/z)	Elem. comp.	Pathways	Identity
2	24.416	559.1776 [M+Na]$^+$	−2.68	C$_{26}$H$_{32}$O$_{12}$	375.1443	C$_{20}$H$_{23}$O$_7^+$	[M+H−Glc]$^+$	nortracheloside
					357.1341	C$_{20}$H$_{21}$O$_6^+$	[M+H−Glc−H$_2$O]$^+$	
					329.1397	C$_{19}$H$_{21}$O$_5^+$	[M+H−Glc−H$_2$O−CO]$^+$	
					311.1290	C$_{19}$H$_{19}$O$_4^+$	[M+H−Glc−2H$_2$O−CO]$^+$	
					233.0822	C$_{11}$H$_{13}$O$_4^+$	[C+H]$^+$	
					189.0927	C$_{12}$H$_{13}$O$_2^+$	[C+H−CO$_2$]$^+$	
					137.0620	C$_8$H$_9$O$_2^+$	[A]$^+$/[A']$^+$	
3	25.296	559.1802 [M+Na]$^+$	1.97	C$_{26}$H$_{32}$O$_{12}$	375.1459	C$_{20}$H$_{23}$O$_7^+$	[M+H−Glc]$^+$	nortrachelogenin 8'-O-β-D-glucoside
					357.1356	C$_{20}$H$_{21}$O$_6^+$	[M+H−Glc−H$_2$O]$^+$	
					329.1408	C$_{19}$H$_{21}$O$_5^+$	[M+H−Glc−H$_2$O−CO]$^+$	
					311.1293	C$_{19}$H$_{19}$O$_4^+$	[M+H−Glc−2H$_2$O−CO]$^+$	
					233.0830	C$_{11}$H$_{13}$O$_4^+$	[C+H]$^+$	
					189.0930	C$_{12}$H$_{13}$O$_2^+$	[C+H−CO$_2$]$^+$	
					137.0612	C$_8$H$_9$O$_2^+$	[A]$^+$/[A']$^+$	
4	26.719	705.2364 [M+Na]$^+$	−0.99	C$_{32}$H$_{42}$O$_{16}$	543.1852	C$_{26}$H$_{32}$O$_{11}$Na$^+$	[M+Na−Glc]$^+$	matairesinol 4'-O-β-gentiobioside
					381.1320	C$_{20}$H$_{22}$O$_6$Na$^+$	[M+Na−2Glc]$^+$	
					159.0423	C$_8$H$_8$O$_2$Na$^+$	[A+Na]$^+$/[A'+Na]$^+$	
					137.0620	C$_8$H$_9$O$_2^+$	[A]$^+$/[A']$^+$	
5	29.039	730.2937 [M+NH$_4$]$^+$	2.05	C$_{33}$H$_{44}$O$_{17}$	551.2149	C$_{27}$H$_{35}$O$_{12}^+$	[M+H−Glc]$^+$	trachelogenin 4'-O-β-gentiobioside
					389.1616	C$_{21}$H$_{25}$O$_7^+$	[M+H−2Glc]$^+$	
					371.1515	C$_{21}$H$_{23}$O$_6^+$	[M+H−2Glc−H$_2$O]$^+$	
					343.1567	C$_{20}$H$_{23}$O$_5^+$	[M+H−2Glc−H$_2$O−CO]$^+$	
					325.1446	C$_{20}$H$_{21}$O$_4^+$	[M+H−2Glc−2H$_2$O−CO]$^+$	
					247.0984	C$_{14}$H$_{15}$O$_4^+$	[C+H]$^+$	
					203.1091	C$_{13}$H$_{15}$O$_2^+$	[C+H−CO$_2$]$^+$	
					151.0773	C$_9$H$_{11}$O$_2^+$	[A']$^+$	
					137.0615	C$_8$H$_9$O$_2^+$	[A]$^+$	

Peak No.	t_R (min)	Precursor Ion (m/z)	Error (ppm)	Formula	Fragments (m/z)	Elem. comp.	Pathways	Identity
6	32.188	543.1849 [M+Na]$^+$	1.29	C$_{26}$H$_{32}$O$_{11}$	359.1506	C$_{20}$H$_{23}$O$_6^+$	[M+H−Glc]$^+$	matairesinoside
					341.1404	C$_{20}$H$_{21}$O$_5^+$	[M+H−Glc−H$_2$O]$^+$	
					323.1299	C$_{20}$H$_{19}$O$_4^+$	[M+H−Glc−2H$_2$O]$^+$	
					223.0980	C$_{12}$H$_{15}$O$_4^+$	[M+H−Glc−A]$^+$	
					163.0770	C$_{10}$H$_{11}$O$_2^+$	[B]$^+$	
					137.0615	C$_8$H$_9$O$_2^+$	[A]$^+$/[A']$^+$	
7	33.679	573.1940 [M+Na]$^+$	−1.40	C$_{27}$H$_{34}$O$_{12}$	389.1601	C$_{21}$H$_{25}$O$_7^+$	[M+H−Glc]$^+$	tracheloside
					371.1499	C$_{21}$H$_{23}$O$_6^+$	[M+H−Glc−H$_2$O]$^+$	
					343.1549	C$_{20}$H$_{23}$O$_5^+$	[M+H−Glc−H$_2$O−CO]$^+$	
					325.1445	C$_{20}$H$_{21}$O$_4^+$	[M+H−Glc−2H$_2$O−CO]$^+$	
					247.0975	C$_{14}$H$_{15}$O$_4^+$	[C+H]$^+$	
					203.1078	C$_{13}$H$_{15}$O$_2^+$	[C+H−CO$_2$]$^+$	
					151.0768	C$_9$H$_{11}$O$_2^+$	[A']$^+$	
					137.0610	C$_8$H$_9$O$_2^+$	[A]$^+$	
8	34.517	714.2968 [M+NH$_4$]$^+$	−0.70	C$_{33}$H$_{44}$O$_{16}$	373.1658	C$_{21}$H$_{25}$O$_6^+$	[M+H−2Glc]$^+$	arctigenin 4'-O-β-gentiobioside
					355.1570	C$_{21}$H$_{23}$O$_5^+$	[M+H−2Glc−H$_2$O]$^+$	
					237.1138	C$_{13}$H$_{17}$O$_4^+$	[M+H−2Glc−A]$^+$	
					137.0598	C$_8$H$_9$O$_2^+$	[A]$^+$	
9	37.687	397.1275 [M+Na]$^+$	3.02	C$_{20}$H$_{22}$O$_7$	357.1343	C$_{20}$H$_{21}$O$_6^+$	[M+H−H$_2$O]$^+$	nortrachelogenin
					329.1394	C$_{19}$H$_{21}$O$_5^+$	[M+H−H$_2$O−CO]$^+$	
					311.1288	C$_{19}$H$_{19}$O$_4^+$	[M+H−2H$_2$O−CO]$^+$	
					233.0819	C$_{13}$H$_{13}$O$_4^+$	[C+H]$^+$/[C'+H]$^+$	
					189.0923	C$_{12}$H$_{13}$O$_2^+$	[C+H−CO$_2$]$^+$	
					175.0764	C$_{11}$H$_{11}$O$_2^+$	[C+H−CO−HCOH]$^+$	
					137.0609	C$_8$H$_9$O$_2^+$	[A]$^+$/[A']$^+$	

Peak No.	t_R (min)	Precursor Ion (m/z)	Error (ppm)	Formula	Fragments (m/z)	Elem. comp.	Pathways	Identity
10	40.708	557.1996 [M+Na]⁺	−0.54	$C_{27}H_{34}O_{11}$	373.1651	$C_{21}H_{25}O_6^+$	[M+H−Glc]⁺	arctiin
					355.1548	$C_{21}H_{23}O_5^+$	[M+H−Glc−H₂O]⁺	
					337.1438	$C_{21}H_{21}O_4^+$	[M+H−Glc−2H₂O]⁺	
					295.1339	$C_{19}H_{19}O_3^+$	[M+H−Glc−H₂O−2HCOH]⁺	
					237.1130	$C_{13}H_{17}O_4^+$	[M+H−Glc−A]⁺	
					177.0911	$C_{11}H_{13}O_2^+$	[B]⁺	
					151.0765	$C_9H_{11}O_2^+$	[A']⁺	
					137.0603	$C_8H_9O_2^+$	[A]⁺	
11	47.006	359.1510 [M+H]⁺	4.18	$C_{20}H_{22}O_6$	341.1396	$C_{20}H_{21}O_5^+$	[M+H−H₂O]⁺	matairesinol
					323.1295	$C_{20}H_{19}O_4^+$	[M+H−2H₂O]⁺	
					305.1178	$C_{20}H_{17}O_3^+$	[M+H−3H₂O]⁺	
					291.1021	$C_{19}H_{15}O_3^+$	[M+H−2H₂O−CH₃OH]⁺	
					263.1074	$C_{18}H_{15}O_2^+$	[M+H−2H₂O−CO−CH₃OH]⁺	
					231.0814	$C_{17}H_{11}O^+$	[M+H−2H₂O−2CH₃OH−CO]⁺	
					223.0968	$C_{12}H_{15}O_4^+$	[M+H−A]⁺	
					163.0758	$C_{10}H_{11}O_2^+$	[B]⁺	
					137.0603	$C_8H_9O_2^+$	[A]⁺/[A']⁺	
					131.0495	$C_9H_7O^+$	[B−CH₃OH]⁺	
12	48.446	411.1426 [M+Na]⁺	1.46	$C_{21}H_{24}O_7$	371.1496	$C_{21}H_{23}O_6^+$	[M+H−H₂O]⁺	trachelogenin
					343.1549	$C_{20}H_{23}O_5^+$	[M+H−H₂O−CO]⁺	
					325.1443	$C_{20}H_{21}O_4^+$	[M+H−2H₂O−CO]⁺	
					247.0970	$C_{14}H_{15}O_4^+$	[C+H]⁺	
					233.0820	$C_{13}H_{13}O_4^+$	[C'+H]⁺	
					203.1073	$C_{13}H_{15}O_2^+$	[C+H−CO₂]⁺	
					189.0912	$C_{12}H_{13}O_2^+$	[C'+H−CO₂]⁺	
					151.0760	$C_9H_{11}O_2^+$	[A']⁺	
					137.0604	$C_8H_9O_2^+$	[A]⁺	

Peak No.	t_R (min)	Precursor Ion (m/z)	Error (ppm)	Formula	Fragments (m/z)	Elem. comp.	Pathways	Identity
13	48.855	441.1526 [M+Na]⁺	0.23	$C_{22}H_{26}O_9$				5-methoxytrachelogenin
14	52.955	395.1463 [M+Na]⁺	−1.01	$C_{21}H_{24}O_6$	159.0408	$C_8H_8O_2Na^+$	[A+Na]⁺	arctigenin

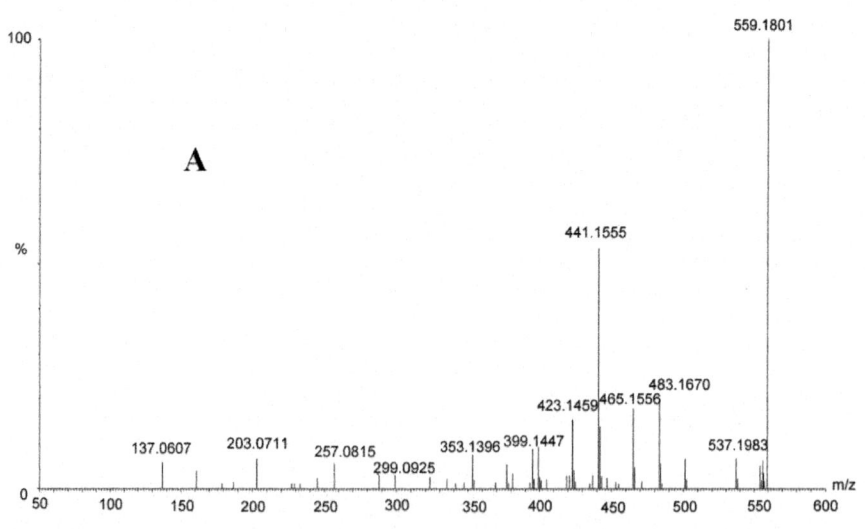

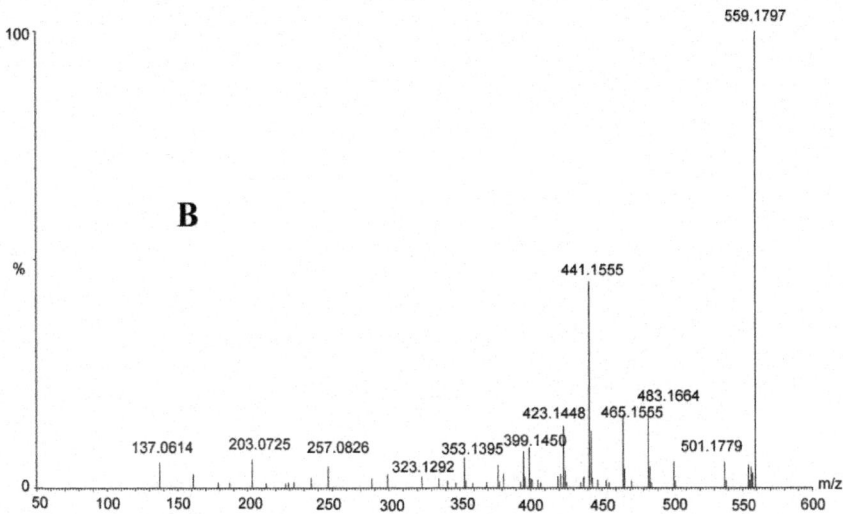

Figure S1. MS/MS spectrum of nortrachelogenin 5'-C-β-D-glucoside (1) in the standard sample (A) and in the sample extracted from Caulis Trachelospermi (B), respectively.

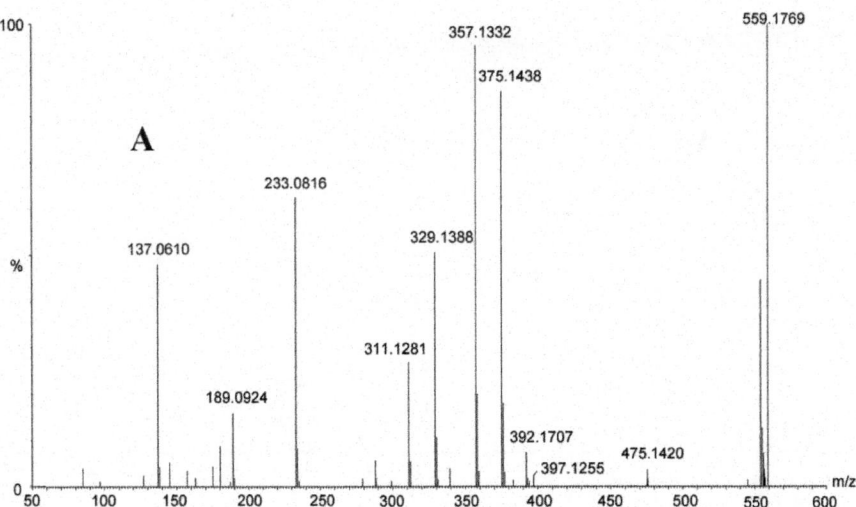

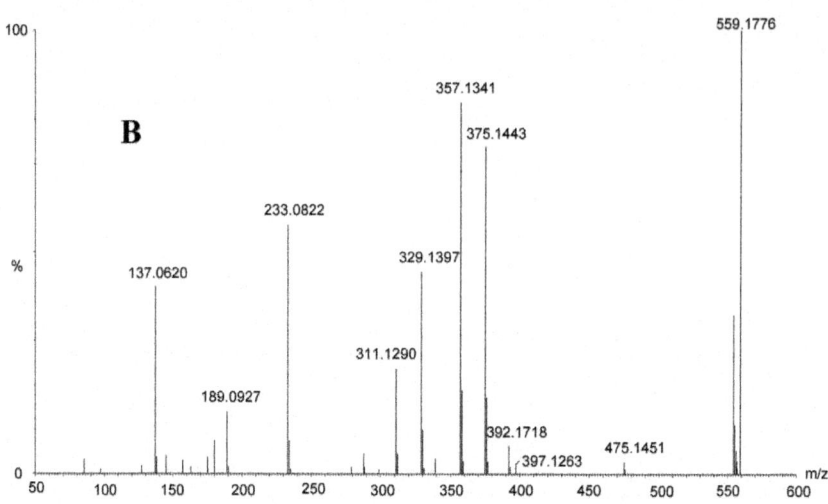

Figure S2. MS/MS spectrum of nortracheloside (2) in the standard sample (A) and in the sample extracted from Caulis Trachelospermi (B), respectively.

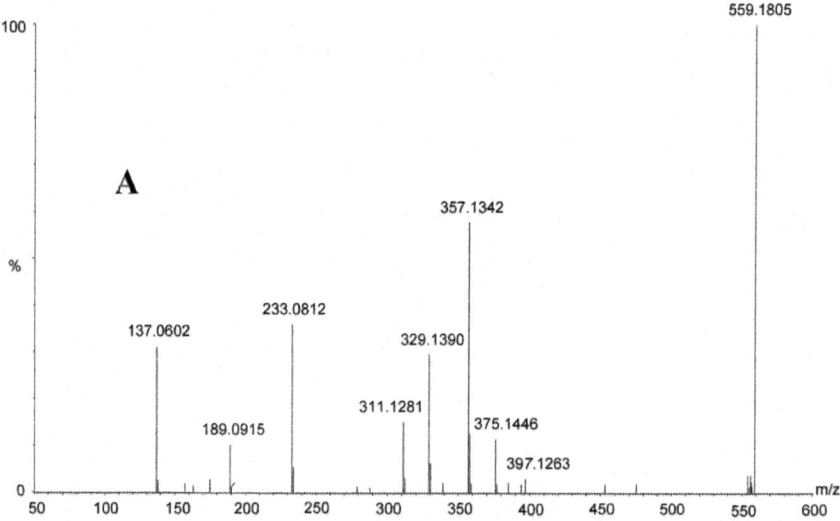

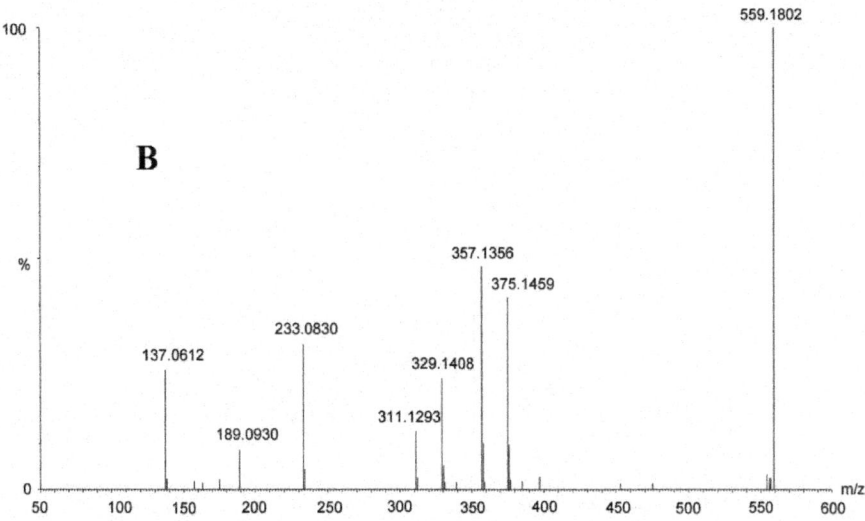

Figure S3. MS/MS spectrum of nortrachelogenin 8'-O-β-D-glucoside (3) in the standard sample (A) and in the sample extracted from Caulis Trachelospermi (B), respectively.

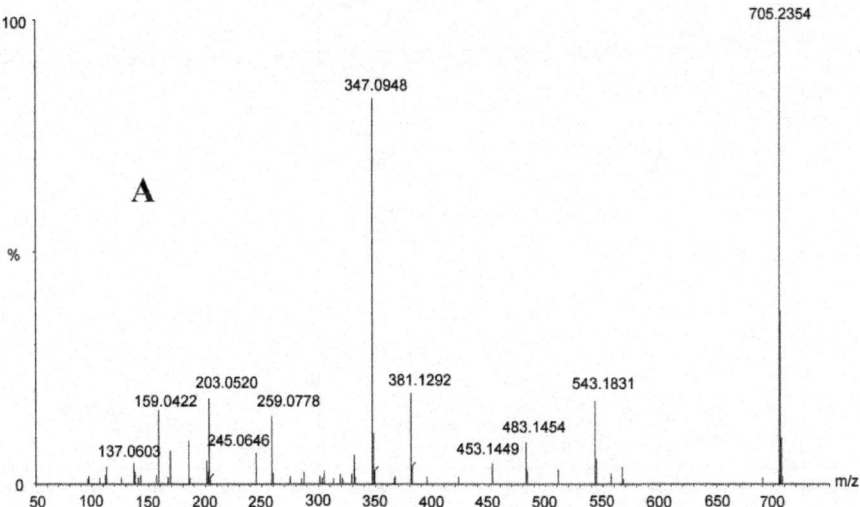

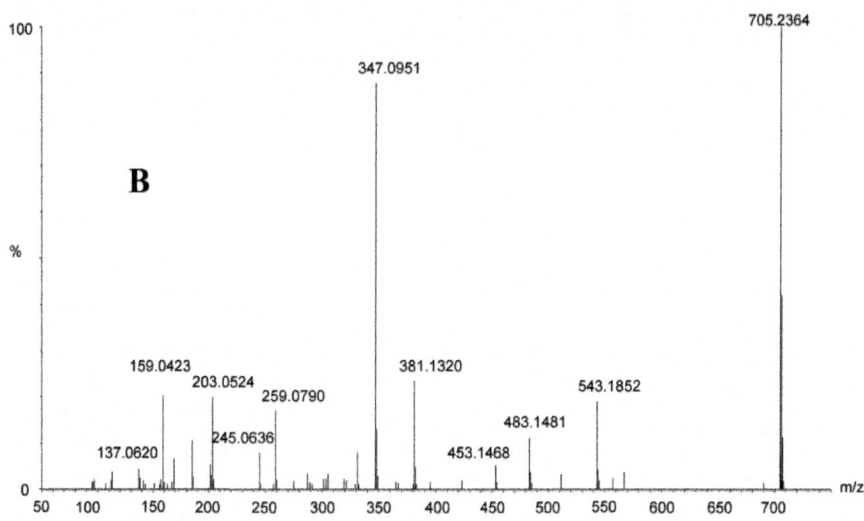

Figure S4. MS/MS spectrum of matairesinol 4'-O-β-gentiobioside (4) in the standard sample (A) and in the sample extracted from Caulis Trachelospermi (B), respectively.

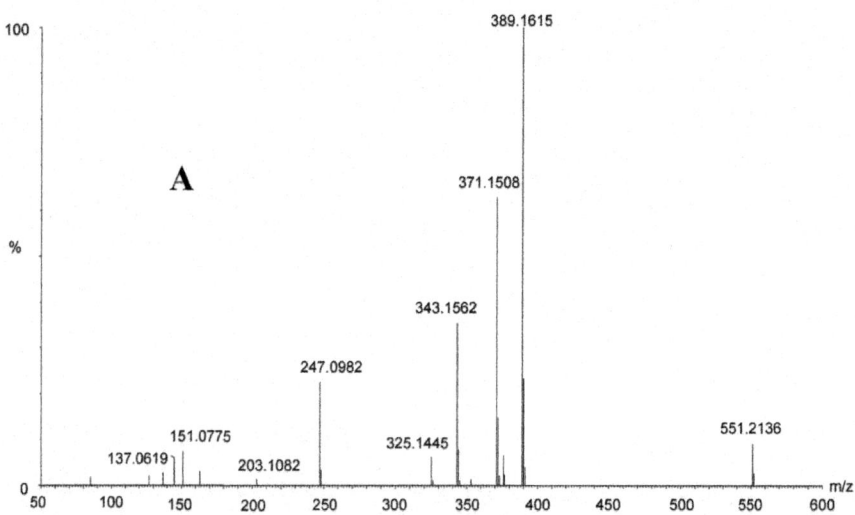

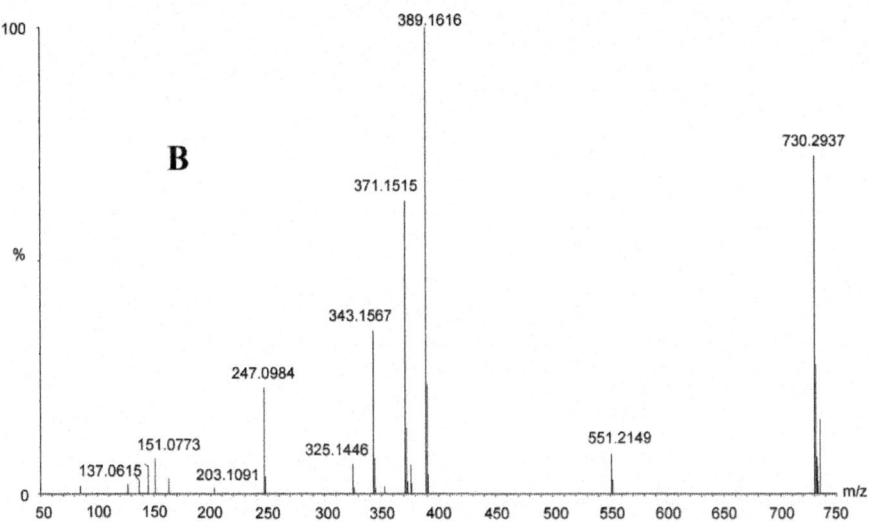

Figure S5. MS/MS spectrum of trachelogenin 4'-O-β-gentiobioside (5) in the standard sample (A) and in the sample extracted from Caulis Trachelospermi (B), respectively.

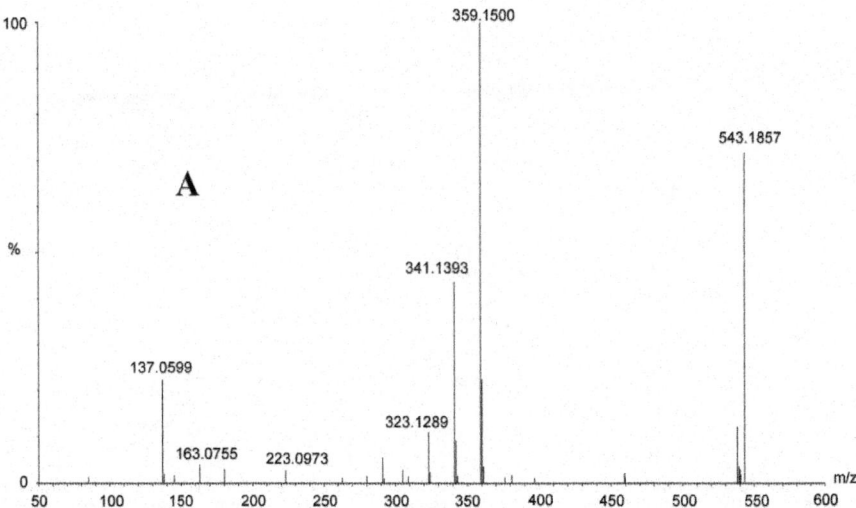

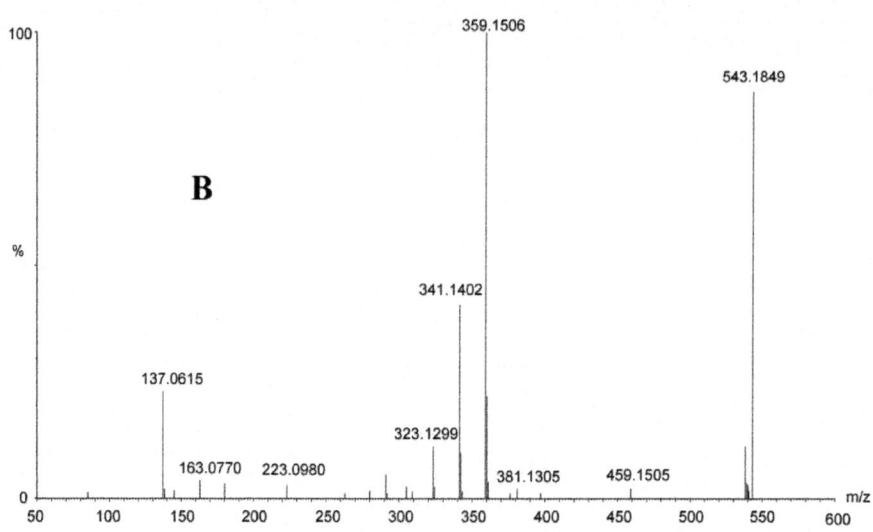

Figure S6. MS/MS spectrum of matairesinoside (6) in the standard sample (A) and in the sample extracted from Caulis Trachelospermi (B), respectively.

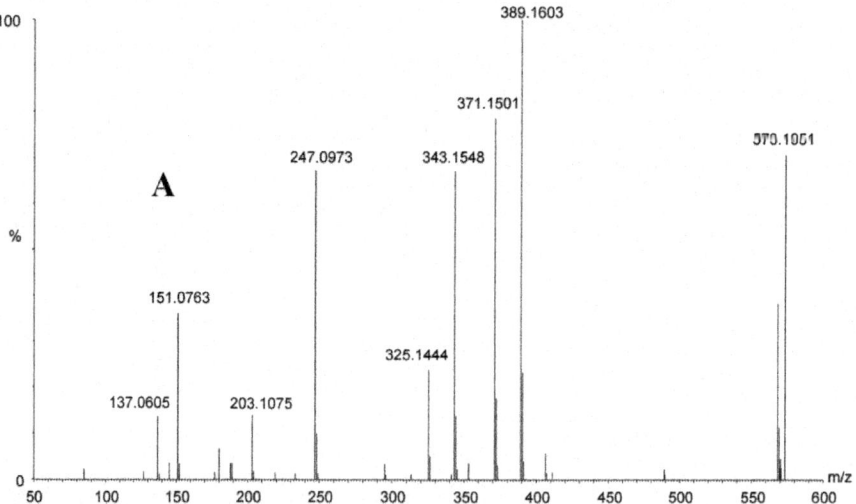

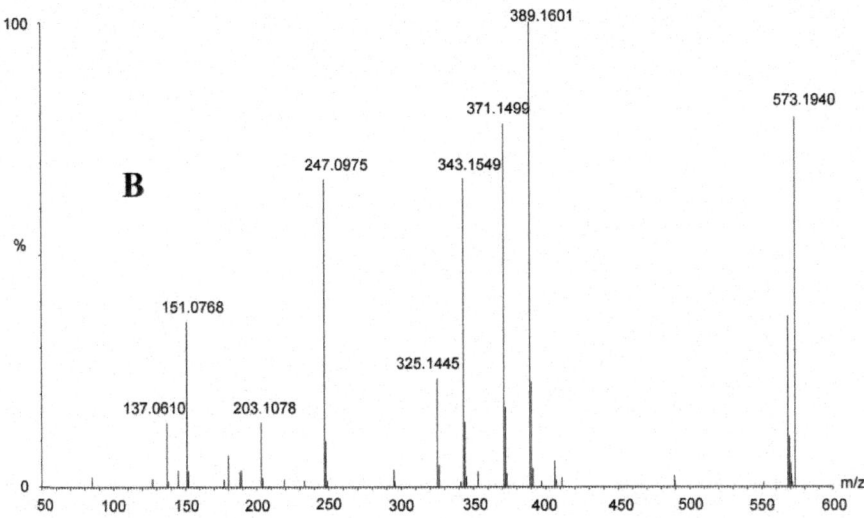

Figure S7. MS/MS spectrum of tracheloside (7) in the standard sample (A) and in the sample extracted from Caulis Trachelospermi (B), respectively.

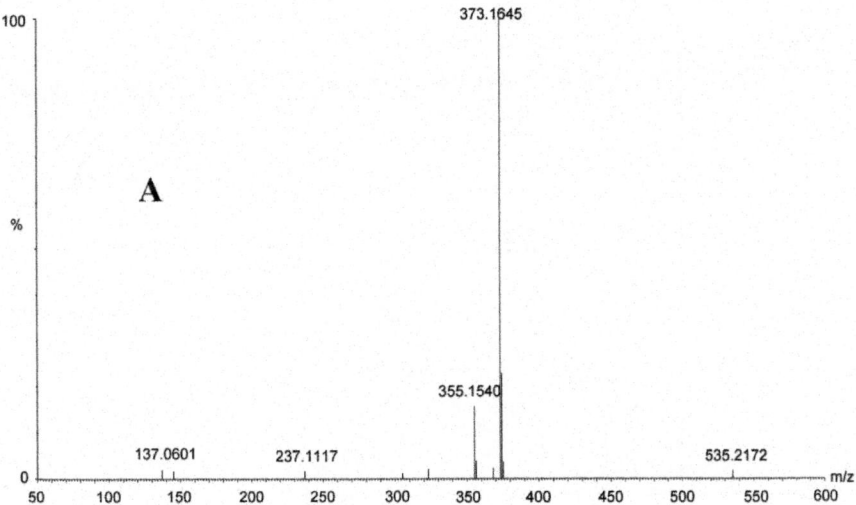

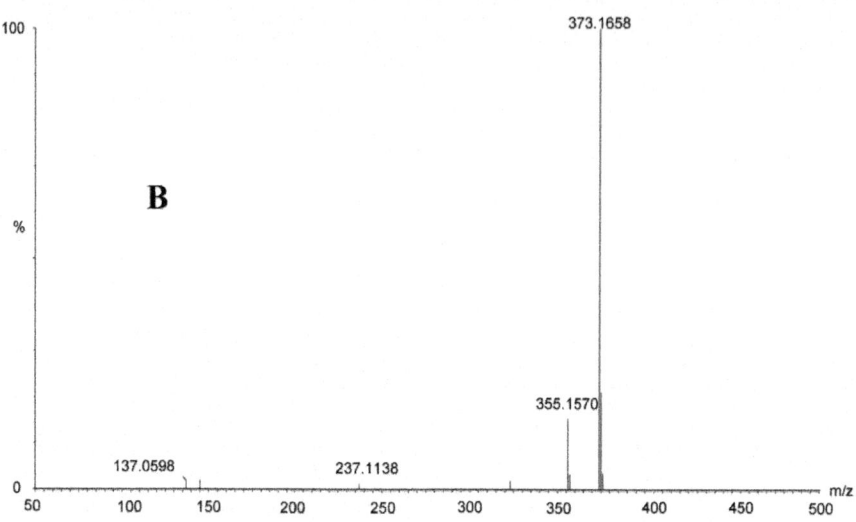

Figure S8. MS/MS spectrum of arctigenin 4'-O-β-gentiobioside (8) in the standard sample (A) and in the sample extracted from Caulis Trachelospermi (B), respectively.

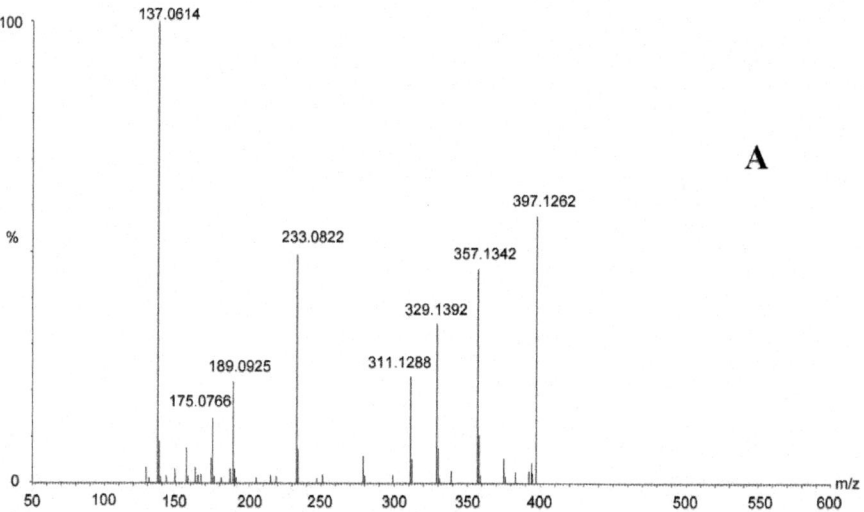

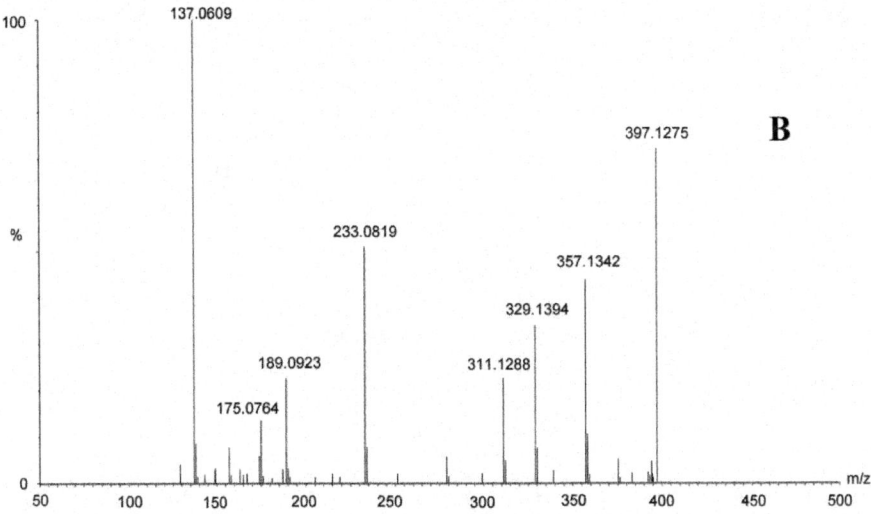

Figure S9. MS/MS spectrum of nortrachelogenin (9) in the standard sample (A) and in the sample extracted from Caulis Trachelospermi (B), respectively.

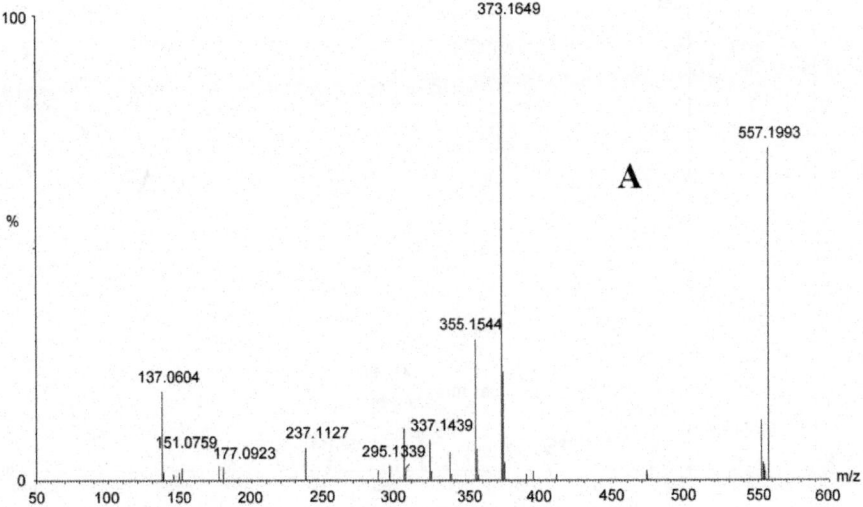

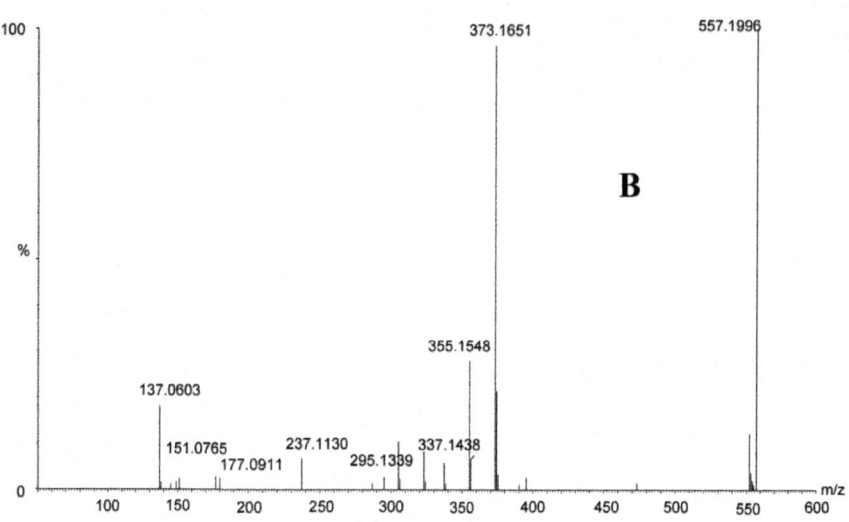

Figure S10. MS/MS spectrum of arctiin (10) in the standard sample (A) and in the sample extracted from Caulis Trachelospermi (B), respectively.

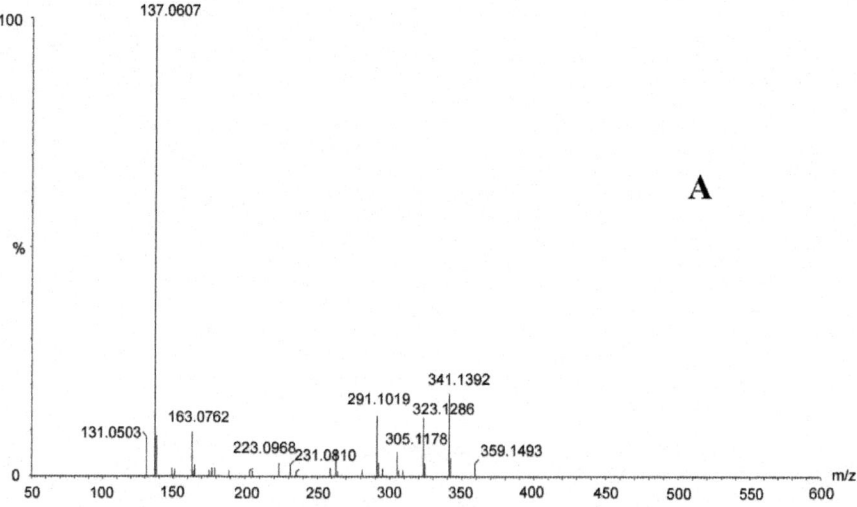

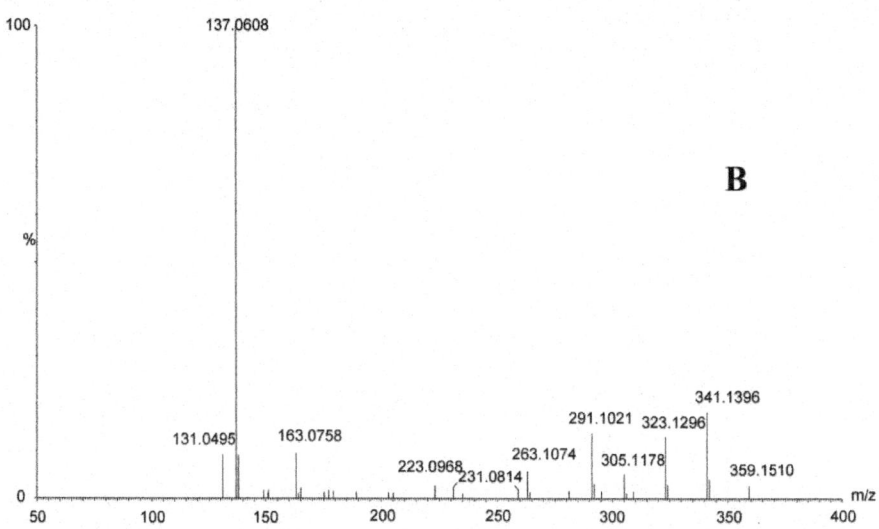

Figure S11. MS/MS spectrum of matairesinol (11) in the standard sample (A) and in the sample extracted from Caulis Trachelospermi (B), respectively.

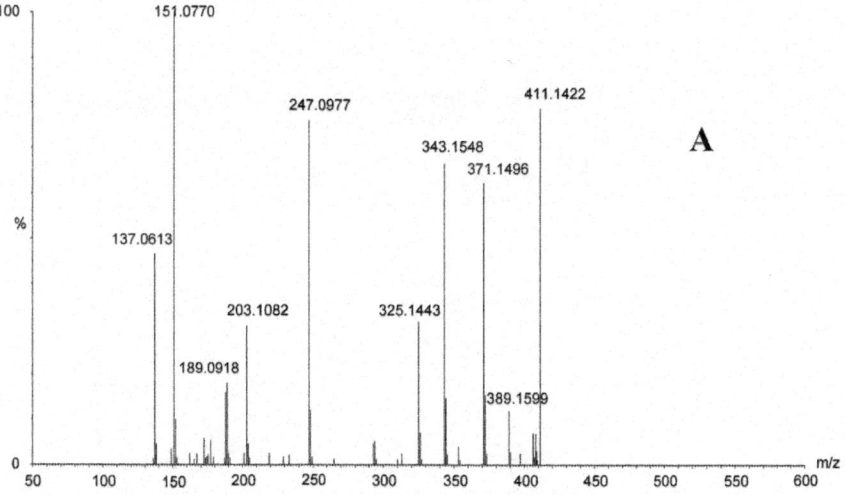

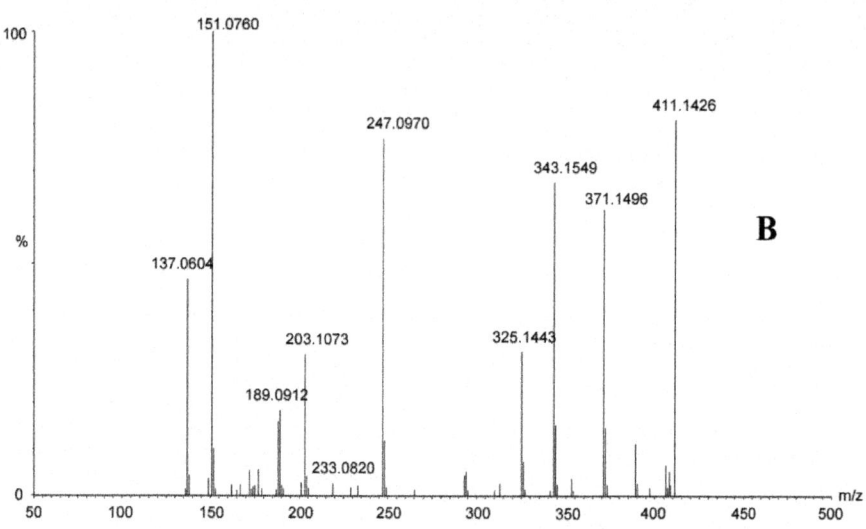

Figure S12. MS/MS spectrum of trachelogenin (12) in the standard sample (A) and in the sample extracted from Caulis Trachelospermi (B), respectively.

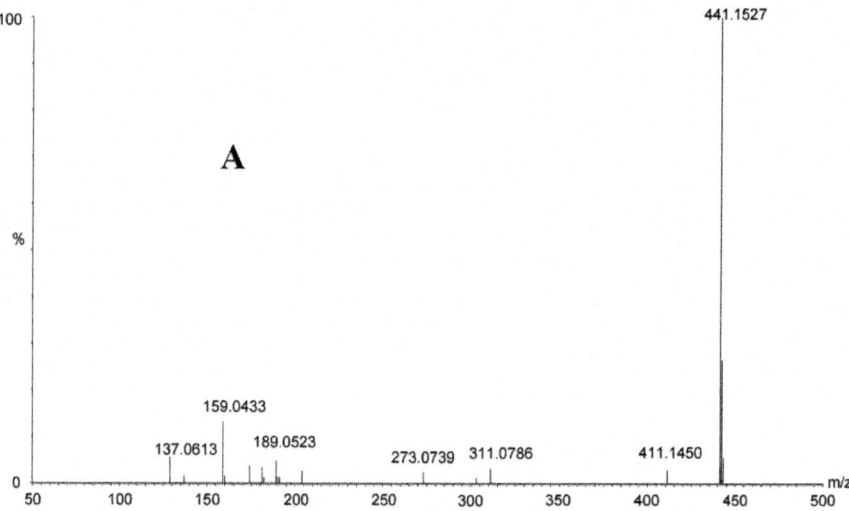

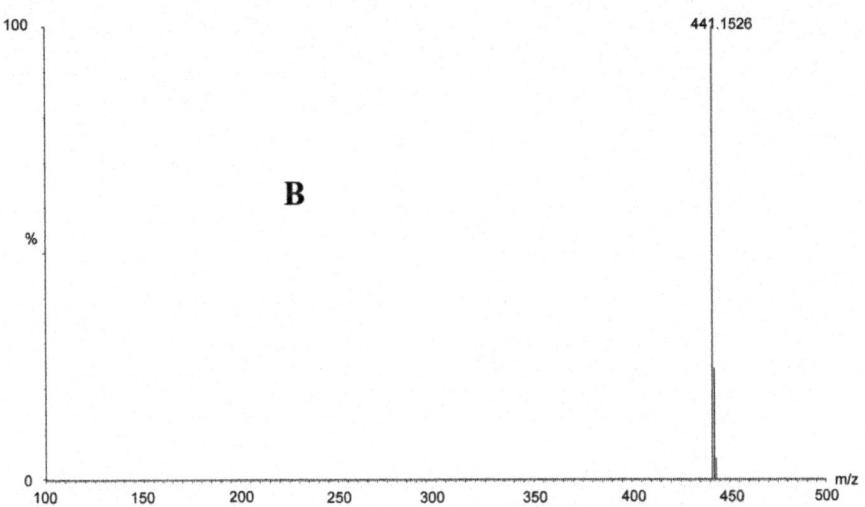

Figure S13. MS/MS spectrum of 5-methoxytrachelogenin (13) in the standard sample (A) and in the sample extracted from Caulis Trachelospermi (B), respectively.

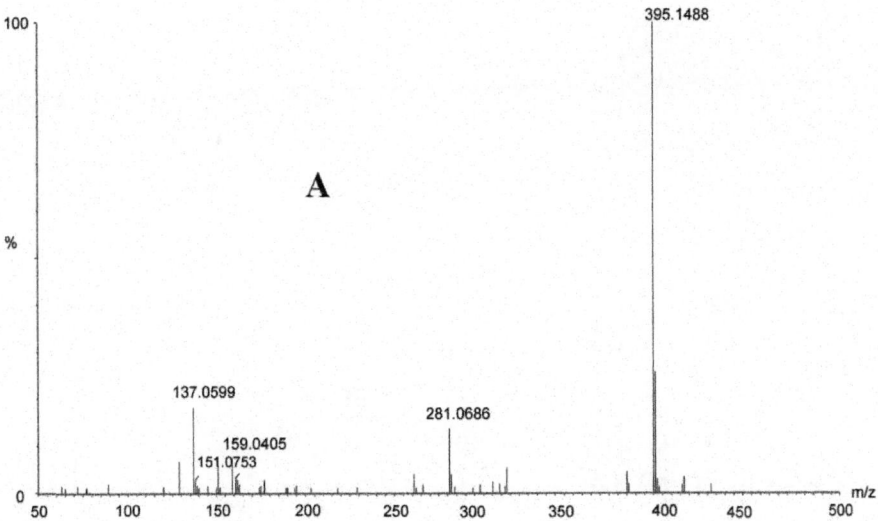

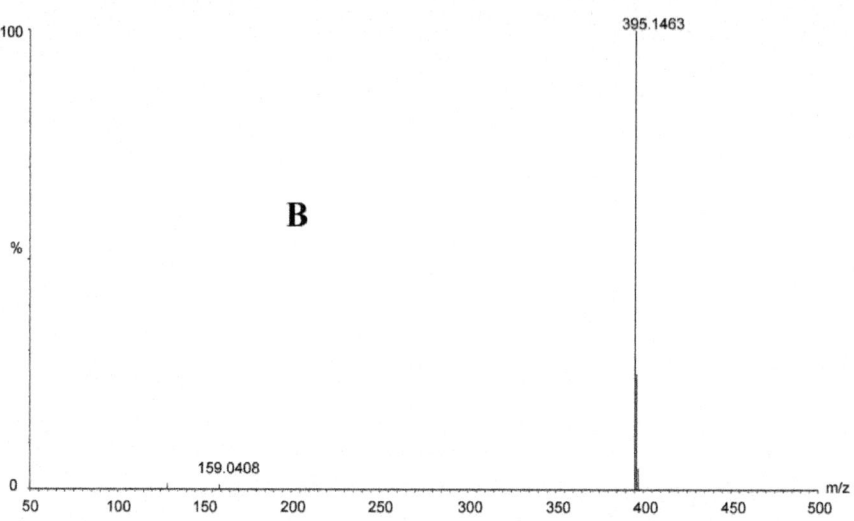

Figure S14. MS/MS spectrum of arctigenin (14) in the standard sample (A) and in the sample extracted from Caulis Trachelospermi (B), respectively.

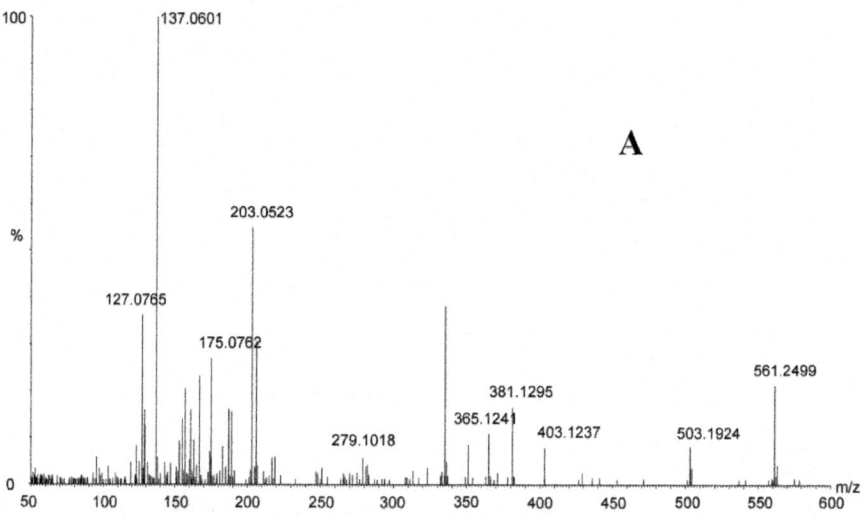

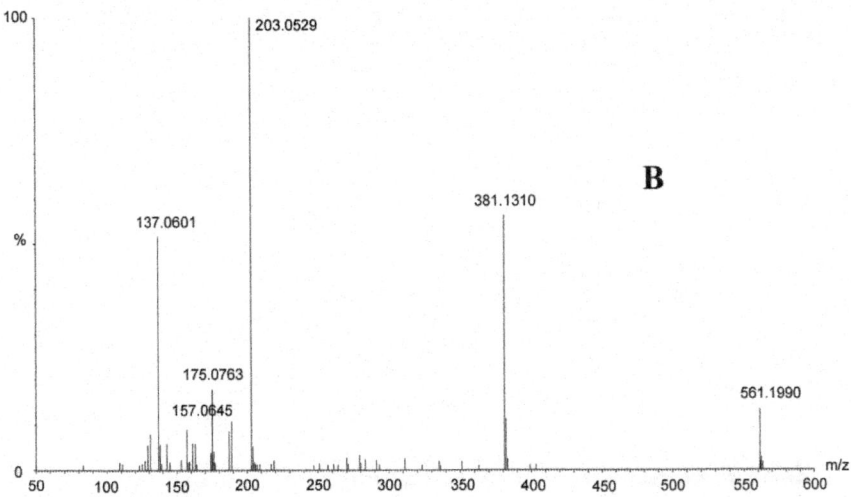

Figure S15. MS/MS spectrum of tanegoside A (17) in the standard sample (A) and in the sample extracted from Caulis Trachelospermi (B), respectively.

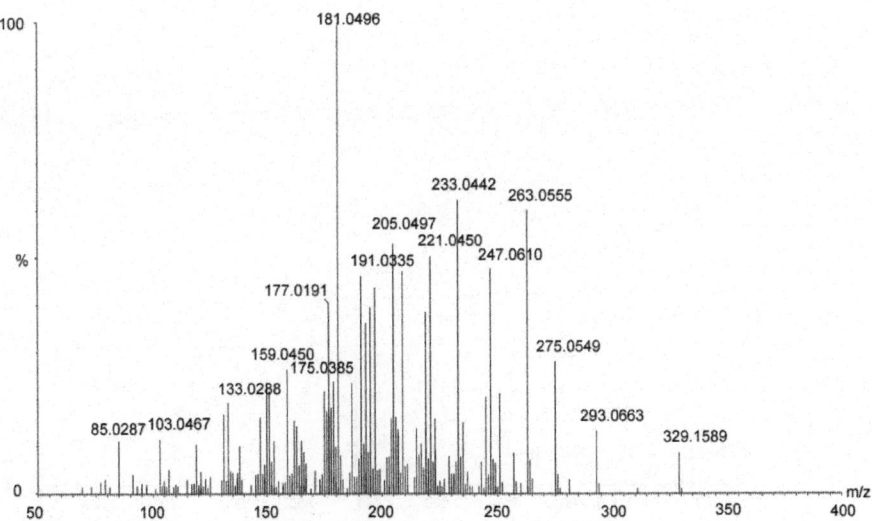

Figure S16. MS/MS spectrum of bergenin (15) in the sample extracted from Caulis Trachelospermi.

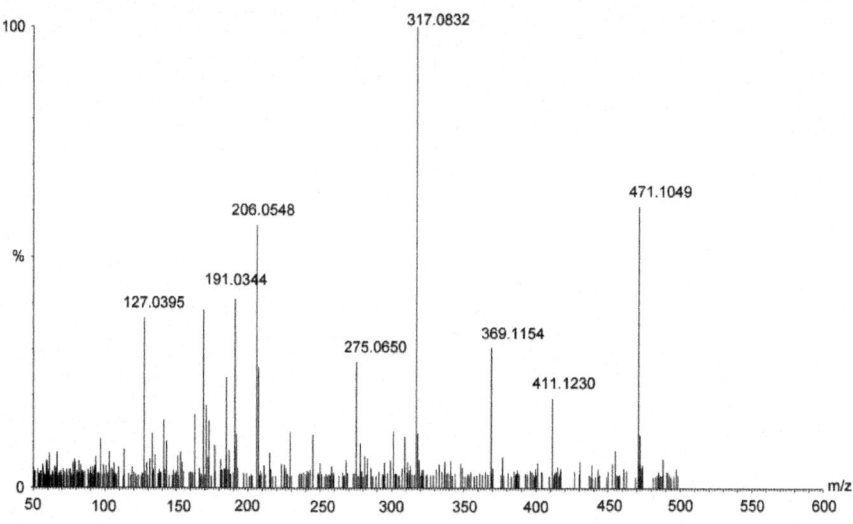

Figure S17. MS/MS spectrum of kelampayoside A (16) in the sample extracted from Caulis Trachelospermi.

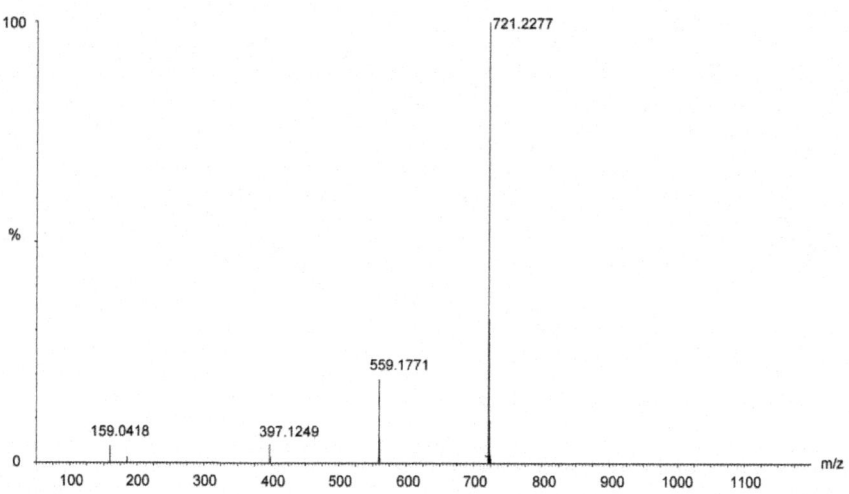

Figure S18. MS/MS spectrum of nortrachelogenin 4, 4'-di-O-β-D-glucoside (18) in the sample extracted from Caulis Trachelospermi.

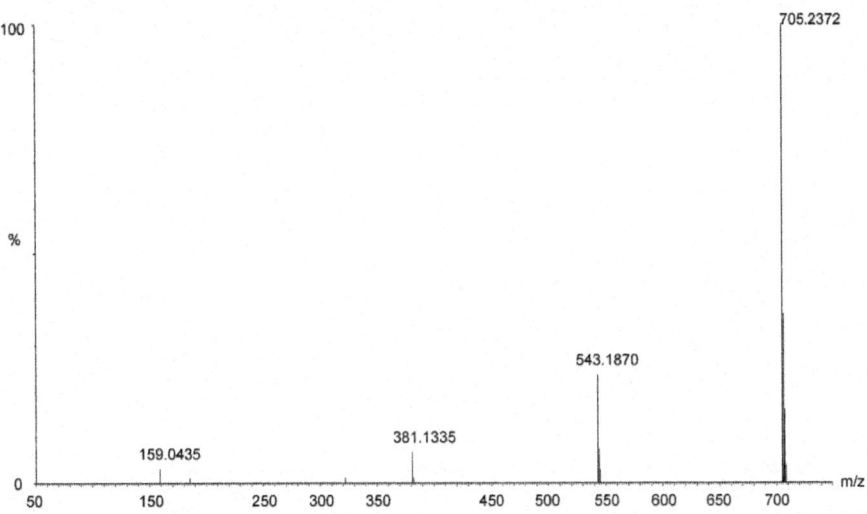

Figure S19. MS/MS spectrum of matairesinol 4, 4'-di-O-β-D-glucoside (19) in the sample extracted from Caulis Trachelospermi.

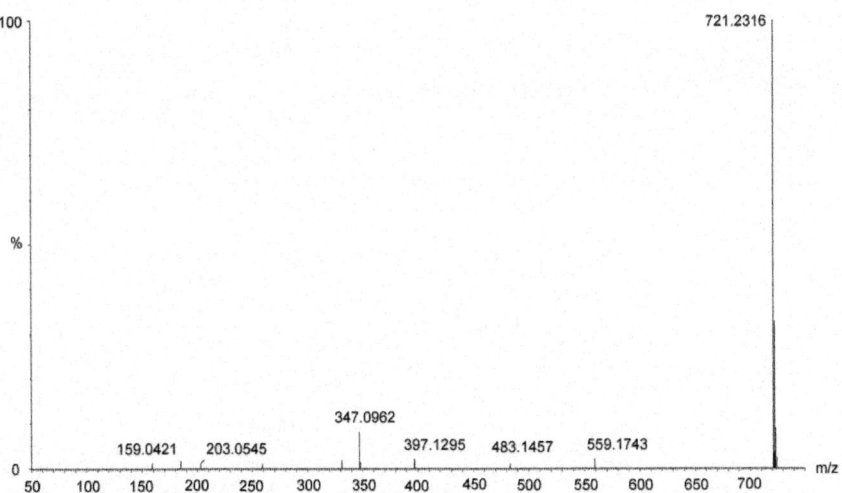

Figure S20. MS/MS spectrum of nortrachelogenin 4'-O-β-gentiobioside (20) in the sample extracted from Caulis Trachelospermi.

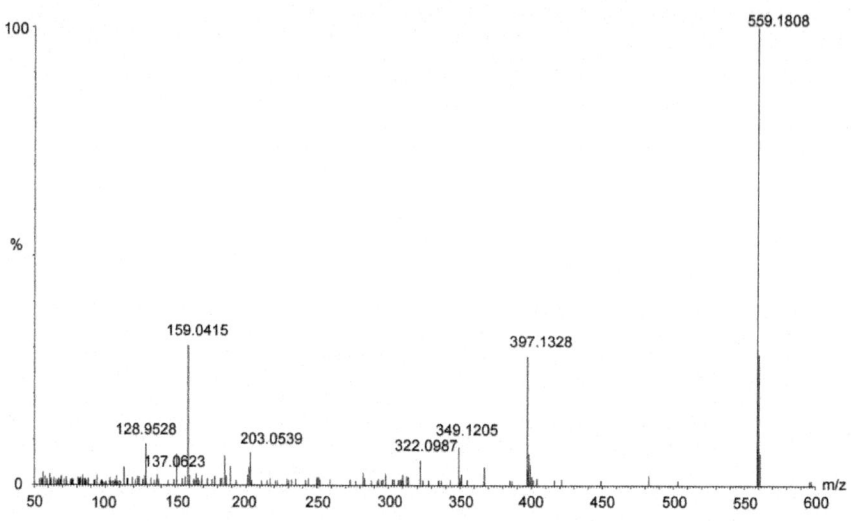

Figure S21. MS/MS spectrum of nortrachelogenin 4-O-β-D-glucoside (21) in the sample extracted from Caulis Trachelospermi.

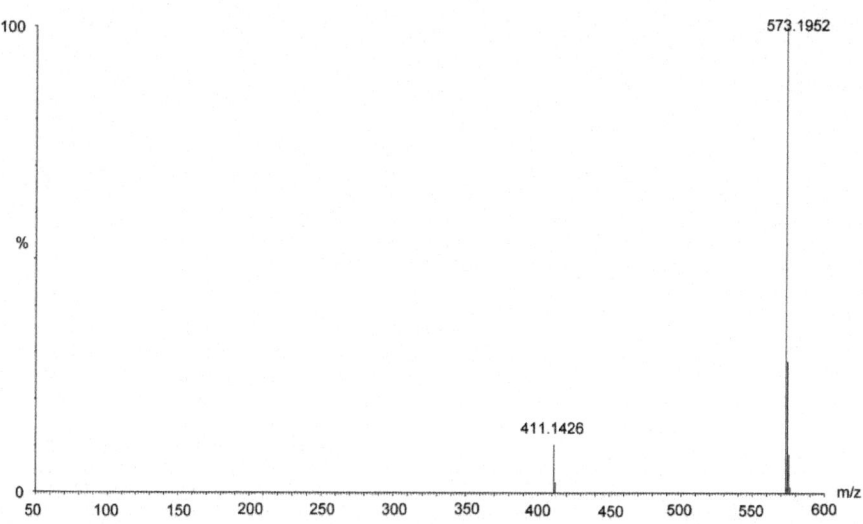

Figure S22. MS/MS spectrum of 4-demethyltraxillaside (22) in the sample extracted from Caulis Trachelospermi.

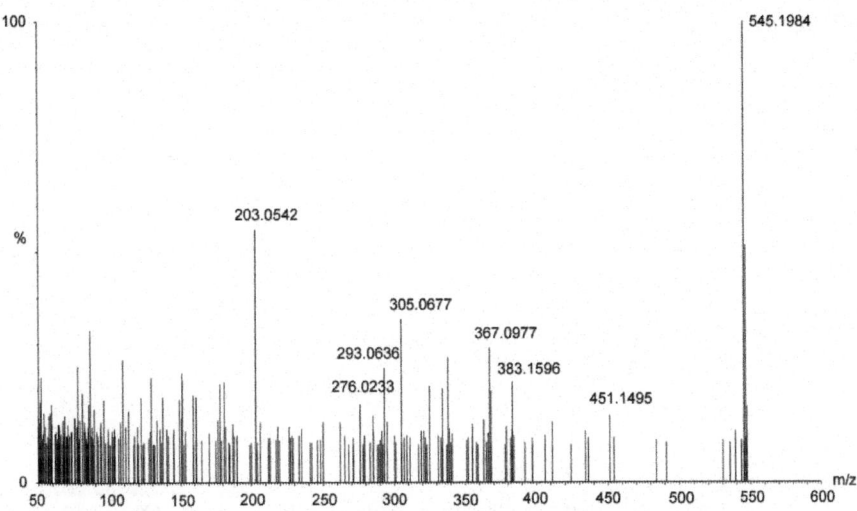

Figure S23. MS/MS spectrum of dihydrodehydrodiconiferyl alcohol-9-O-β-D-glucoside (23) in the sample extracted from Caulis Trachelospermi.

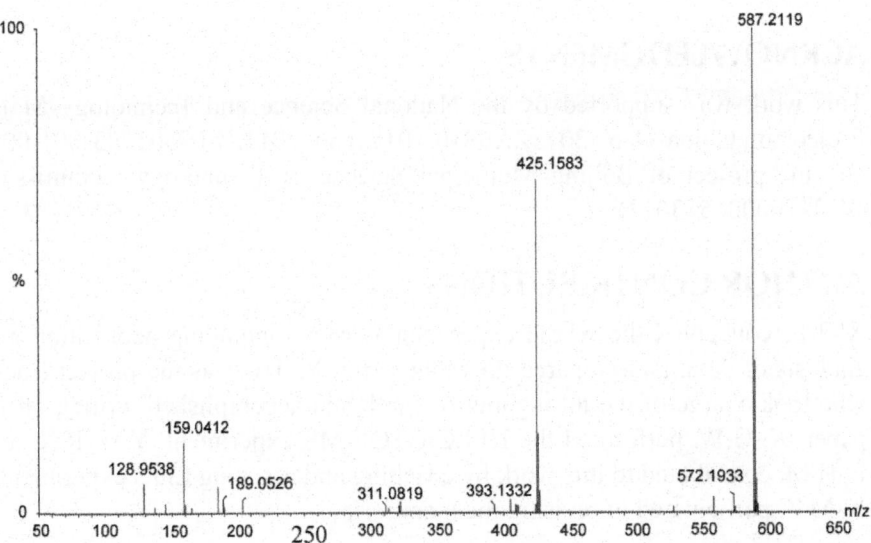

Figure S24. MS/MS spectrum of traxillageside (24) in the sample extracted from Caulis Trachelospermi.

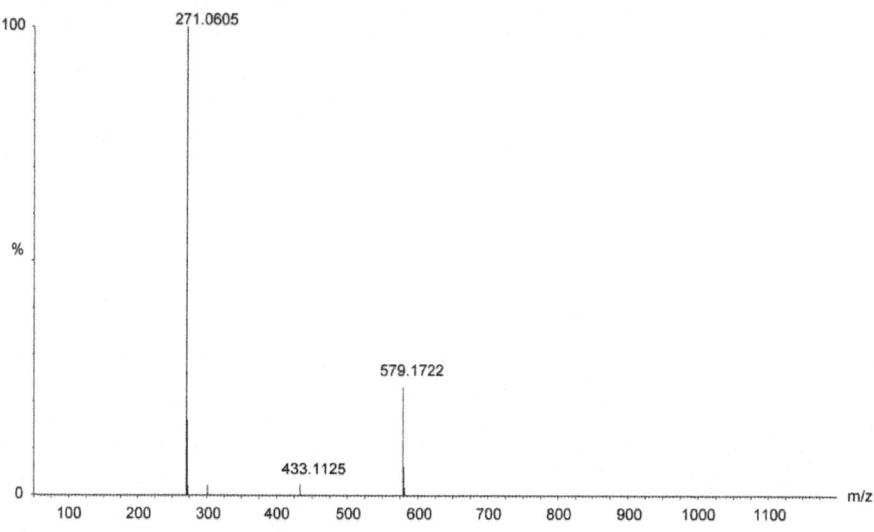

Figure S25. MS/MS spectrum of apigenin 7-O-β-neospheroside (25) in the sample extracted from Caulis Trachelospermi.

ACKNOWLEDGMENTS

This work was supported by the National Science and Technology Major Project of China (No. 2013ZX09102-018 and 2012ZX09301003-001-004) and the project of Beijing Municipal Science & Technology Commission (Z131100006513013).

AUTHOR CONTRIBUTIONS

X.-T.L. conceived the whole experimental work, including acquisition and analysis of data, and prepared the manuscript. N.-J.Y. was the project leader who took charge of the throughout research and accomplished writing of the paper. X.-G.W. performed the HPLC-QTOF-MS experiment. Y.Y., R.X. and F.-H.M. contributed to this work by coaching and assisting Liu's experiments. Y.-M.Z. participated in revising the manuscript.

REFERENCES

1. Committee of pharmacopoeia of the People's Republic of China, *Pharmacopoeia of the People's Republic of China (First Part)*; China Medical Science Technology Press: Beijing, China, 2010; p. 252.

2. Nishibe, S.; Han, Y.M. Chemical constituents from *Trachelosperomum*

jasminoides and its anticancer activity. *World Phytomed.* **2002**, *17*, 57–58.

3. Lee, M.H.; Lee, J.M.; Jun, S.H.; Ha, C.G.; Lee, S.-H.; Kim, N.M.; Lee, J.H.; Ko, N.Y.; Mun, S.H.; Park, S.H.; *et al. In vitro*and *in vivo* anti-inflammatory action of the ethanol extract of Trachelospermi Caulis. *J. Pharm. Pharmacol.* **2007**, *59*, 123–130.

4. Tan, X.Q.; Chen, H.S.; Liu, R.H.; Tan, C.H.; Xu, C.L.; Xuan, W.D.; Zhang, W.D. Lignans from *Trachelospermum jasminoides*. *Planta Med.* **2005**, *71*, 93–95.

5. Tan, X.Q.; Chen, H.S.; Zhou, M.; Zhang, Y. Triterpenoids from canes with leaves of *Trachelospermum jasminoides*. *Chin. Tradit. Herb. Drugs* **2006**, *37*, 171–174.

6. Tan, X.Q.; Guo, L.J.; Chen, H.S.; Wu, L.S.; Kong, F.F. Study on the flavonoids constituents of *Trachelospermum jasminoides*. *J. Chin. Med. Mater.* **2010**, *33*, 58–60.

7. Jing, L.; Yu, N.J.; Li, Y.S.; Fu, L.; Zhao, Y.M. Novel lignans from the stems and leaves of *Trachelospermum jasminoides*.*Chin. Chem. Lett.* **2011**, *22*, 1075–1077.

8. Zhu, C.C.; Jing, L.; Yu, N.J.; Yang, X.D.; Zhao, Y.M. A new lignan and active compounds inhibiting NF- B signaling pathway from Caulis Trachelospermi. *Acta Pharm. Sin. B* **2013**, *3*, 109–112.

9. Liu, X.T.; Wang, Z.X.; Yang, Y.; Wang, L.; Sun, R.F.; Zhao, Y.M.; Yu, N.J. Active components with inhibitory activities on IFN-γ/STAT1 and IL-6/STAT3 signaling pathways from Caulis Trachelospermi. *Molecules* **2014**, *19*, 11560–11571.

10. Li, X.X.; Wan, L.L.; Zhu, J.H.; Li, Y.; Zhang, J.P.; Guo, C. Determination of flavonoid aglycones in *Trachelospermum jasminoides* (Lindl.) Lem by HPLC. *China Pharm.* **2008**, *19*, 436–437.

11. Tan, X.Q.; Guo, L.J.; Kong, F.F. Determination of trachelogenin in *Trachelospermum jasminoides* (*Lindl.*) Lem. by HPLC.*Anhui Med. Pharm. J.* **2011**, *15*, 308–309.

12. Guo, L.; Tan, X.; Lu, P.; Wu, L.; Kong, F. Determination of salicylic acid in *Trachelospermum jasminoides* (*Lindl.*) Lem. by HPLC. *China Pharm.* **2009**, *12*, 1779–1781.

13. Kong, M.; Zhang, J.; Yao, N.; Li, Y.; Jiang, C.H.; Gao, M.; Fang, Z.J.; Liang, J.Y. Simultaneous determination of five compounds in *Trachelospermum jasminodes* by HPLC. *Chin. J. Exp. Tradit. Med. Formulae* **2013**, *19*, 93–96.

14. Zhu, S.; Tan, X.; Chen, H.; Lei, Y. Determination the content of tracheloside in *Trachelospermum jasminoides* (*Lindl.*) Lem. by RP-HPLC. *China Pharm.* **2005**, *8*, 1008–1009.

15. Fujimoto, T.; Nose, M.; Takeda, T.; Ogihara, Y.; Nishibe, S. Quantitative analysis of lignan components in Chinese crude drugs "Zihualuoshi" and "Luoshiteng". *Jpn. J. Pharmacogn.* **1993**, *47*, 218–221.

16. Li, Q.; Wang, Z.M.; Fu, X.T. Study on quality standards of trachelospermum jasminoides. *Northwest Pharm. J.* **2010**, *25*, 431–432.

17. Liu, M.P.; Yu, N.J.; Zhao, J.; Xu, B.; Zhao, Y.M. Quantitative Analysis of Total Lignans in the Lignan Extract from Trachelospermum jasminoides by Ultraviolet Spectrophotomey. *Med. Pharm. J. Chin. PLA* **2010**, *26*, 162–164.

18. Zhou, Q.; Fu, H.Y.; Bai, J.W.; Li, L. Determination of the Total Flavonoids in Chinese Starjasmine Stem by Spectrophotometry. *Lishizhen Med. Mater. Med. Res.* **2007**, *18*, 351–352.

19. Liu, Y.Q.; Yu, N.J.; Yang, X.D.; Zhao, Y.M. Study on HPLC fingerprint of *Trachelospermum jasminoides*. *China J. Chin. Mater. Med.* **2009**, *34*, 727–730.

20. Ferrer, I.; García-Reyes, J.F.; Mezcua, M.; Thurman, E.M.; Fernández-Alba, A.R. Multi-residue pesticide analysis in fruits and vegetables by liquid chromatography-time-of-flight mass spectrometry. *J. Chromatogr. A* **2005**, *1082*, 81–90.

21. Xu, S.Y.; Ye, M.L.; Xu, D.K.; Li, X.; Pan, C.S.; Zou, H.F. Matrix with high salt tolerance for the analysis of peptide and protein samples by desorption/ionization time-of-flight mass spectrometry. *Anal. Chem.* **2006**, *78*, 2593–2599.

22. Zhang, J.L.; Li, P.; Li, H.J.; Jiang, Y.; Ren, M.T.; Liu, Y. Development and validation of a liquid chromatography/electrospray ionization time-of-flight mass spectrometry method for relative and absolute quantification of steroidal alkaloids in Fritillaria species. *J. Chromatogr. A* **2008**, *1177*, 126–137.

23. Wu, H.; Guo, J.; Chen, S.; Liu, X.; Zhou, Y.; Zhang, X.; Xu, X. Recent developments in qualitative and quantitative analysis of phytochemical constituents and their metabolites using liquid chromatography-mass spectrometry. *J. Pharm. Biomed. Anal.* **2013**, *72*, 267–291.

24. Schmidt, T.J.; Alfermann, A.W.; Fuss, E. High-performance liquid chromatography/mass spectrometric identification of dibenzylbutyrolactone type lignans: insights into electrospray ionization tandem mass spectrometric fragmentation of lign-7-eno-9,9'-lactones

and application to the lignans of *Linum usitatissimum* L. (Common Flax). *Rapid Commun. Mass Spectrom.* **2008**, *22*, 3642–3650.

25. Schmidt, T.J.; Hemmati, S.; Fuss, E.; Alfermann, A.W. A combined HPLC-UV and HPLC-MS method for the identification of lignans and its application to the lignans of *Linum usitatissimum* L. and *L. bienne* mill. *Phytochem. Anal.* **2006**, *17*, 299–311.

26. Jing, L.; Yu, N.J.; Zhao, Y.M.; Li, Y.S. Trace chemical constituents contained in *Trachelospermum jasminoides* and structure identification. *China J. Chin. Mater. Med.* **2012**, *37*, 1581–1585.

27. Yuan, Q.S.; Yu, N.J.; Zhao, Y.M.; Xu, B.; Yao, Z.W. Chemical constituents from *Trachelospermum jasminoides*. *Chin. Tradit. Herb. Drugs* **2010**, *41*, 179–181.

28. Yu, N.J.; Zhao, Y.M.; Ren, F.X. Total Lignans Extract from Caulis Trachelospermi, Its Extraction Method, and the Medicinal Usage of the Extract and Its Active Constituents. CN 200510093357.X, 26 August 2005.

Chapter 2

QUANTITATIVE PROTEOMIC ANALYSIS OF THE HFQ-REGULON IN SINORHIZOBIUM MELILOTI 2011

Patricio Sobrero[1], Jan-Philip Schlu¨ter[2,4], Ulrike Lanner[3], Andreas Schlosser[3,5], Anke Becker[2,4], Claudio Valverde[1]

[1] Laboratorio de Bioquı´mica, Microbiologı´a e Interacciones Biolo´gicas en el Suelo, Departamento de Ciencia y Tecnologı´a, Universidad Nacional de Quilmes, Buenos Aires, Argentina

[2] Institute of Biology III, Faculty of Biology, University of Freiburg, Freiburg, Germany

[3] Core Facility Proteomics, Center for Biological Systems Analysis (ZBSA), Freiburg, Germany

[4] LOEWE Center for Synthetic Microbiology, Marburg, Germany

[5] Rudolf Virchow Center, University of Wuerzburg, Wuerzburg, Germany

ABSTRACT

Riboregulation stands for RNA-based control of gene expression. In bacteria, small non-coding RNAs (sRNAs) are a major class of riboregulatory elements, most of which act at the post-transcriptional level by base-pairing target mRNA genes. The RNA chaperone Hfq facilitates antisense interactions between target mRNAs and regulatory sRNAs, thus influencing mRNA stability and/or translation rate. In the α-proteobacterium *Sinorhizobium meliloti* strain 2011, the identification and detection of multiple sRNAs genes and the broadly pleitropic phenotype associated to the absence of a functional Hfq protein both support the existence of riboregulatory circuits controlling gene expression to ensure the fitness of this bacterium in both free living and symbiotic conditions. In order to identify target mRNAs subject to Hfq-dependent riboregulation, we have compared the proteome of an *hfq* mutant and the wild type *S. meliloti* by quantitative proteomics following protein labelling with ^{15}N. Among 2139 univocally identified proteins, a total of 195 proteins showed a differential abundance between the Hfq mutant and the wild type strain; 65 proteins accumulated $\geq$2-fold whereas 130 were downregulated ($\leq$0.5-fold) in the absence of Hfq. This profound proteomic impact implies a major role for Hfq on regulation of diverse physiological processes in *S. meliloti*, from transport of small molecules to homeostasis of iron and nitrogen. Changes in the cellular levels of proteins involved in transport of nucleotides, peptides and amino acids, and in iron homeostasis, were confirmed with phenotypic assays. These results represent the first quantitative proteomic analysis in *S. meliloti*. The comparative analysis of the *hfq* mutant proteome allowed identification of novel strongly Hfq-regulated genes in *S. meliloti*.

INTRODUCTION

In bacteria, regulation of gene expression based on small non-coding RNA molecules (sRNAs) has emerged and consolidated as fundamental post-transcriptional checkpoints, usually involved in the response to environmental stress to maintain cell homeostasis [1]. In particular, trans-encoded sRNAs have the ability to establish imperfect complementary interactions with one or more mRNAs, thus affecting their translation rate and/or stability [2]. The activity of the RNA chaperone Hfq is often required to facilitate such interaction *in vivo* [3], [4]. Originally discovered as a host factor necessary for the replication of the coliphage Qβ [5], Hfq has been recognized as a key player in riboregulatory processes and a limiting factor for sRNA action [6]. The absence of this RNA-binding protein alters the steady concentration and/ or stability of cellular sRNAs, leading to an abrogate response in the regulation of gene expression under certain conditions [7], [8].

The biological role of Hfq has been explored in diverse eubacterial taxons, namely α-proteobacteria [9], [10], β-proteobacteria [11], [12], γ-proteobacteria [13], [14] and Gram positives [15]. In the α-proteobacterium *Sinorhizobium meliloti* strain 2011, the absence of *hfq*promotes a pleiotropic phenotype [16], outlining the critical role of this protein in *S. meliloti*physiology. This microorganism is of particular interest because of its endosymbiotic association with the roots of legumes belonging to the *Medicago-Melilotus-Trigonella* tribe, in which these bacteria induce and colonize root nodule organs specifically devoted to nitrogen fixation [17], [18]. Some of the phenotypes observed in the *hfq* minus background of strain 2011 have been described in the related *S. meliloti* strain 1021, whose proteomic profile has been explored by 2D-PAGE [19]. Indeed, an *hfq* mutant serves to identify possible target mRNAs subject to direct or indirect Hfq-dependent riboregulatory processes [13], [20], [21].

Since its introduction in 2002 [22], the metabolic labelling strategy that uses stable isotopes in the growth medium has shown its potential for the determination of highly accurate quantitative proteomic profiles when coupled with high performance liquid chromatography and mass spectrometry [23], [24]. Still, this technical approach remains mostly unexplored in bacteria: only a few projects considered metabolic labelling with stable isotopes coupled with LC/MS [25],[26], [27] compared with the huge number of proteomic studies based on comparative 2D-PAGE profiles. In most cases, Hfq-dependent riboregulation is not an "all or nothing" process, and contributes to the fine tuning of target gene expression leading to modest regulatory factors. In this context, and due to its documented accuracy and sensitivity, high-throughput quantitative proteomic approaches that make use of metabolic labelling with stable isotopes, provide an excellent platform to quantify mild changes in gene expression at the protein level[24]. We here present the results of the first comparative quantitative proteomic analysis in the α-proteobacterial legume symbiont *S. meliloti* that made use of stable protein labelling with ^{15}N. The study aimed to quantify the impact of knocking out the RNA chaperone Hfq on the proteome of *S. meliloti* strain 2011.

MATERIALS AND METHODS

Bacterial Strains and Culture Conditions

In this work, *S. meliloti* 2011 [28] was used as the wild type strain. Strain 20PS01 is an isogenic mutant that bears an in-frame deletion of 63 bp within the central region of the *hfq* gene [16]. *S. meliloti* was cultured at 28°C in tryptone-yeast extract (TY; in g l^{-1}: tryptone, 5; yeast extract, 3, CaCl$_2$, 0.7) or in MOPS-buffered defined medium (MDM) [29]. When required, streptomycin was added to the growth medium at 400 μg/ml. For cytotoxicity growth assays, 5-fluorouracil (Sigma-Aldrich, USA), sodium glufosinate (BASTA, Bayer Crop Science, Argentina) or Bialaphos (Toku-E, USA) were added to the growth medium at the concentrations indicated in the text. In all cases, growth was monitored by measuring OD$_{600}$ in cultures shaken at 120 rpm. For each test, three independent cultures were analysed and the experiments were repeated twice with similar results.

^{15}N Isotopic Labelling of *S. meliloti* Cellular Proteins, and Preparation of Protein Samples

S. meliloti strains 2011 and 20PS01 were grown in 100 ml of MDM medium containing 0.1% w/v of ^{14}NH$_4$Cl, or ^{15}NH$_4$Cl, as the only nitrogen source. Cells were harvested at exponential phase by centrifugation at 5000×*g* for 10 min at

4°C. Proteins from harvested cells were separated into cytoplasmic, periplasmic and membrane fractions, as described in [30]. Cells were resuspended in 2 ml of TEX buffer [50 mM Tris/HCl (pH 8.0), 3 mM EDTA, 0.1% Triton X-100], incubated on ice for 45 min, centrifuged and washed with 2 ml of TEX buffer. The supernatants of both centrifugation steps were pooled, and kept as the 'periplasmic fraction'. The remaining pellet was resuspended in 4 ml of 10 mM $MgCl_2$, 50 mg ml^{-1} DNase A, 50 mg ml^{-1} RNase I, 20 mM Tris/HCl, pH 8.0, and incubated on ice for 30 min. After three passages through a French press at 20.000 psi, the resulting extract was freed of unbroken cells by centrifugation at 2000×g for 2 min. Broken cells were centrifuged at 160.000×g for 1.5 h at 4°C, yielding the supernatant as cytoplasmic fraction and the pellet as membrane fraction. Total protein content was quantified with the Bradford method [31]. The quality of protein fractions was verified by SDS-PAGE [32].

Quantitative LC-MS Proteomic Analysis

Protein separation and digestion.

The three fractions were reduced with DTT (50 mM, 10 min, 70°C), alkylated with iodoacetamide (120 mM, 20 min, room temperature) and separated by SDS-PAGE (Invitrogen, NuPAGE 4–12% Bis-Tris gels). The gels were stained with Simply Blue (Invitrogen). Each lane was cut into 24 bands. For in-gel digestion the excised gel bands were destained with 30% ACN, shrunk with 100% ACN, and dried in a Vacuum Concentrator (Concentrator 5301, Eppendorf, Hamburg, Germany). Digests with trypsin were performed overnight at 37°C in 0.05 M NH_4HCO_3 (pH 8.0). About 0.1 µg of protease was used for one gel band. Peptides were extracted from the gel slices with 5% formic acid.

LC-MS/MS Analysis

LC-MS/MS analysis were performed on a Q-TOF mass spectrometer (Agilent 6520, Agilent Technologies) coupled to an 1200 Agilent nanoflow system via a HPLC-Chip cube ESI interface. Peptides were separated on a HPLC-chip with an analytical column of 75 µm i.d. and 150 mm length and a 40-nL trap column, both packed with Zorbax 300SB C-18 (5 µm particle size). Peptides were elutes with a linear acetonitrile gradient with 1%/min at flow rate of 300 nL/min (starting with 3% of acetonitrile).

The Q-TOF was operated in the 2 Ghz extended dynamic range mode. MS/MS analyses were performed using data-dependent acquisition mode. After a MS scan (2 spectra/s), a maximum of three peptides were selected for MS/

MS (2 spectra/s). Singly charged precursor ions were excluded from selection. Internal calibration was applied using two reference masses.

Protein Identification

Mascot Distiller 2.3 was used for raw data processing and quantitation, essentially with standard settings for the Agilent Q-Tof. Mascot Server 2.3 was used for database searching with the following parameters: peptide mass tolerance: 20 ppm, MS/MS mass tolerance: 0.05 Da, enzyme: "trypsin" with 2 uncleaved sites allowed, variable modifications: Carbamidomethyl (C), Gln->pyroGlu (N-term. Q), oxidation (M). A custom made database containing all UniProt entries for the taxonomy *Sinorhizobium meliloti* (Swiss-Prot and TrEMBL) was used. All protein sequences of this target database were reversed, and the resulting decoy database was concatenated with the target database. On average the calculated protein false discovery rate (FDR) was 2%. The resulting lists of identified proteins were filtered by excluding all protein identifications with only one peptide.

Protein Quantitation

The individual L/H ratios for all peptides and the L/H ratios for all proteins have been calculated using Mascot Distiller Quantitation Toolbox (Matrix Science, London, UK). 98 atom % ^{15}N incorporation was used to calculate the theoretical isotopic patterns of the heavy peptides. Only peptides with a Mascot peptide score equal or higher as the Mascot homology threshold have been used for quantitation. Proteins with less than 2 (or 3) peptides were excluded from quantitation. Mass time matches have been allowed, so that the identification of one peptide version (heavy or light) is sufficient for L/H ratio calculation. In addition, the following quality filters have been applied for excluding low quality data from quantitation: correlation threshold (matched rho) 0.7 and standard error threshold (elution profile correlation threshold) 0.1. Protein L/H ratios have been calculated as the median of the corresponding peptide ratios.

Glutamine Synthetase Activity

Total GS activity was measured by the incorporation of hydroxylamine to γ-glutamyl hydroxamate from *S. meliloti* cell extract as described in [33], [34]. The GS unit was defined as the amount of enzyme that catalyses the production of 1 μmol γ-glutamyl hydroxamate per min. The cellular protein content was estimated with the Bradford Method [31].

Siderophore Production

Siderophore production was measured in the supernatant of conditioned cultures in MDM medium with different iron supply, colorimetrically with the CAS method [35].

Intracellular Iron Content

Cellular iron content was estimated with the commercial Liquid Fer-color AA kit (Wiener lab, Argentina). An equivalent of 20 OD_{600} of *S. meliloti* cells were harvested at exponential phase from cultures grown in MDM medium. Cell pellets were resuspended in 1 ml of saline solution (NaCl 0.85% w/v) and lysed by sonication. 100 µl of the cell extracts were used to estimate total protein content with the Bradford method, and 200 µl of the clarified supernatants were used for iron colorimetry.

RESULTS AND DISCUSSION

The Absence of *hfq* Promotes Profound Changes in the *S. meliloti* Proteome

In this study, we aimed to quantitatively characterize changes in the LC/MS protein profile of an *hfq* mutant derived from *S. meliloti* strain 2011 [16], after metabolic labelling of cellular proteins with stable ^{15}N. In strain 2011, the absence of *hfq* generates a pleiotropic phenotype [16], which could be a consequence of a major proteome remodelling. This was evident as a growth deficiency in the defined MDM medium (Figure 1). For this reason, appropriate cell densities were chosen to ensure that cultures of wild type and *hfq* mutant cells were at a comparable growth phase (Figure 1). The quantitative proteomic analysis was carried out on protein extracts from exponential phase cultures growing in MDM medium containing $^{15}NH_4Cl$ as the only source of N. The efficiency of ^{15}N incorporation was examined in the wild type strain 2011. After 6 generations in MDM medium with 0.1% (w/v) $^{15}NH_4Cl$ all identified proteins showed an incorporation of 98% ^{15}N.

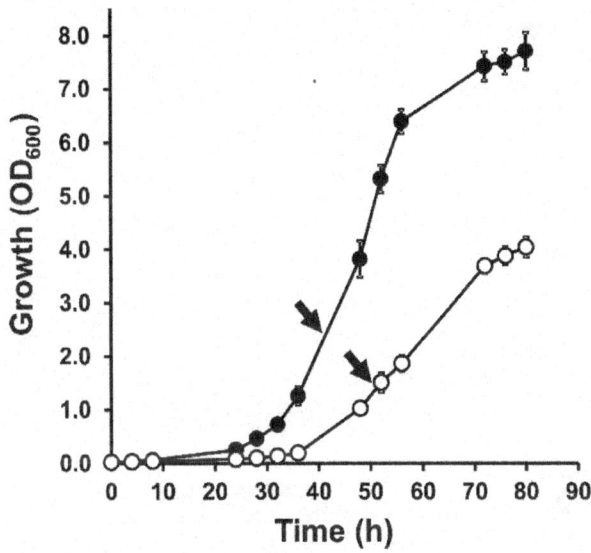

Figure 1. The absence of Hfq results in deficient growth in MDM defined medium. (•), wild type strain 2011; (○), Δ*hfq* mutant strain 20PS01. Each curve represents the average from three different cultures ± SD. The experiment was repeated twice, with essentially the same results. Arrows point to the growth stage in which cells were harvested for comparative quantitative analysis of ^{15}N-labelled proteins.

In order to maximize the chances to identify regulated proteins, we carried out a subcellular protein fractionation, resulting in three different comparative protein profiles: cytoplasmic-, membrane- and periplasm-enriched fractions [30]. We could identify a total of 2139 unique proteins among the three fractions, which represented 34% of all the protein coding genes of *S. meliloti* 1021 genome [36]. In the cytoplasmic-enriched fraction, 833 proteins were quantified out of 1183 identified proteins. In the membrane-enriched fraction, 932 proteins were quantified out of 1276 identified proteins. Finally, in the periplasm-enriched fraction, 969 unique proteins were quantified out of 1258 identified proteins. Three biological replicates were analyzed. In two of these replicates, the wild type strain *S. meliloti* 2011 was grown in the heavy nitrogen source, whereas the *hfq* mutant strain was grown in ^{15}N in the third replicate. The high degree of correlation in the relative abundance of the quantified proteins discards any effect of the heavy nitrogen source on the proteomic profile. An important number of unique proteins was identified and quantified in only one fraction, supporting the subcellular fractionation strategy (Figure 2).

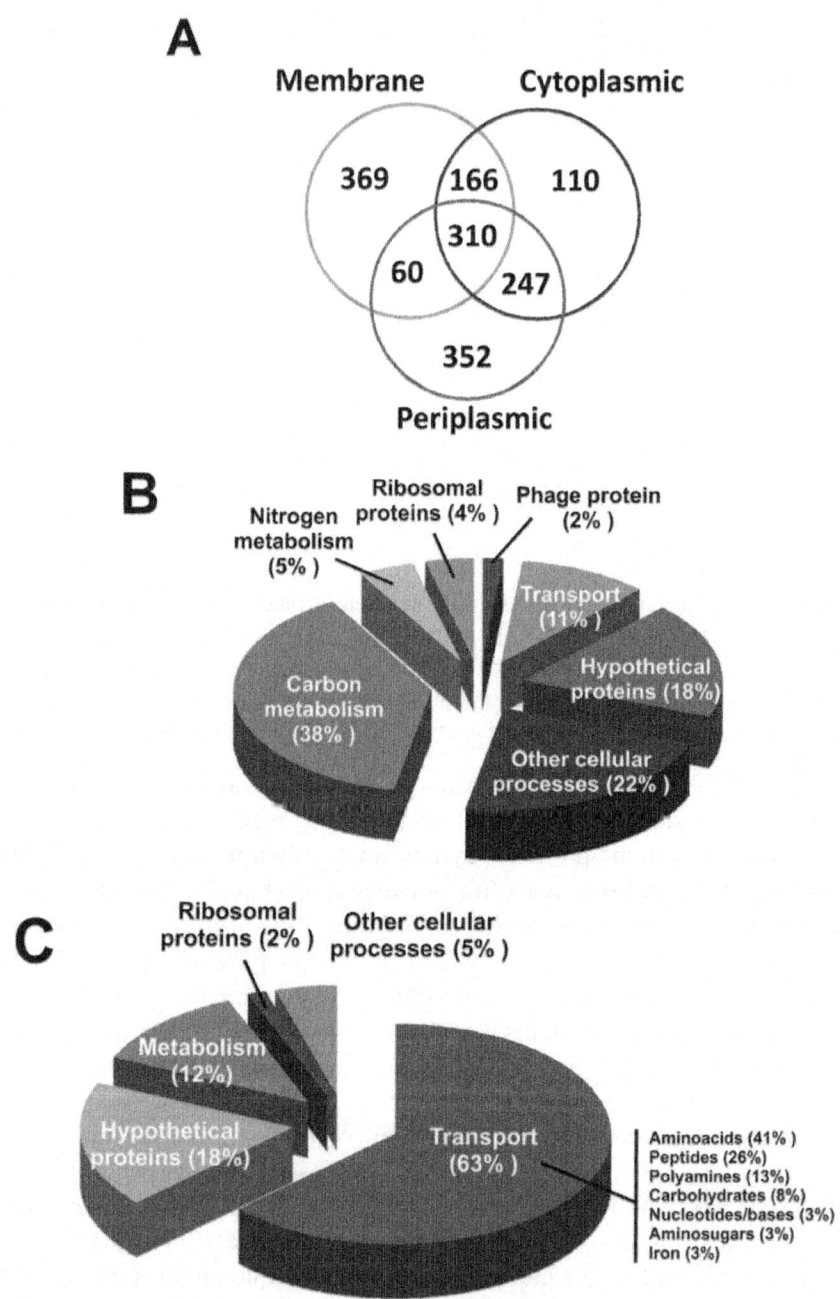

Figure 2. Summary of the quantitative LC/MS proteomic analysis of ^{15}N labelled proteins. A) Representative Venn diagram of identified and quantified unique proteins in

each of the subcellular fractions. B) Functional distribution of positively Hfq-regulated proteins. C) Functional distribution of negatively Hfq-regulated proteins.

The relative abundance of each protein was expressed as average L/H ratios, which correspond to the relative abundance of a unique protein in the Δ*hfq* mutant strain 20PS01 (L) over that of the wild type strain 2011 (H). Thus, the quantitative proteomic analysis of ^{15}N-labelled proteins allowed us to classify quantified proteins in three different groups: 1) proteins positively controlled by Hfq (those with L/H ratios <0.5); 2) proteins negatively regulated by Hfq (those with L/H ratios >2); 3) proteins whose relative abundance remained stable in the absence of Hfq (0.5< L/H <2) (Figure S1). The total number of proteins comprised into groups 1 and 2, raised to 195 unique polypeptides. Thus, 3% of the *S. meliloti* 2011 protein coding potential was detected as, directly or indirectly, controlled by Hfq in exponential growth phase in MDM medium. This figure most probably underestimates the real magnitude of the Hfq regulon in *S. meliloti*, as a large proportion of the encoded proteins were either expressed at levels below the detection limit or not expressed under the tested experimental condition. Previous studies on the impact of *hfq* deletion on global gene expression in *S. meliloti* revealed that, depending on the strain and growing condition, from 2.7% to 7% of the transcriptome [37], [38], and from 0.5% to 0.9% of the proteome [38], [39]were misregulated in the absence of Hfq. Thus, the results of our quantitative proteomic comparative study markedly enlarged the Hfq regulon of *S. meliloti*.

Importantly, the *hfq* in-frame internal deletion carried by strain 20PS01 was confirmed since the Hfq cellular levels were dramatically reduced (L/H=0.05±0.01 in the membrane fraction and an average of 0.07 in two replicates of the cytoplasmic fraction). As for *E. coli*, in which a relevant fraction of the Hfq pool is closely associated to the cell membrane [40], the presence of Hfq in the membrane-enriched fraction may be partly due to its participation in riboregulatory mechanisms to control expression of the large set of membrane and secreted proteins of *S. meliloti* [41]. The level of the immediately downstream encoded gene product HflX was barely affected (L/H=0.72) by the *hfq* deletion. Nevertheless, HflX could only be identified and quantified in one replicate of the membrane fraction. Even though, we can safely assume that the *hfq* mutation is not polar. Thus, it is expected that most, if not all, riboregulatory processes that require Hfq will be affected in the *hfq* mutant 20PS01.

Proteins Positively Regulated by Hfq

In this section, we focus on unique proteins with L/H ratios <0.5. The majority of the positively controlled proteins (74%) are expressed from the chromosome of *S. meliloti*. In terms of biological functions, 38% are involved in carbon metabolism, such as alcohol dehydrogenase (SMa1296, L/H=0.46±0.15 in the cytoplasmic fraction and 0.32±0.12 in the periplasm) or*smoS*, a putative manitol 2-dehydrogenase (SMc04093, L/H=0.42±0.16 in the cytoplasmic fraction). Hypothetical proteins represented 18% of the proteins downregulated in the *hfq* mutant (Figure 2). In comparison with two recent studies that characterized the *S. meliloti*proteome in the absence of *hfq* by 2D-PAGE in strains 2011 [38] and 1021 [39], our quantitative proteomic analysis revealed 120 unique proteins with differential accumulation, which were not previously identified. The overlap of positively Hfq-controlled proteins was limited to only 2 proteins with Torres Quesada *et al.* (SMb20895, SMc04093) [38] and 8 proteins with Barra-Bily *et al.* (SMb21549, SMc01834, SMc02344, SMc02501, SMc02514, SMc02692, SMc02788 y SMc03157) [39].

A remarkable finding was that of the phage protein p077, as one of the least abundant protein in the *hfq* mutant. p077 has been reported as a putative recombinase of the lysogenic phage 16-3 (UniprotKB RM163_077) [42]. More than 2% of *S. meliloti* 1021 genome represents mobile elements, such as phages or insertion sequences [36]. However, neither p077 nor other phage genes have been identified in the genome sequence of *S. meliloti* strain 1021 [36]. So, this mobile element could be specific of *S. meliloti* 2011.

The cobalamine-dependent ribonucleotide reductase NrdJ (SMc01237) was another strongly downregulated protein (L/H=0.09, an average of the quantification in two biological replicates of the cytoplasmic fraction. Moreover, CobW (SMc04304), directly involved in the biosynthesis of cobalamine [43], also showed a reduced accumulation in the *hfq* mutant (L/H=0.23, average from two replicates of the membrane fraction). As the biosynthesis of cobalamine is essential for the symbiosis between *S. meliloti* 1021 and *Medicago sativa* [44], [45], the observed depression of the cobalamine metabolism does not seem to block the symbiotic competence of the *hfq* mutant [16]. Interestingly, the corresponding transcripts of both detected proteins appeared downregulated only 2-fold in an *hfq* mutant [37], suggesting a major translational regulation by Hfq of *nrdJ* and *cobW* expression.

The SMc00986 protein was quantified among the strongest downregulated proteins, comparable to Hfq itself, although its transcript showed a modest dependence on Hfq [37]. SMc00986 encodes a hypothetical protein with six DUF2117 domains, which are highly conserved domains in eubacterial proteins of unknown function. Interestingly, SMc00986 was found to be controlled by the iron master regulator RirA [46], as well as by CbrA [47] (a regulator of several envelope proteins in *S. meliloti*), and to be accumulated in the absence of the outer membrane protein TolC [48].

Structural proteins of the *S. meliloti* flagella were moderately repressed in the *hfq* mutant background. In the insoluble fraction enriched in membrane associated proteins, the relative abundance of FlaA, FlaB, FlaC and FlaD reached *ca.* 0.6 (L/H=0.65±0.21, 0.60±0.28, 0.55±0.35 and 0.64±0.23, respectively). In the periplasmic fraction, the L/H ratio of FlaC was 0.93±0.4, whereas FlaA, FlaB and FlaD showed averages near 0.65 in two replicates of this fraction. In agreement with our observations, flagellar proteins were not detected as regulated by Hfq in previous proteomic studies [38], [39], and the corresponding transcripts were detected as barely downregulated in one of the two transcriptomic comparative analyses of *hfq* mutants[37]. These results suggest a modest contribution of Hfq in the control of flagellar structural genes. In turn, the reduced motility of *hfq* mutant cells derived from both *S. meliloti* 1021 and 2011 strains [16], [37] could not be attributed to major changes in the relative abundance of these proteins.

Proteins Negatively Regulated by Hfq

Proteins whose L/H ratios resulted ≥2 were identified as negatively controlled by Hfq. In this group, 71% of the upregulated proteins derived from the *S. meliloti* chromosome (Figure 2), being 63% of them involved in transport of small molecules, notably L-amino acids or peptides (Figure 2). Among 65 quantified proteins with differential overexpression, 52 hits were only identified in this work. Only a few upregulated proteins matched those identified by 2D-PAGE in previous studies: 6 proteins shared with Barra-Bily *et al.* [39] (SMc00140, SMc00242, SMc00770, SMc00784, SMc02884 and SMc03786), 3 proteins shared with Torres-Quesada *et al.* [38] (SMc02121, SMc02171, SMc02259), and 4 proteins that were common to all three proteomic studies (SMc00786, SMc01525, SMc01946, SMc02118). Thus, our quantitative proteomic analysis confirmed those hits and significantly expanded the list of candidate genes possibly subject to Hfq-dependent negative riboregulation.

As observed for proteins positively regulated by Hfq, the degree of overlap among proteomic studies is markedly low. Possible reasons for such apparent lack of consistency include differences in growth conditions (complex TY medium [38], defined GAS medium [39], or defined MDM medium in this study), growth phase of sampled cells (exponential [38], [39] or stationary [39]), strain (1021 [39] or 2011 [38]), mutant construction (*lacZ::accC1* insertion [39], plasmid insertion [38], or internal deletion [16]), and finally, the resolution power of the analytical technique (2D-SDS-PAGE followed by MS identification [38], [39], or LC-MS/MS analysis of ^{15}N-labeled proteins in this study). However, the number of misregulated proteins detected in our work shows a greater degree of overlap with misregulated transcripts detected in microarray-based analyses of *hfq* mutants: 37 out of 130 polypeptides positively controlled by Hfq were reported as downregulated transcripts in *hfq* mutants in previous studies, whereas 26 out of 65 negatively controlled polypeptides were reported as upregulated transcripts in *hfq* mutants [37], [38]. Such correlations suggest that at least part of the Hfq-controlled genes is subject to direct riboregulatory mechanisms coupled to changes in mRNA stability, as reported for certain *E. coli* sRNAs [7], [49], [50]. By contrast, the level of mRNAs like SMc00986, SMc01237, SMc00786 and SMc02884 was proportionally less influenced by Hfq (1.8, 2.5, 1.5 and 1.6-fold, respectively) [37] than the corresponding proteins in our study (11.1, 20.0, 11.2 and 9.7-fold, respectively), indicating that the expression of these genes is mainly subject to translational Hfq-dependent regulation with less or no significant impact on mRNA stability.

It cannot be discarded that some of the observed changes in the *hfq* mutant could be also due to transcriptional control and the affected genes would thus represent indirect targets. The number of misregulated transcriptional regulatory proteins identified in the *hfq* mutant was very low, probably due to their typical low cellular concentration. However, for the two misregulated transcriptional regulators detected in our study (AniA and FrcR), we found that also one of the proteins under their control were co-regulated: AniA (SMc03880) and PhbB (SMc03879) [51] were upregulated *ca.* 2.8-fold in the *hfq* mutant, whereas FrcR (SMc02172) and FrcA (SMc02169) [52] were downregulated *ca.* 3-fold in the Δ*hfq* strain. Thus, part of the *S. meliloti* Hfq regulon may be a consequence of direct Hfq-dependent riboregulation of mRNAs encoding transcriptional regulators, resulting in indirect control on the subrogated mRNAs.

Hfq Controls Uptake of Nitrogenous Compounds and N Metabolism

Several components of ABC transport systems possibly involved in uptake of oligopeptides were accumulated in the *hfq* mutant: DppA1 (SMc00786), DppA2 (SMc01525), DppD2 (SMc01528) and DppF2 (SMc01529) are part of dipeptide transporters [53]; OppA (SMb21196) has been identified as a periplasmic binding component of an ABC transport system for di- and tripeptides [53]; all proteins encoded by the *aap* operon (*aapJQMP*; SMc02121, SMc02120, SMc02119, SMc02118), constitute a high affinity transport system for L-amino acids [41]; and several genes of the *liv* operon (*livHMGFK*; SMc01946,

SMc01948, SMc01949, SMc01950, SMc01951) [41], are involved in the high affinity transport of leucine, valine and isoleucine. Several of the proteins encoded by these operons related to uptake of amino acid and small peptides were also previously reported to be negatively controlled by Hfq at either the protein or mRNA level [37], [38], [39] , confirming the strong influence of Hfq on the flow of amino acids in *S. meliloti*. These findings point to the existence of Hfq-dependent sRNAs in *S. meliloti* targeting the mRNAs encoding those transporter proteins. sRNAs that control multiple transport systems have been identified in *Salmonella enterica* and*E. coli* (GcvB and RybB) [54], [55] and in the α-proteobacteria *Agrobacterium tumefaciens*(AbcR1) [56] and *Brucella abortus* (AbcR1 and AbcR2) [57]. GcvB and RybB homolog genes seem to be confined to Enterobacteriaceae (Rfam [58]), whereas AbcR1 and AbcR2 homologs have been identified in *S. meliloti* as the non-coding RNA transcripts SmrC15 and SmrC16 [59],[60], [61]. As suggested by computational predictions of mRNA targets, SmrC15 and SmrC16 may fulfill the role of a multi-target sRNA controlling a number of proteins involved in small molecule transport. However, neither Dpp, Opp or Aap components have been identified *in silico* as SmrC15/C16 targets. Conversely, the computational search of putative sRNAs encoded in *S. meliloti* intergenic regions which would be able to bind the mRNA leader of the *opp*, *liv*, *aap*, *dpp1* and *dpp2* operons did not identify the SmrC15/C16 sRNA genes nor revealed a single putative sRNA having all these operons as common targets. Altogether, these computational analyses suggest that, most probably, several sRNAs (including SmrC15/C16) may be responsible for the strong Hfq-dependent riboregulation imposed to multiple oligopeptide and amino acid transporters in *S. meliloti*.

In order to get an *in vivo* correlation with the observed quantitative proteomic alterations on oligopeptide transporters, we tested the sensitivity of the *hfq* mutant strain 20PS01 to the antibacterial tripeptide Bialaphos. This herbicidal and antibiotic compound produced by some*Streptomyces* strains [62], presents two alanines and the glutamate toxic analogue phosphinotricine. Strain 20PS01 showed an increased sensitivity to Bialaphos (Figure 3); a concentration of 100 µg/ml inhibited the growth of the *hfq* mutant, whereas a concentration 2.5-fold higher was required to obtain the same degree of growth inhibition in the wild type strain (Figure 3). Most probably, the increased sensitivity to Bialaphos could be explained by the overexpression of oligopeptide transport systems such as Opp. In addition, the *hfq* mutant presented an increased sensitivity to ammonium glufosinate, an L-glutamate toxic analogue (Figure 3). As for Bialaphos, this sensitivity could be a direct effect of the accumulation of several proteins involved in the transport of L-amino acids, like the amino acid permease (Aap) transport system or other putative amino acid transporters (SMc02259, SMb20706). It is worth pointing out that the tight control of amino acid transporter expression by Hfq has also been recognized in the pea symbiont *Rhizobium leguminosarum* [63]. In this related α-proteobacterium, a number of spontaneous second site suppressor mutants were found to arise from a *gltB* mutant, which is unable to carry out *de novo* amino acids synthesis and shows reduced amino acid transport via Aap and Bra systems. Strikingly, 12 independent spontaneous mutants that regained growth on glutamate, had unique point mutations mapping within *hfq*, and showed strongly increased expression of several amino acid transport systems (Bra, Opp, App) [63]. Thus, negative control of amino acid and oligopeptide transport by Hfq seems to be a conserved feature among plant symbiotic rhizobia.

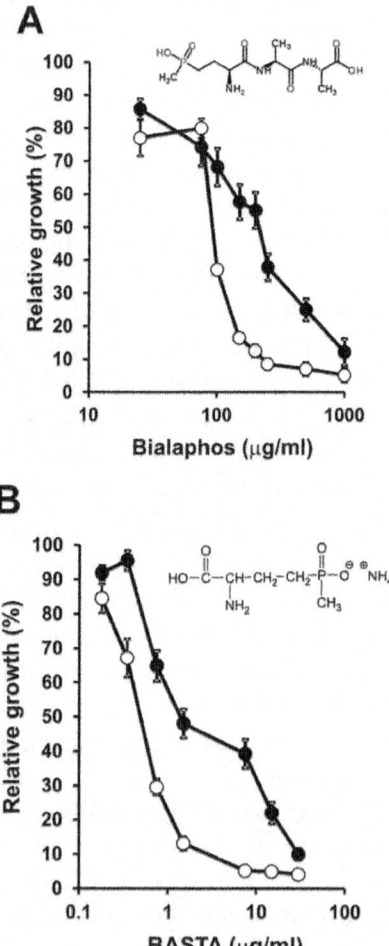

Figure 3. Bialaphos and sodium glufosinate sensitivity assay. In both cases, bacterial growth was estimated by OD_{600} measurements after 72 h in the presence of different

concentrations of the toxic tripeptide Bialaphos (A) or sodium glufosinate (B). (•), wild type strain 2011; (○), Δ*hfq* mutant strain 20PS01. Data represent the growth measured for each *S. meliloti* strain in the presence of a given concentration of the toxic reagent relative to the growth in the absence of the chemical. Each bar shows average values for n=3 replicate cultures ± SD. The experiment was repeated twice with similar results. The chemical structure of both chemicals is shown within each panel.

On the other hand, our analysis identified and quantified a group of transport proteins possibly involved in the uptake of polyamines, such as putrescine, agmatine and spermidine (SMa0392, SMb20284, SMc00770, SMc01652 and SMc01966). Among these, only PotF (SMc00770) has been previously identified as part of the Hfq regulon [39], although it was found to be repressed in an *hfq* mutant background in exponential phase in a defined medium[39]. Polyamines are key components involved in several physiological processes, such as transcription, mRNA stability, translation and oxidative stress resistance [64]. None of these proteins or putative polyamine transport systems has been studied in *S. meliloti* or other α-proteobacteria.

Altogether, our quantitative proteomic analysis revealed a major role of Hfq in the control of nitrogen metabolism of *S. meliloti* 2011 growing on ammonium as N source. Hence, changes in regulatory proteins that govern the nitrogen stress response could contribute to the observed changes in the *hfq* mutant background. The PII protein GlnB (SMc00947) was identified in our analysis but its level did not change significantly in the absence of *hfq*, whereas the PII protein GlnK and the regulators NtrB/C could not be identified in our study. On the other hand, the relative abundance of glutamine synthetase I (GlnAI or SMc00948) remained unaltered in the*hfq* mutant. Instead, GlnAI was upregulated ca. 3-fold in a *S. meliloti hfq* mutant in stationary phase cells [39]. Other enzymes with the same biological function, like GlnII (GlnT - SMc02613), SMc02352 or SMc01594, were not detected in our proteomic analysis of exponential phase cells. A probable glutamine synthetase (SMc00762) presented an L/H ratio of 0.42. Nevertheless, this deficiency would be complemented by GlnAI. The nitrogen assimilatory activity of glutamine synthetase (GS) was measured in order to get a picture of the nitrogen status in the absence of *hfq*. The GS activity in complex TY medium was not affected by the *hfq* deletion. However, in the defined MDM medium containing NH_4Cl (same growth conditions used for the quantitative proteomic analysis), the GS activity of the *hfq*mutant doubled that of the wild type strain. Despite the fact that GlnAI levels remained constant, the higher GS activity could be due to a post-translational activation associated to changes in the intracellular level of GlnD, which in turn activates PII proteins and, finally, GS activity [65], [66], [67].

Table 1. Glutamine synthetase activity of an *S. meliloti* Δ*hfq* **mutant** growing in complex and defined media

Strain (genotype)	Growth medium	
	TY	MDM-NH$_4^+$1% (w/v)
2011 (wild type)	2224.51±355.96 a	3221.69±574.29 a
20PS01 (Δ*hfq*)	2026.49±206.57 a	4741.83±974.29 b

Data correspond to average GS activity (Units/mg protein) of n = 3 replicate cultures ± SD. Different letters indicate statistically significant differences among strains based on pairwise comparisons (Student's *t*-test, $P<0.05$). doi:10.1371/journal.pone.0048494.t001

Hfq Control of Uracil/uridine Uptake

A striking upregulation (>10-fold) was observed for the periplasmic substrate-binding component (SMc01827) from a putative ABC transport involved in the incorporation of uracil and/or uridine (Figure 4). SMc01827 showed the highest L/H ratios in the *hfq* mutant background. In agreement with our finding, almost all members of the operon (Figure 4A) showed elevated mRNA levels in comparative microarray analyses of a *S. meliloti* 1021 and its isogenic Δ*hfq* mutant [37], [38]. These annotations, not yet functionally characterized in *S. meliloti*, seem to constitute the sole active transport system involved in the specific incorporation of uracil and/or uridine to be encoded in the *S. meliloti* genome. The accumulation of these proteins were verified *in vivo* by the incorporation of a cytotoxic uracil analogue 5-fluorouracil (5-FU) [68]. The *hfq* mutant presented a higher sensitivity to 5-FU (Figure 4). 5–10 ng per ml were sufficient to provoke growth inhibition of strain 20PS01, whereas in the wild type strain a comparable effect required 600 ng per ml of 5-FU (Figure 4). The accumulation of this ABC transport system would facilitate the uptake of 5-FU and reduce the tolerance to this agent. In order to discard an increased sensitivity of the *hfq* mutant to 5-FU rather than a more pronounced accumulation of the chemical, direct measurement of uracil uptake rate would be required. To our knowledge, we report for the first time evidences of Hfq-dependent regulation of uracil transport in bacteria.

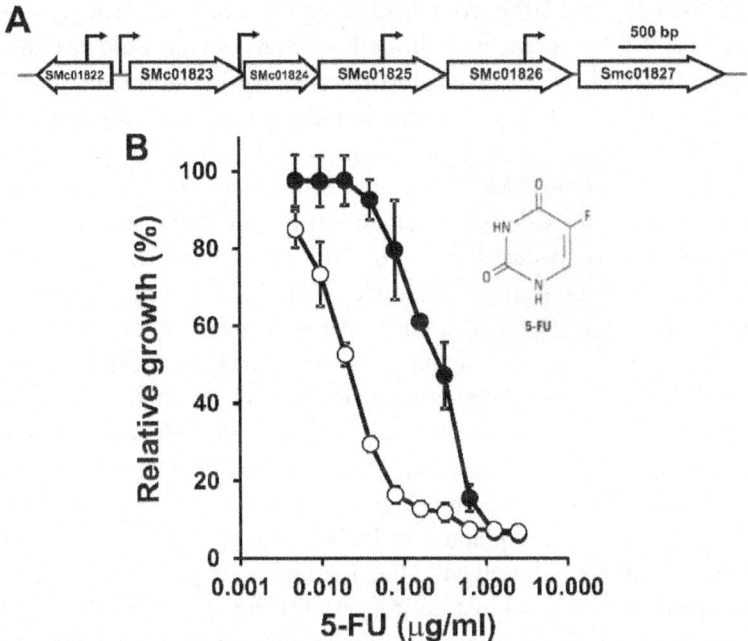

Figure 4. 5-fluorouracil sensitivity assay. A) Schematic representation of the chromosomal *S. meliloti* 1021 locus encompassing the annotation SMc01827. The arrows indicate possible transcription start sites (J.-P. Schlüter & A. Becker, personal communication). B) Bacterial growth was estimated by OD_{600} measurements after 96 h in the presence of different concentrations of the uracil analogue. (•), wild type strain 2011; (○), Δ*hfq* mutant strain 20PS01. Values represent the relative growth of *S. meliloti* 2011 and 20PS01 in presence and absence of the toxic analogue. Each bar shows average values for n=3 replicate cultures ± SD. The experiment was repeated twice with similar results. The chemical structure of the uracil analogue is shown.

hfq Influences Iron Homeostasis in *S. meliloti* 2011

Among the differentially expressed proteins identified in the *hfq* mutant, there was a set of proteins involved in the transport and storage of iron. Notably, the putative iron-storage bacterioferritin (Bfr or SMc03786) of *S. meliloti* 2011 accumulated in the Δ*hfq* strain (L/H=2.41±0.23 in the cytoplasmic fraction, and L/H=3.14±0.91 in the membrane fraction). In strain 1021, the opposite effect was reported: Bfr was downregulated in the *hfq* mutant[39]. The reasons for this discrepancy are unknown. Bacterioferritins have an important role in the homeostasis of iron [69], although its contribution to the control of cellular iron concentration has not been explored yet in *S. meliloti*. The intracellular iron content of the Δ*hfq* strain was 259.0±68.1 µmoles per mg of total protein, whereas the wild type strain contained 28.6±4.9 µmoles of iron per mg of total protein. This 9-fold increment in the cellular iron content could be explained by the higher availability of bacterioferritin detected in our quantitative analysis of the *S. meliloti* 2011 *hfq* mutant. Another evidence of an exacerbated iron accumulation response in the Δ*hfq* mutant was the upregulation of the putative Fe^{+3} transporters FbpA (SMc00784) and SMc01605. Higher than normal iron levels could lead to oxidative stress due to Fenton chemistry in the presence of reactive oxygen species like H_2O_2 [70]. The sensitivity of the *S. meliloti* *hfq* mutant was studied in the presence of H_2O_2, and we found an important growth inhibition in MDM medium under iron sufficient conditions (37 µM) (Figure 5). In line with the higher sensitivity to H_2O_2, the KatB peroxidase/catalase (SMa2379) was downregulated in the Δ*hfq* mutant (average L/H=0.30, in two replicates of the membrane fraction). Similar results were reported for *S. meliloti* 1021 [39]. Other catalase/peroxidase proteins could not be identified in the protein profiles. Thus, the higher sensitivity to oxidative stress in the absence of *hfq* could be explained, at least partially, by the inability to activate expression of catalases involved in the detoxification process and, possibly, by a higher than normal level of free iron. As for the sensitivity to toxic analogs of amino acids and uracil (Figures 3 and 4), we cannot rule out that the higher sensitivity of the *hfq* mutant to H_2O_2 could be partially due to its slower growth rate (Figure 1). However, the observed phenotypes (Figures 3, 4, 5) are consistent with the quantitated changes in the cellular level of proteins directed involved in such phenotypes.

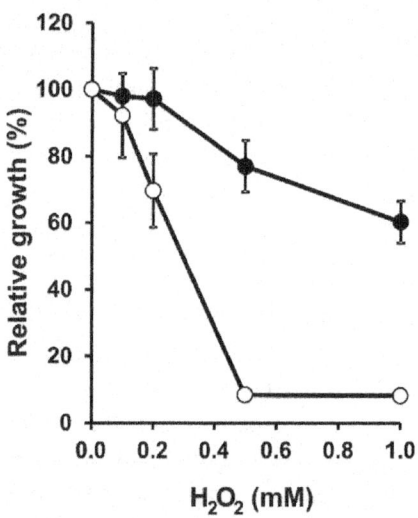

Figure 5. Hfq contributes to tolerance to oxidative stress. Bacterial growth in MDM was estimated by OD_{600} measurements after 72 h in the presence of increasing

concentrations of H_2O_2. (•), wild type strain 2011; (○), Δhfq mutant strain 20PS01. Values represent the relative growth of *S. meliloti* 2011 and 20PS01 in presence and absence of the oxidative stress agent. Each bar shows average values for n=3 replicate cultures ± SD. The experiment was repeated twice with similar results.

Considering that the iron concentration in the growth medium used for our proteomic analysis corresponded to an iron sufficient condition [46], the results led us to hypothesize that the genetic regulatory mechanisms operating on iron homeostasis genes are exacerbated in the absence of *hfq*. Under iron sufficient conditions, the Δhfq mutant accumulates much more iron than the wild type strain. Thus, we studied the production of secreted iron-chelating compounds (siderophores) under different iron supplies in the wild type and Δhfq mutant strains. As expected, both strains responded to lower iron availability by increasing the production of siderophores (Table 2). However, in the absence of *hfq*, more siderophores were produced in iron-limiting conditions and less iron-chelators were secreted in iron-sufficient conditions, with respect to the wild type strain (Table 2). Again, this pattern of siderophore response is consistent with a misregulated iron cellular level in the absence of Hfq.

Table 2. Hfq is required for normal siderophore production

Strain (genotype)	Total iron concentration in the growth medium	
	0.37 μM	37 μM
2011 (wild type)	0.57±0.03 a	0.11±0.01 a
20PS01 (Δhfq)	0.74±0.04 b	0.06±0.01 b

The results are presented as siderophore relative units measured by the CAS interference assay [35]. Values represent the average of three independent cultures ± SD. Different letters indicate statistically significant differences among strains based on pairwise comparisons (Student's *t*-test, $P<0.05$.).
doi:10.1371/journal.pone.0048494.t002

Altogether, these results reveal a critical role of *hfq* as a fine-tuning regulatory factor of iron homeostasis in *S. meliloti* 2011. This could be executed through a direct Hfq-dependent regulation of an iron-related global regulator like Fur, or, as in γ-proteobacteria, through Fur-regulated sRNAs of the RyhB family [71]. However, unlike γ-proteobacteria, the *S. meliloti* Fur homolog is dedicated to control Mn^{+2} uptake [72], whereas regulation of iron cellular levels relies on the transcriptional regulator RirA [46], and possibly on the yet uncharacterized iron response regulator Irr [73], [74]. Neither RirA nor

Irr were detected as Hfq targets in this and previous studies [37], [38], [39]. Yet, sRNAs under RirA or Irr control may contribute to iron homeostasis in *S. meliloti*. As there is no experimental evidence for the expression of RyhB-like sRNAs in *S. meliloti*, we scanned intergenic regions for the presence of putative RirA binding sites [74] located upstream annotated non-coding RNAs [75], following a similar strategy to that reported for identification of the *P. aeruginosa* Fur-regulated PrrF1 and PrrF2 sRNAs [76]. The search identified three annotated small transcripts just downstream putative RirA binding sites. All three hits corresponded to transcripts detected by RNA pyrosequencing as cis-regulatory mRNA leaders linked to iron-metabolism genes. Target mRNA searches for these annotated non-coding transcripts did not reveal obvious iron-related mRNAs. Thus, the identification of bona-fide RyhB homolog(s) in *S. meliloti* remains an open task.

CONCLUSIONS

The results described here represent the first quantitative proteomic approach in the α-proteobacterial legume symbiont *S. meliloti*, following stable protein labelling of cell cultures with ^{15}N. The choice of the technical approach is relevant because: 1) post-transcriptional riboregulatory mechanisms may result in changes in protein levels without being reflected in changes in mRNA levels and thus escape transcriptomic profiling; 2) subtle Hfq-dependent control of gene expression requires sensitive techniques like quantitative proteomics of stable isotope labelled proteins. The outcome of the comparative protein profile in the absence of *hfq* significantly expands the repertoire of cellular processes influenced by the RNA chaperone Hfq in *S. meliloti*. Overall, nearly 200 polypeptides have been found to be misregulated in the Hfq mutant, including several polypeptides recently identified by 2D-PAGE [38], [39]. We reveal novel Hfq targets involved in the transport of small molecules, like peptides and uracil, and in iron homeostasis, and confirmed the associated phenotypes by *in vivo* assays. These results lead us to hypothesize that *S. meliloti* expresses sRNAs dedicated to regulate those cellular processes with the assistance of Hfq, like the γ-proteobacterial sRNAs GcvB and RybB, and the α-proteobacterial AbcR1 and AbcR2 sRNAs, that controls multiple ABC uptake systems [56],[57], [77], [78], and the regulatory sRNA RyhB that control iron homeostasis [7], [71]. The absence of *hfq* also affected the expression of genes encoding proteins involved in other cellular processes, like cell division and adaptation to heat shock. The corresponding genes or operons may be subject to either direct or indirect control by Hfq. A comparative expression analysis of translational reporter fusions to the putative target genes would help to confirm a direct control by Hfq.

SUPPORTING INFORMATION

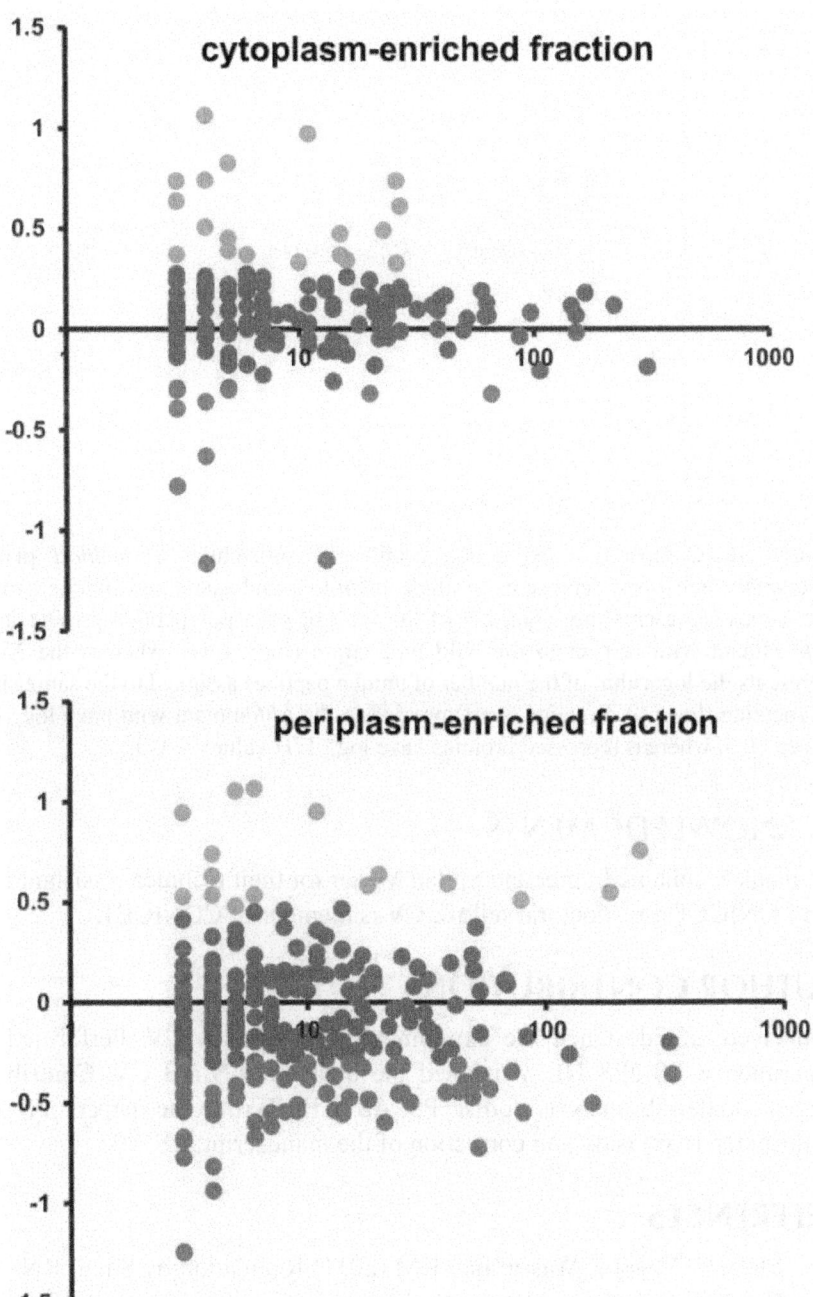

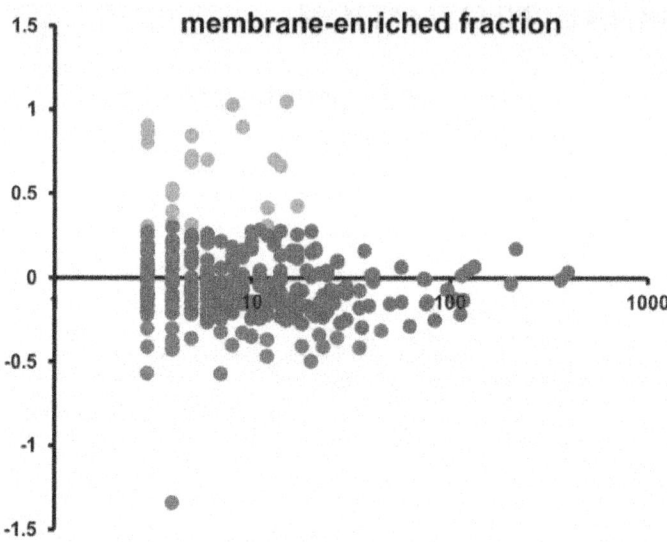

Figure S1. Comparative expression profiles of subcellular *S. meliloti* **protein fractions.** Each point represents a single identified and quantified unique protein. The Y-axis represents the logarithm of the ratio of each polypeptide present in the Δ*hfq* mutant with respect to the wild type strain ($\log_{10}$ L/H), whereas the X-axis represents the logarithm of the number of unique peptides assigned to the same single polypeptide ($\log_{10}$ #). Proteins overexpressed in the Δ*hfq*mutant with have $\log_{10}$ L/H values >0.3, whereas repressed proteins have $\log_{10}$ L/H values <−0.3.

ACKNOWLEDGMENTS

We thank Stephanie Lamer and Stefan Meyer for their technical assistance. PS is a CONICET post-doctoral fellow. CV is member of CONICET.

AUTHOR CONTRIBUTIONS

Conceived and designed the experiments: PS AS AB CV. Performed the experiments: PS JPS UL. Analyzed the data: PS AS AB CV. Contributed reagents/materials/analysis tools: PS AS AB. Wrote the paper: PS CV. Contributed to revision and correction of the manuscript: AS AB.

REFERENCES

1.   Storz G, Vogel J, Wassarman KM (2011) Regulation by Small RNAs in Bacteria: Expanding Frontiers. Mol Cell 43: 880–891. doi: 10.1016/j. molcel.2011.08.022

2.   Beisel CL, Storz G (2010) Base pairing small RNAs and their roles in global regulatory networks. FEMS Microbiol Rev 34: 866–882. doi: 10.1111/j.1574-6976.2010.00241.x

3.   Sobrero P, Valverde C (2012) The bacterial protein Hfq: much more than a mere RNA-binding factor. Crit Rev Microbiol.

4.   Vogel J, Luisi BF (2011) Hfq and its constellation of RNA. Nat Rev Microbiol 9: 578–589. doi: 10.1038/nrmicro2615

5.   Franze de Fernandez MT, Eoyang L, August JT (1968) Factor fraction required for the synthesis of bacteriophage Qbeta-RNA. Nature 219: 588–590. doi: 10.1038/219588a0

6.   Adamson DN, Lim HN (2011) Essential requirements for robust signaling in Hfq dependent small RNA networks. PLoS Comput Biol 7: e1002138. doi: 10.1371/journal.pcbi.1002138

7.   Masse E, Escorcia FE, Gottesman S (2003) Coupled degradation of a small regulatory RNA and its mRNA targets in Escherichia coli. Genes Dev 17: 2374–2383. doi: 10.1101/gad.1127103

8.   Morita T, Maki K, Aiba H (2005) RNase E-based ribonucleoprotein complexes: mechanical basis of mRNA destabilization mediated by bacterial noncoding RNAs. Genes Dev 19: 2176–2186. doi: 10.1101/gad.1330405

9.   Robertson GT, Roop RM Jr (1999) The Brucella abortus host factor I (HF-I) protein contributes to stress resistance during stationary phase and is a major determinant of virulence in mice. Mol Microbiol 34: 690–700. doi: 10.1046/j.1365-2958.1999.01629.x

10.   Kaminski PA, Desnoues N, Elmerich C (1994) The expression of nifA in Azorhizobium caulinodans requires a gene product homologous to Escherichia coli HF-I, an RNA-binding protein involved in the replication of phage Q beta RNA. Proc Natl Acad Sci U S A 91: 4663–4667. doi: 10.1073/pnas.91.11.4663

11.   Sittka A, Pfeiffer V, Tedin K, Vogel J (2007) The RNA chaperone Hfq is essential for the virulence of Salmonella typhimurium. Mol Microbiol 63: 193–217. doi: 10.1111/j.1365-2958.2006.05489.x

12.   Tsui HC, Leung HC, Winkler ME (1994) Characterization of broadly pleiotropic phenotypes caused by an hfq insertion mutation in Escherichia coli K-12. Mol Microbiol 13: 35–49. doi: 10.1111/j.1365-2958.1994.tb00400.x

13.   Dietrich M, Munke R, Gottschald M, Ziska E, Boettcher JP, et al. (2009) The effect of hfq on global gene expression and virulence in Neisseria gonorrhoeae. FEBS J 276: 5507–5520. doi: 10.1111/j.1742-

4658.2009.07234.x

14. Sousa SA, Ramos CG, Moreira LM, Leitao JH (2010) The hfq gene is required for stress resistance and full virulence of Burkholderia cepacia to the nematode Caenorhabditis elegans. Microbiology 156: 896–908. doi: 10.1099/mic.0.035139-0

15. Christiansen JK, Larsen MH, Ingmer H, Sogaard-Andersen L, Kallipolitis BH (2004) The RNA-binding protein Hfq of Listeria monocytogenes: role in stress tolerance and virulence. J Bacteriol 186: 3355–3362. doi: 10.1128/jb.186.11.3355-3362.2004

16. Sobrero P, Valverde C (2011) Evidences of autoregulation of hfq expression in Sinorhizobium meliloti strain 2011. Arch Microbiol 193: 629–639. doi: 10.1007/s00203-011-0701-1

17. Gage DJ (2004) Infection and invasion of roots by symbiotic, nitrogen-fixing rhizobia during nodulation of temperate legumes. Microbiol Mol Biol Rev 68: 280–300. doi: 10.1128/mmbr.68.2.280-300.2004

18. Jones KM, Kobayashi H, Davies BW, Taga ME, Walker GC (2007) How rhizobial symbionts invade plants: the Sinorhizobium-Medicago model. Nat Rev Microbiol 5: 619–633. doi: 10.1038/nrmicro1705

19. Barra-Bily L, Pandey SP, Trautwetter A, Blanco C, Walker GC (2010) The Sinorhizobium meliloti RNA chaperone Hfq mediates symbiosis of S. meliloti and alfalfa. J Bacteriol 192: 1710–1718. doi: 10.1128/jb.01427-09

20. Zhang A, Wassarman KM, Rosenow C, Tjaden BC, Storz G, et al. (2003) Global analysis of small RNA and mRNA targets of Hfq. Mol Microbiol 50: 1111–1124. doi: 10.1046/j.1365-2958.2003.03734.x

21. Ansong C, Yoon H, Porwollik S, Mottaz-Brewer H, Petritis BO, et al. (2009) Global systems-level analysis of Hfq and SmpB deletion mutants in Salmonella: implications for virulence and global protein translation. PLoS One 4: e4809. doi: 10.1371/journal.pone.0004809

22. Ong SE, Blagoev B, Kratchmarova I, Kristensen DB, Steen H, et al. (2002) Stable isotope labeling by amino acids in cell culture, SILAC, as a simple and accurate approach to expression proteomics. Mol Cell Proteomics 1: 376–386. doi: 10.1074/mcp.m200025-mcp200

23. Emadali A, Gallagher-Gambarelli M (2009) [Quantitative proteomics by SILAC: practicalities and perspectives for an evolving approach]. Med Sci (Paris) 25: 835–842. doi: 10.1051/medsci/20092510835

24. Otto A, Bernhardt J, Hecker M, Becher D (2012) Global relative and absolute quantitation in microbial proteomics. Curr Opin Microbiol.

25. Soufi B, Kumar C, Gnad F, Mann M, Mijakovic I, et al. (2010) Stable isotope labeling by amino acids in cell culture (SILAC) applied to quantitative proteomics of Bacillus subtilis. J Proteome Res 9: 3638–3646. doi: 10.1021/pr100150w

26. Auweter SD, Bhavsar AP, de Hoog CL, Li Y, Chan YA, et al. (2011) Quantitative mass spectrometry catalogues Salmonella pathogenicity island-2 effectors and identifies their cognate host binding partners. J Biol Chem 286: 24023–24035. doi: 10.1074/jbc.m111.224600

27. Vogels MW, van Balkom BW, Heck AJ, de Haan CA, Rottier PJ, et al. (2011) Quantitative proteomic identification of host factors involved in the Salmonella typhimurium infection cycle. Proteomics 11: 4477–4491. doi: 10.1002/pmic.201100224

28. Meade HM, Signer ER (1977) Genetic mapping of Rhizobium meliloti. Proc Natl Acad Sci U S A 74: 2076–2078. doi: 10.1073/pnas.74.5.2076

29. McIntosh M, Krol E, Becker A (2008) Competitive and cooperative effects in quorum-sensing-regulated galactoglucan biosynthesis in Sinorhizobium meliloti. J Bacteriol 190: 5308–5317. doi: 10.1128/jb.00063-08

30. Eggenhofer E, Haslbeck M, Scharf B (2004) MotE serves as a new chaperone specific for the periplasmic motility protein, MotC, in Sinorhizobium meliloti. Mol Microbiol 52: 701–712. doi: 10.1111/j.1365-2958.2004.04022.x

31. Bradford MM (1976) A rapid and sensitive method for the quantitation of microgram quantities of protein utilizing the principle of protein-dye binding. Anal Biochem 72: 248–254. doi: 10.1006/abio.1976.9999

32. Laemmli UK (1970) Cleavage of structural proteins during the assembly of the head of bacteriophage T4. Nature 227: 680–685. doi: 10.1038/227680a0

33. Bender RA, Janssen KA, Resnick AD, Blumenberg M, Foor F, et al. (1977) Biochemical parameters of glutamine synthetase from Klebsiella aerogenes. J Bacteriol 129: 1001–1009.

34. Somerville JE, Kahn ML (1983) Cloning of the glutamine synthetase I gene from Rhizobium meliloti. J Bacteriol 156: 168–176.

35. Schwyn B, Neilands JB (1987) Universal chemical assay for the detection and determination of siderophores. Anal Biochem 160: 47–56. doi: 10.1016/0003-2697(87)90612-9

36. Galibert F, Finan TM, Long SR, Puhler A, Abola P, et al. (2001) The composite genome of the legume symbiont Sinorhizobium meliloti.

Science 293: 668–672. doi: 10.1126/science.1060966

37. Gao M, Barnett MJ, Long SR, Teplitski M (2010) Role of the Sinorhizobium meliloti global regulator Hfq in gene regulation and symbiosis. Mol Plant Microbe Interact 23: 355–365. doi: 10.1094/mpmi-23-4-0355

38. Torres-Quesada O, Oruezabal RI, Peregrina A, Jofre E, Lloret J, et al. (2010) The Sinorhizobium meliloti RNA chaperone Hfq influences central carbon metabolism and the symbiotic interaction with alfalfa. BMC Microbiol 10: 71. doi: 10.1186/1471-2180-10-71

39. Barra-Bily L, Fontenelle C, Jan G, Flechard M, Trautwetter A, et al. (2010) Proteomic alterations explain phenotypic changes in Sinorhizobium meliloti lacking the RNA chaperone Hfq. J Bacteriol 192: 1719–1729. doi: 10.1128/jb.01429-09

40. Diestra E, Cayrol B, Arluison V, Risco C (2009) Cellular electron microscopy imaging reveals the localization of the Hfq protein close to the bacterial membrane. PLoS One 4: e8301. doi: 10.1371/journal.pone.0008301

41. Mauchline TH, Fowler JE, East AK, Sartor AL, Zaheer R, et al. (2006) Mapping the Sinorhizobium meliloti 1021 solute-binding protein-dependent transportome. Proc Natl Acad Sci U S A 103: 17933–17938. doi: 10.1073/pnas.0606673103

42. Deak V, Lukacs R, Buzas Z, Palvolgyi A, Papp PP, et al. (2010) Identification of tail genes in the temperate phage 16–3 of Sinorhizobium meliloti 41. J Bacteriol 192: 1617–1623. doi: 10.1128/jb.01335-09

43. Rodionov DA, Vitreschak AG, Mironov AA, Gelfand MS (2003) Comparative genomics of the vitamin B12 metabolism and regulation in prokaryotes. J Biol Chem 278: 41148–41159. doi: 10.1074/jbc.m305837200

44. Campbell GR, Taga ME, Mistry K, Lloret J, Anderson PJ, et al. (2006) Sinorhizobium meliloti bluB is necessary for production of 5,6-dimethylbenzimidazole, the lower ligand of B12. Proc Natl Acad Sci U S A 103: 4634–4639. doi: 10.1073/pnas.0509384103

45. Taga ME, Walker GC (2010) Sinorhizobium meliloti requires a cobalamin-dependent ribonucleotide reductase for symbiosis with its plant host. Mol Plant Microbe Interact 23: 1643–1654. doi: 10.1094/mpmi-07-10-0151

46. Chao TC, Buhrmester J, Hansmeier N, Puhler A, Weidner S (2005) Role of the regulatory gene rirA in the transcriptional response of Sinorhizobium meliloti to iron limitation. Appl Environ Microbiol 71: 5969–5982. doi: 10.1128/aem.71.10.5969-5982.2005

47.   Gibson KE, Barnett MJ, Toman CJ, Long SR, Walker GC (2007) The symbiosis regulator CbrA modulates a complex regulatory network affecting the flagellar apparatus and cell envelope proteins. J Bacteriol 189: 3591–3602. doi: 10.1128/jb.01834-06

48.   Santos MR, Cosme AM, Becker JD, Medeiros JM, Mata MF, et al. (2010) Absence of functional TolC protein causes increased stress response gene expression in Sinorhizobium meliloti. BMC Microbiol 10: 180. doi: 10.1186/1471-2180-10-180

49.   Ikeda Y, Yagi M, Morita T, Aiba H (2011) Hfq binding at RhlB-recognition region of RNase E is crucial for the rapid degradation of target mRNAs mediated by sRNAs in Escherichia coli. Mol Microbiol 79: 419–432. doi: 10.1111/j.1365-2958.2010.07454.x

50.   Prevost K, Desnoyers G, Jacques JF, Lavoie F, Masse E (2011) Small RNA-induced mRNA degradation achieved through both translation block and activated cleavage. Genes Dev 25: 385–396. doi: 10.1101/gad.2001711

51.   Povolo S, Casella S (2000) A critical role for aniA in energy-carbon flux and symbiotic nitrogen fixation in Sinorhizobium meliloti. Arch Microbiol 174: 42–49. doi: 10.1007/s002030000171

52.   Lambert A, Osteras M, Mandon K, Poggi MC, Le Rudulier D (2001) Fructose uptake in Sinorhizobium meliloti is mediated by a high-affinity ATP-binding cassette transport system. J Bacteriol 183: 4709–4717. doi: 10.1128/jb.183.16.4709-4717.2001

53.   Nogales J, Munoz S, Olivares J, Sanjuan J (2009) Genetic characterization of oligopeptide uptake systems in Sinorhizobium meliloti. FEMS Microbiol Lett 293: 177–187. doi: 10.1111/j.1574-6968.2009.01527.x

54.   Sharma CM, Darfeuille F, Plantinga TH, Vogel J (2007) A small RNA regulates multiple ABC transporter mRNAs by targeting C/A-rich elements inside and upstream of ribosome-binding sites. Genes Dev 21: 2804–2817. doi: 10.1101/gad.447207

55.   Urbanowski ML, Stauffer LT, Stauffer GV (2000) The gcvB gene encodes a small untranslated RNA involved in expression of the dipeptide and oligopeptide transport systems in Escherichia coli. Mol Microbiol 37: 856–868. doi: 10.1046/j.1365-2958.2000.02051.x

56.   Wilms I, Voss B, Hess WR, Leichert LI, Narberhaus F (2011) Small RNA-mediated control of the Agrobacterium tumefaciens GABA binding protein. Mol Microbiol 80: 492–506. doi: 10.1111/j.1365-2958.2011.07589.x

57.   Caswell CC, Gaines JM, Ciborowski P, Smith D, Borchers CH, et al.

(2012) Identification of two small regulatory RNAs linked to virulence in Brucella abortus 2308. Mol Microbiol 85: 345–360. doi: 10.1111/j.1365-2958.2012.08117.x

58. Gardner PP, Daub J, Tate J, Moore BL, Osuch IH, et al. (2011) Rfam: Wikipedia, clans and the "decimal" release. Nucleic Acids Res 39: D141–145. doi: 10.1093/nar/gkq1129

59. del Val C, Rivas E, Torres-Quesada O, Toro N, Jimenez-Zurdo JI (2007) Identification of differentially expressed small non-coding RNAs in the legume endosymbiont Sinorhizobium meliloti by comparative genomics. Mol Microbiol 66: 1080–1091. doi: 10.1111/j.1365-2958.2007.05978.x

60. del Val C, Romero-Zaliz R, Torres-Quesada O, Peregrina A, Toro N, et al. (2012) A survey of sRNA families in alpha-proteobacteria. RNA Biol 9: 119–129. doi: 10.4161/rna.18643

61. Valverde C, Livny J, Schluter JP, Reinkensmeier J, Becker A, et al. (2008) Prediction of Sinorhizobium meliloti sRNA genes and experimental detection in strain 2011. BMC Genomics 9: 416. doi: 10.1186/1471-2164-9-416

62. Schinko E, Schad K, Eys S, Keller U, Wohlleben W (2009) Phosphinothricin-tripeptide biosynthesis: an original version of bacterial secondary metabolism? Phytochemistry 70: 1787–1800. doi: 10.1016/j.phytochem.2009.09.002

63. Mulley G, White JP, Karunakaran R, Prell J, Bourdes A, et al. (2011) Mutation of GOGAT prevents pea bacteroid formation and N2 fixation by globally downregulating transport of organic nitrogen sources. Mol Microbiol 80: 149–167. doi: 10.1111/j.1365-2958.2011.07565.x

64. Shah P, Swiatlo E (2008) A multifaceted role for polyamines in bacterial pathogens. Mol Microbiol 68: 4–16. doi: 10.1111/j.1365-2958.2008.06126.x

65. Yurgel SN, Kahn ML (2008) A mutant GlnD nitrogen sensor protein leads to a nitrogen-fixing but ineffective Sinorhizobium meliloti symbiosis with alfalfa. Proc Natl Acad Sci U S A 105: 18958–18963. doi: 10.1073/pnas.0808048105

66. Yurgel SN, Rice J, Kahn M (2011) Nitrogen metabolism in S. meliloti?alfalfa symbiosis: Dissecting the role of GlnD and PII proteins. Mol Plant Microbe Interact.

67. Yurgel SN, Rice J, Mulder M, Kahn ML (2010) GlnB/GlnK PII proteins and regulation of the Sinorhizobium meliloti Rm1021 nitrogen stress response and symbiotic function. J Bacteriol 192: 2473–2481. doi: 10.1128/jb.01657-09

68. Tomasz A, Borek E (1960) The Mechanism of Bacterial Fragility Produced by 5-Fluorouracil: The Accumulation of Cell Wall Precursors. Proc Natl Acad Sci U S A 46: 324–327. doi: 10.1073/pnas.46.3.324

69. Carrondo MA (2003) Ferritins, iron uptake and storage from the bacterioferritin viewpoint. EMBO J 22: 1959–1968. doi: 10.1093/emboj/cdg215

70. Cornelis P, Wei Q, Andrews SC, Vinckx T (2011) Iron homeostasis and management of oxidative stress response in bacteria. Metallomics 3: 540–549. doi: 10.1039/c1mt00022e

71. Salvail H, Masse E (2012) Regulating iron storage and metabolism with RNA: an overview of posttranscriptional controls of intracellular iron homeostasis. Wiley Interdiscip Rev RNA 3: 26–36. doi: 10.1002/wrna.102

72. Platero R, Peixoto L, O'Brian MR, Fabiano E (2004) Fur is involved in manganese-dependent regulation of mntA (sitA) expression in Sinorhizobium meliloti. Appl Environ Microbiol 70: 4349–4355. doi: 10.1128/aem.70.7.4349-4355.2004

73. Johnston AW, Todd JD, Curson AR, Lei S, Nikolaidou-Katsaridou N, et al. (2007) Living without Fur: the subtlety and complexity of iron-responsive gene regulation in the symbiotic bacterium Rhizobium and other alpha-proteobacteria. Biometals 20: 501–511. doi: 10.1007/s10534-007-9085-8

74. Rodionov DA, Gelfand MS, Todd JD, Curson AR, Johnston AW (2006) Computational reconstruction of iron- and manganese-responsive transcriptional networks in alpha-proteobacteria. PLoS Comput Biol 2: e163. doi: 10.1371/journal.pcbi.0020163.eor

75. Schluter JP, Reinkensmeier J, Daschkey S, Evguenieva-Hackenberg E, Janssen S, et al. (2010) A genome-wide survey of sRNAs in the symbiotic nitrogen-fixing alpha-proteobacterium Sinorhizobium meliloti. BMC Genomics 11: 245. doi: 10.1186/1471-2164-11-245

76. Wilderman PJ, Sowa NA, FitzGerald DJ, FitzGerald PC, Gottesman S, et al. (2004) Identification of tandem duplicate regulatory small RNAs in Pseudomonas aeruginosa involved in iron homeostasis. Proc Natl Acad Sci U S A 101: 9792–9797. doi: 10.1073/pnas.0403423101

77. Sharma CM, Papenfort K, Pernitzsch SR, Mollenkopf HJ, Hinton JC, et al. (2011) Pervasive post-transcriptional control of genes involved in amino acid metabolism by the Hfq-dependent GcvB small RNA. Mol Microbiol 81: 1144–1165. doi: 10.1111/j.1365-2958.2011.07751.x

78. Balbontin R, Fiorini F, Figueroa-Bossi N, Casadesus J, Bossi L (2011)

Recognition of heptameric seed sequence underlies multi-target regulation by RybB small RNA in Salmonella enterica. Mol Microbiol 78: 380–394. doi: 10.1111/j.1365-2958.2010.07342.x

Chapter 3

DYNAMICS OF ACTIN WAVES ON PATTERNED SUBSTRATES: A QUANTITATIVE ANALYSIS OF CIRCULAR DORSAL RUFFLES

Erik Bernitt[1], Cheng Gee Koh[2] , Nir Gov[3] , Hans-Günther Döbereiner[1]

[1] Institut für Biophysik, Universität Bremen, Bremen, Germany

[2] School of Biological Sciences, Nanyang Technological University, Singapore, Singapore

[3] Department of Chemical Physics, Weizmann Institute of Science, Rehovot, Israel

ABSTRACT

Circular Dorsal Ruffles (CDRs) have been known for decades, but the mechanism that organizes these actin waves remains unclear. In this article we systematically analyze the dynamics of CDRs on fibroblasts with respect to characteristics of current models of actin waves. We studied CDRs on heterogeneously shaped cells and on cells that we forced into disk-like morphology. We show that CDRs exhibit phenomena such as periodic cycles of formation, spiral patterns, and mutual wave annihilations that are in accord with an active medium description of CDRs. On cells of controlled morphologies, CDRs exhibit extremely regular patterns of repeated wave formation and propagation, whereas on random-shaped cells the dynamics seem to be dominated by the limited availability of a reactive species. We show that theoretical models of reaction-diffusion type incorporating conserved species capture partially the behavior we observe in our data.

INTRODUCTION

The polymerization of the structural protein actin from its monomeric to its filamentous state accounts for fundamental aspects of the dynamic cell shape [1]. Changes of the cell morphology can therefore be mainly understood by deciphering the underlying actin machinery. In the organization of actin dynamics, waves of protein activity play a central role and have become a field of intense research recently [2]. The prominence of waves for cell morphodynamics and motility seems to be a conserved property among different cell lines and organisms. Popular model systems include *Dictyostelium discoideum*, human neutrophils, and rat tumor mast cells to name but a few [3–5]. The subject of this work is a type of actin wave termed Circular Dorsal Ruffles (CDRs). These are actin-based structures of vertical extension that form and propagate as soliton-like waves at the dorsal side of a large number of cell types such as fibroblasts, epithelial cells, macrophages, glial cells, and aortic smooth muscle cells [6–10]. CDRs were named as such based on their often ring-like appearance and the fact that the membrane at CDRs has ruffle-like morphologies, as visible in scanning electron micrographs, much like peripheral ruffles at the cell edge [11, 12]. Their biological function is still debated, but CDRs clearly serve endocytotic purposes [7]. The fact that several pathogens are known to hijack CDRs as a mechanism of gate opening through the cell membrane brought CDRs to the center of interest of the medical community [12]. CDRs have been comprehensively characterized in terms of protein composition [7, 8], but the mechanism that orchestrates the protein interplay in CDRs remains puzzling.

In this work, we present a detailed study on CDR phenomenology and quantitative analysis of wavefront dynamics that allows us to characterize several properties of the mechanism underlying CDRs. In this respect, we visually identified and quantitatively confirmed features characteristic of current theoretical models of actin waves. Comprehensive theories for the description of actin gels from fundamental properties have been formulated [13–15]. However, most models for description of actin waves focus on a limited number of molecular key-players. A recent review by Allard and Mogilner classifies theories underlying actin wave description into models in which actin is an autocatalytic species that promotes its own growth and into models in which actin only responds to fields of regulatory proteins [2]. The latter are further sub-divided into models of classical reaction diffusion type and into models that also contain morpho- or mechanosensitive species. As an example of a reaction diffusion system, the interactions between the Scar/WAVE complex and actin have been shown to organize into actin waves. In these, Scar/WAVE triggers actin polymerization, but is in turn inactivated by polymerized actin [4]. Actin autocatalysis, supported by the Arp2/3 complex, and the treadmilling properties of actin filaments, on the other side, provide an intrinsic propagator of actin waves as has been shown in *D. discoideum* [3]. In these cells, no Scar/WAVE was required for propagating actin waves. In the organization of actin waves into patterns that allow cells to function, coupling of the actin wave machinery and oscillations of species such as calcium seems to play important roles [5].

For the description of CDRs, there are currently two models, one based on a reaction-diffusion system and the other based on morphosensitive actin factors. The model by Zeng et al relies on an antagonistic reaction-diffusion scheme between the rhoGTPases Rho and Rac [16]. The activity states of these two proteins thereby compete for the organization of actin into either stress fibers or into a meshwork as found in lamellipodia and CDRs. In contrast, Peleg et al consider a system that incorporates membrane-bound proteins that are curvature-sensitive effectors of actin [17]. The curvature of the membrane thus plays a crucial role in this model.

Even though these two studies show that both models support the formation and propagation of actin waves, a comparison to the detailed dynamics of CDRs obtained by live cell imaging is missing. With this work, we show that CDRs on fibroblast cells cover a much wider range of behavior than currently described in the literature on CDRs; among them states of periodic expansion and contraction ("breathing" modes), spiral waves, stalled wavefronts close to cell edges, and mutual wave annihilation upon collision of wavefronts. We analyze the dynamics of CDRs on cells of natural morphology and on cells that we forced into disk-like morphology with centered nuclei. On the latter we find CDRs to propagate laterally between cell nucleus and cell edge while keeping constant sizes. Comparing the data of cells of uncontrolled morphologies and cells of disk-like morphology reveals that CDR size influences their dynamics and that CDRs exhibit very uniform patterns of formation and propagation when their size is fixed. When reviewed in the light of current models of actin waves, CDR dynamics can be understood—without knowledge of molecular details—as waves in an active medium where a limited resource or some other constraint has a large influence.

RESULTS

CDRs Exhibit Behaviors Well Known from Reaction Diffusions Systems with a Strong Dependency on Local Cell Morphology

We were interested whether CDRs exhibit, besides formation of expanding rings, other phenomena known to occur in reaction-diffusion systems. Therefore, we performed long-term experiments using a cell line (NIH 3T3 X2) that spontaneously forms CDRs when cultured in cell medium containing Fetal Bovine Serum (FBS). We confirmed that the CDRs formed under these conditions are identical in appearance to CDRs formed in response to Platelet-

Derived Growth Factor (PDGF) stimulation. The use of PDGF for CDR stimulation is the standard method for studies on CDRs [6–8]. In contrast, using NIH 3T3 X2 cells allowed us to keep the biochemical conditions of the medium fixed in our experiments, refraining from the disturbance of the biochemical surroundings by addition of PDGF. We further confirmed that minima of phase contrast micrographs of CDRs corresponded to maxima in actin concentration. The live-imaging data of cells exhibiting CDRs were then scanned systematically for phenomena characteristic for active media.

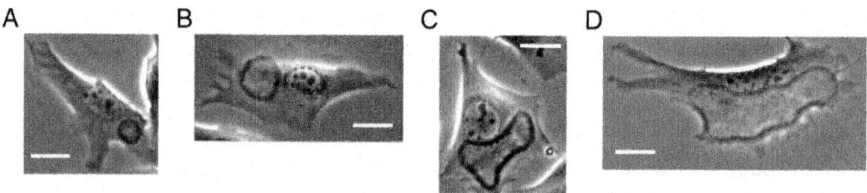

Figure 1. Effects of cell size and morphology on CDR morphology and dynamics. CDRs usually avoid the nucleus region (encircled in red). Cell edge and nucleus therefore define a bounded region available for CDR propagation, which limits the maximal size CDRs can attain. In the panels *A-D* the size of this region is increasing from left to right. The isotropy of CDRs decreases with increasing CDR size, while the tendency to mimic cell morphology increases with CDR size. CDRs in small regions (*A*) cannot extend much, typically forming oscillatory reappearing objects of high isotropy. All scale bars correspond to 25 μm.

We found CDRs to grow outward as closed soliton-like structures originating from points. The width of CDRs was typically 5 μm whereas the shape, size, and lifetime strongly depended on cell morphology (compare Fig. 1A–D). We generally found CDRs to avoid the region of the cell nucleus; we rarely saw CDRs crossing it. CDRs also, trivially, could not propagate beyond the cell edge. CDRs at maximal extension therefore tended to take shapes that reflected the cell morphology (Fig. 1C&D). Since fibroblasts are random-shaped, we consequently found CDRs to largely vary in form, typically strongly differing from circular geometry. The area available for growth and propagation of a CDR, bounded by the nucleus and the cell edge, crucially influenced its dynamics. In the following we describe the phenomena we observed together with the typical size of the region where CDRs emerged.

In small regions of order twice the characteristic CDR diameter, where there is no space for CDR propagation, we found CDRs to re-appear in a periodical manner at the same position (Fig. 2A&B). For these CDRs, the phase of expansion was very short, immediately followed by CDR closure. In sequences of several CDR reappearances, we observed a well-defined frequency of firing (Fig. 2 B&C) that did, however, vary among different cells. The recovery times between two CDRs could be as short as one minute. We also observed repeated formations of larger CDRs on regions of greater spatial extent. However, in such large regions the number of CDRs formed was usually smaller and the frequency lower than for CDRs on small regions (Fig. 2D–G). In no case did we observe concentric wave trains; a succeeding CDR would only form if the preceding CDR either closed or decayed.

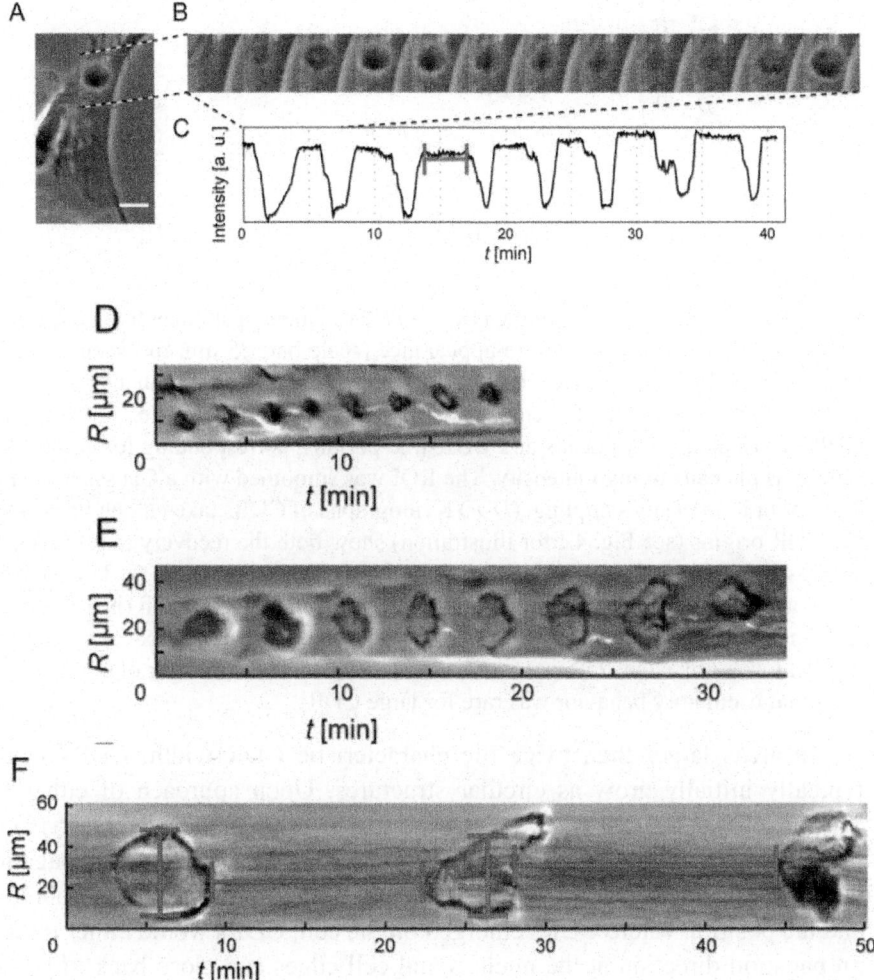

G

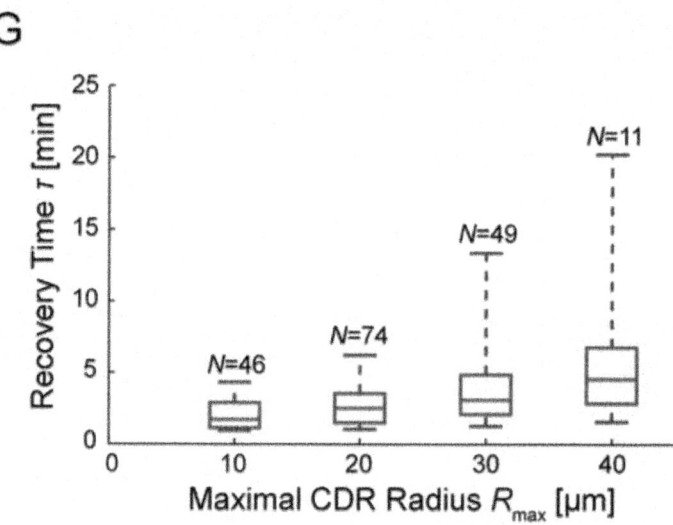

Figure 2. Oscillatory reappearing CDRs. (*A*) CDRs under spatial confinement exhibit oscillatory patterns of pulsating re-appearance (scale bar: 25 μm, full sequence). (*B*) Stills from the region of interest highlighted red in the time-lapse sequence *A* (Δ*t* = 36 s). (*C*) A plot of the minimal intensity value of the ROI in *A* as a function of time shows CDR events as negative peaks and CDR-free periods, corresponding to the recovery time τ, as plateaus of high intensity. The ROI was smoothed with a Gaussian with = 2 μm prior to intensity sampling. (*D-F*) Kymographs of CDRs taken along lines crossing CDR origins (see Fig. 4*A* for illustration) show both the recovery time τ between successive events and their radial extension R_{max} (cells not shown). (*G*) The recovery times increase with CDR size. The data was binned in R_{max}-direction (box width: 10 μm) and plotted as boxes with whiskers (red lines: median, upper box edge: 75th percentile, lower box edge: 25th percentile). *N* values denote the number of observations. Note that oscillatory behavior was rare for large CDRs.

In areas larger than twice the characteristic CDR width, CDRs would typically initially grow as circular structures. Upon approach of either the cell nucleus or the cell edge, CDR shapes typically gained asymmetry. With increasing size and asymmetry, we observed a slow-down in the propagation velocity. The subsequent dynamics depended on the overall cell morphology and the position where CDRs emerged on the cell. CDRs would either reverse propagation direction at the nucleus and cell edges and close back to points, decay at maximal extension or they would separate into typically arc-shaped parts that continued to propagate (Fig. 3D). CDRs that covered the larger part of the cell surface formed closed structures of high anisotropy with life times of up to several ten minutes. Close to the cell edge we observed a persistent stalling of CDR net movement in some cases (Fig. 1D). This was accompanied by spatial oscillatory fluctuations of CDRs around one fixed position for typically several minutes. We employed fluorescence microscopy to visualize actin dynamics of stalled CDRs and found them to be actin-dense structures surrounded by areas of complete actin depletion (Fig. 3A).

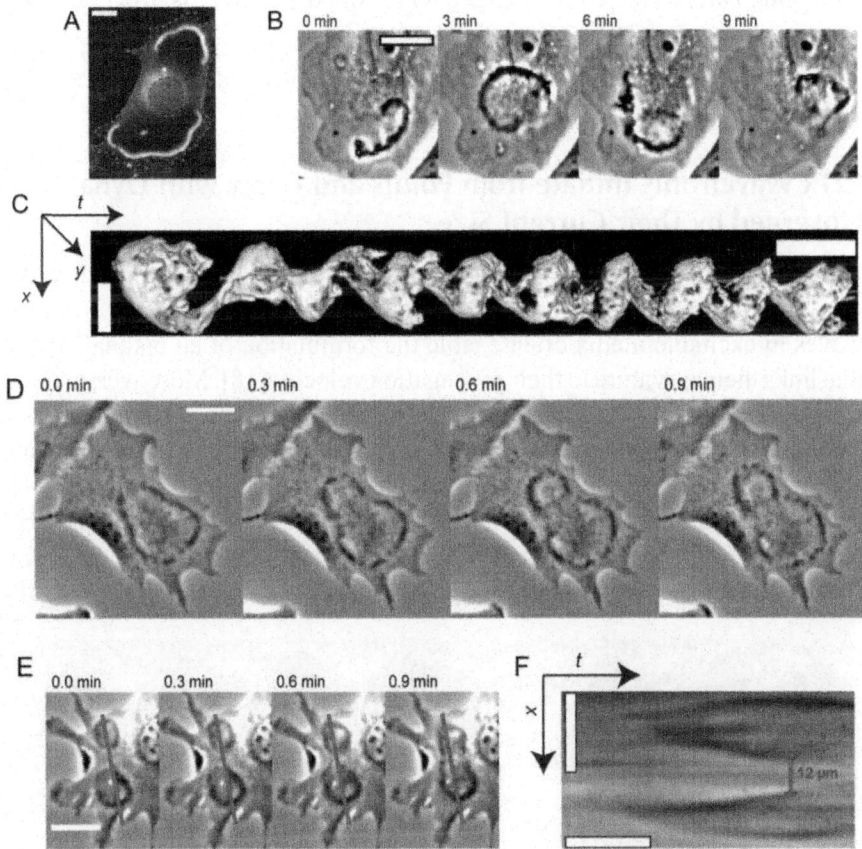

Figure 3. Overview of CDR phenomena. (*A*) CDRs can become trapped close to cell edges; actin staining with pLifeAct–TagGFP2 reveals regions of actin depletion behind wavefronts. Red dashed lines highlight cell edge and nucleus. (*B*) A CDR propagating as a spiral wave. (*C*) Iso-surface visualization of the CDR in *B* as an *x-y-t*-projection. The CDR performed eight full rotations in approximately 70 minutes. (*D*) A CDR dividing into two arc-shaped wavefronts. (*E* and *F*) Time-lapse sequence of two colliding CDRs and the corresponding kymograph respectively. The red lines in *E* mark the position were kymographs where sampled. The CDR wavefronts mutually annihilate each other at a distance of approximately 12 μm before they actually make contact. All spatial scale bars correspond to 25 μm. Temporal scale bar in *C:* 10 minutes, temporal scale bar in *F* 1: minute.

In cases where more than one CDR formed on the same cell simultaneously, we observed fusion of CDRs when the wavefronts of two CDRs happened to collide. The parts of the colliding wavefronts that moved towards each other thereby annihilated at a typical distance of 12 μm from each other. The remaining parts of the CDRs merged and formed one large CDR (Fig. 3E&F).

In rare cases, CDRs could be observed to propagate as spirals (Fig. 3B&C). The periods for one full rotation varied among different cells between 5 and 10 minutes.

CDR Wavefronts Initiate from Points and Grow with Dynamics Governed by Their Current Size

Knowledge of the typical propagation velocities of wavefronts and their dynamics allows for conclusions on the underlying mechanisms [2]. Further, waves in excitable media often enable the formulation of an eikonal equation that links their curvature to their propagation velocity [18]. Moreover, in models in which curvature-sensitive membrane proteins regulate actin dynamics the local curvature of wavefronts is governing the propagation dynamics. In the following, we analyze the wavefront dynamics of CDRs with respect to their local curvature. Our data set consists of CDRs that followed the pattern of opening (centrifugal CDR growth), reversing and closure (centripetal CDR shrinkage), i.e., where no events like separation or fusion could be observed.

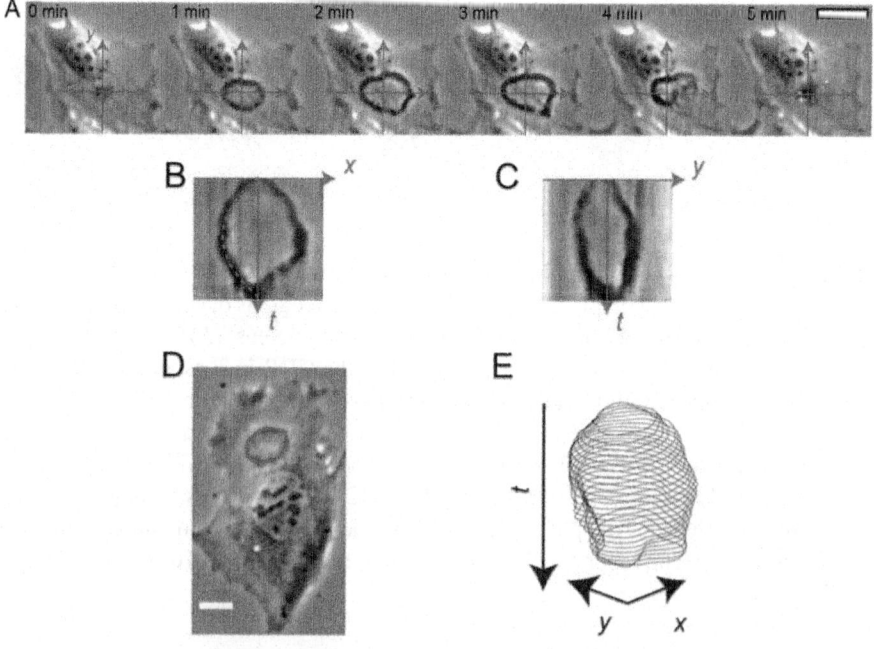

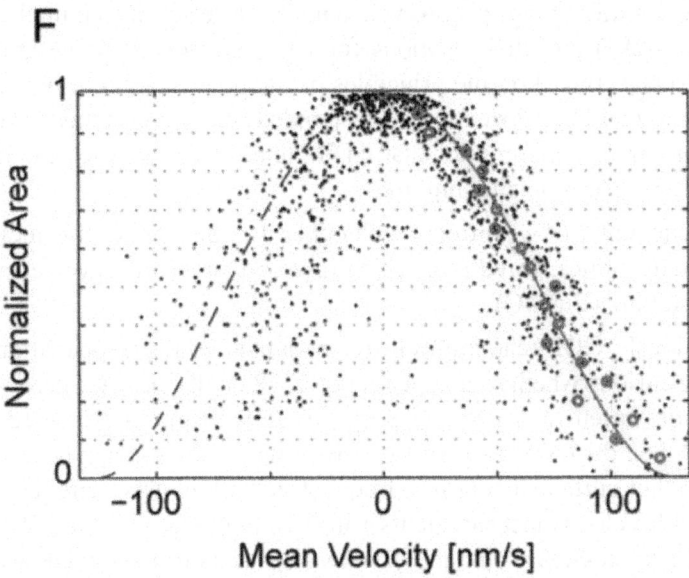

Figure 4. Wavefront dynamics of opening and closing CDRs. (*A-C*) A typical life course of a CDR exhibiting opening and closing. The coordinate system in *A* is the basis for calculation of the kymographs shown in *B* and *C*. Together with time-lapse sequence *A* these kymographs show the dependency of CDR dynamics on cell features. CDR propagation without encounter of obstacles and absence of instability has a parabolic evolution of the CDR radius with time (red dashed line in *B*: empirical parabola fit). In positive *x*-direction, however, the wavefront becomes unstable and partially decays, leading to an asymmetric profile in kymograph *B*. (*D* and*E*) Using active contours, the wavefronts of CDRs can be tracked yielding sets of contours for each CDR. (*F*) The contour mean velocity data of 13 CDRs as a function of the normalized area roughly follow one trajectory. Positive velocities correspond to CDR growth and negative velocities to CDR shrinking. Original data points are shown in black, red circles correspond to average velocities calculated using a box median of width 0.05 in normalized area. The red line is an empirical fit function used for extrapolation to a CDR area of zero. All spatial scale bars correspond to 25 μm.

CDRs originated from points and grew as ring-shaped waves. Upon approach of either the cell nucleus or the cell edge, CDRs typically lost isotropy and mimicked the cell shape to some degree (Fig. 4A). A kymograph analysis revealed that the growth velocity decreased right from the point of CDR formation until CDRs reversed and closed back to points. When propagating unperturbed, without contact with the cell edge or nucleus, CDRs exhibited radius evolutions that were symmetric with respect to the time point of reversal. In a first approximation, this could be described with a parabolic time dependence of the radius (Fig. 4B). Wavefronts with a pronounced halo in phase contrast images, however, tended to decay at maximal extension. See Fig. 4A&B and their captions for a comparison of stable and unstable CDR reversal. Based on the principles of image formation in phase contrast, we assume that halo formation corresponded to a large vertical extension of decaying CDRs. This implies that CDRs tend to decay when the membrane forms high vertical height amplitude.

An encounter of wavefronts with either the cell edge or the nucleus led to disturbed dynamics. Wavefronts either reversed directly or reversed after some time of stalling (Fig. 4C).

Obviously, CDRs can follow different dynamics after propagation reversal. We were curious whether the opening of CDRs still follows a general pattern. Since more often than not, the point of CDR origin and closure were not identical, i.e., the CDR as a whole translocated, a kymograph description was generally not sufficient. Further, frequent deviations from circular shapes of CDRs made velocity measurements using kymographs inaccurate. We thus used the image processing technique of active contours to track CDR wavefronts. We focused on CDRs that had closed shapes of limited complexity, i.e., CDRs similar to those shown in Fig. 1B andFig. 4A. The sets of contours we obtained this way corresponded to the positions of the CDR wavefronts as a function of time (Fig. 4D&E). We calculated the local normal velocity of CDR contours and their local curvature. A cross-correlation analysis yielded no correlation between local curvature and local velocity. We thus asked whether an integral measure such as the area covered by a CDR could govern its dynamics. A possible mechanism to link area and velocity could incorporate, e.g., a limited resource such as an involved protein or the available membrane area. Since the size of both cells and CDRs varied, we calculated the CDR area and normalized it to its maximal area. We then plotted the mean contour velocity as a function of this normalized area (Fig. 4F). The data points corresponding to the opening of CDRs collapse on a polynomial of fourth order as expected from a parabolic evolution of the mean radius in time (see kymographs in Fig. 4B&C). We used this fit function for extrapolation and found a velocity of 0.13 µm/s at small CDR areas, i.e., directly after CDR formation and shortly before its closure. The fact that we find highest wave velocities directly after wave formation might seem to suggest highest velocity for largest curvature.

CDRs Propagate in a Highly Regular Manner on Cells of Controlled Morphology

We found the propagation velocity of CDRs to be a dynamic quantity on naturally shaped cells. Since we observed the encounter of CDRs with the cell edge or nucleus to crucially influence their dynamics, we were curious how CDRs would behave on cells exhibiting a ring-like region for CDR propagation around the nucleus, i.e., cells of disk-like morphology with centered nuclei. Unfortunately, such morphology is not common for fibroblasts. However, on single cells that happened to have this morphology by chance, we found CDRs orbiting the nucleus with constant velocities (Fig. 5). Based on this result, we chose to force cells into disk-like morphologies—using microcontact-printing—to realize the simplest possible boundary geometry. On these cells, the nucleus was situated at their center (Fig. 6A&B). Since CDRs avoided crossing of the nucleus, the resulting path of CDR propagation was quasi one-dimensional with respect to CDR propagation direction and had periodic boundary conditions. On random-shaped cells, CDRs extended over the whole cell surface in some cases. In contrast, on disk-like cells the small region for CDR propagation, i.e., the area between cell nucleus and cell edge, only allowed for waves of a small extent when compared to the cell size (Fig. 6B). CDRs formed spontaneously on cells of controlled morphologies, without requiring stimulation by growth factors. We verified that these structures indeed corresponded to CDRs based on comparison between them and CDRs formed with PDGF stimulation. Both experimental approaches leaded to CDRs with the same morphological properties. In the following we report on a data set of 38 cells. We used kymographs in which we plotted the image intensity along a circle with origin at the cell nucleus versus time to display wave propagation in lateral direction (Fig. 6B&C). We found waves to exhibit remarkably conserved velocities on disk-like cells (Fig. 6C). Wave initiation was followed by expansion of a circular wave that split shortly after formation, which along a line of fixed radius appeared as the propagation of two pulses into both lateral directions. Propagating pulses usually collided head-on with newly formed pulses on their way. Collisions of pulses lead in 79% of the cases to mutual annihilation while in 21% of the cases one pulse survived the collision (Fig. 6C). In other cases CDRs translocated as a whole and therefore appeared in kymographs as two pulses propagating in the same direction, where one pulse closely followed the other.

A

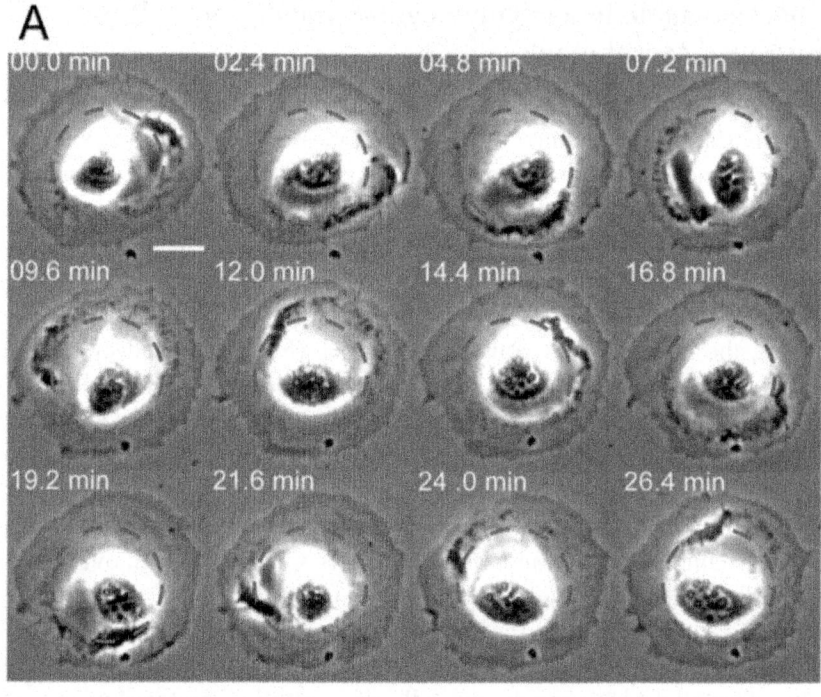

B

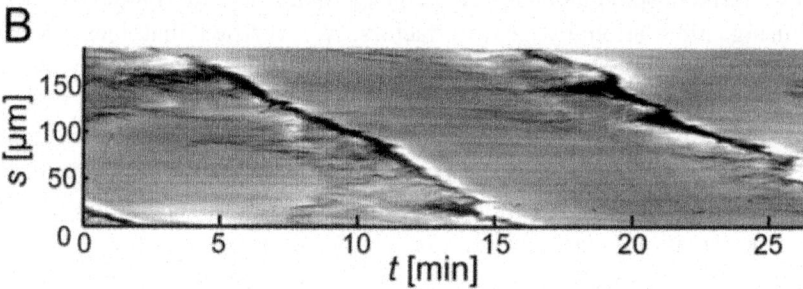

Figure 5. CDR orbiting the nucleus of a cell of disk-morphology. (*A*) Time-lapse sequence of a cell that had disk-like morphology without being plated on a micro protein patch. A CDR propagates between cell edge and cell nucleus, circling the nucleus almost twice. (*B*) Sampling of the image intensity along the arc length of a circle (highlighted red in *A*) and as a function of time gives rise to a circular kymograph. The nearly constant slope in this kymograph indicates a constant lateral velocity ($v = 0.21$ μm/s) of the CDR. Scale bar in *A*: 25 μm.

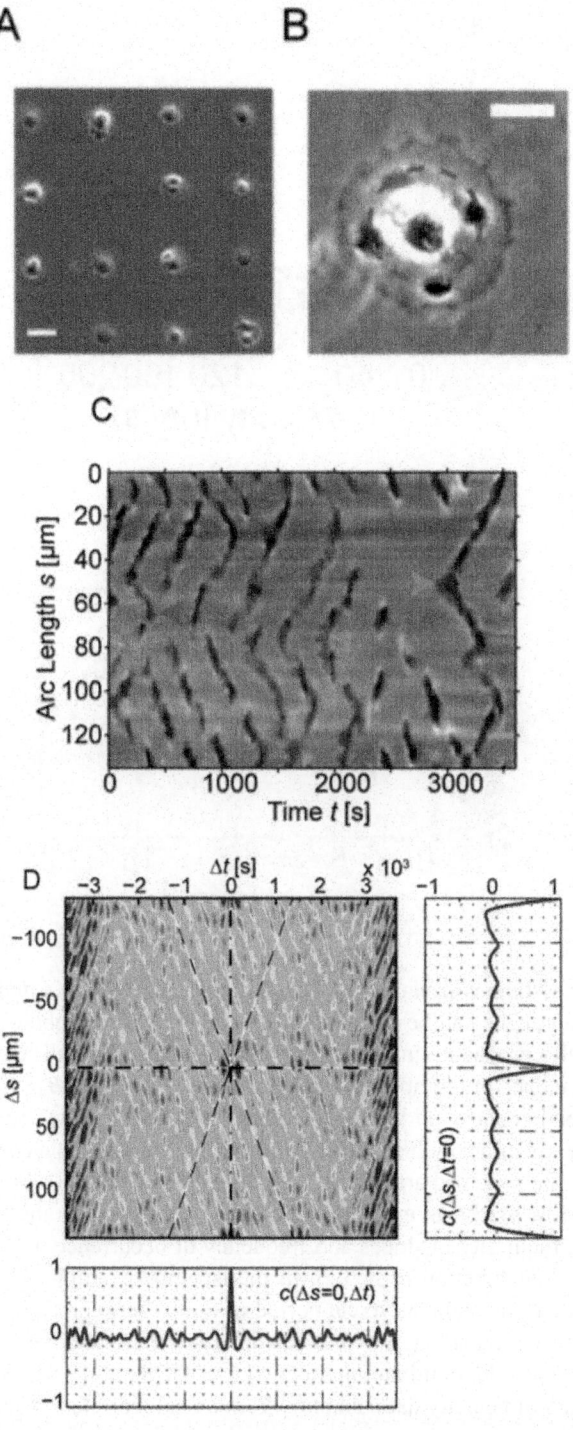

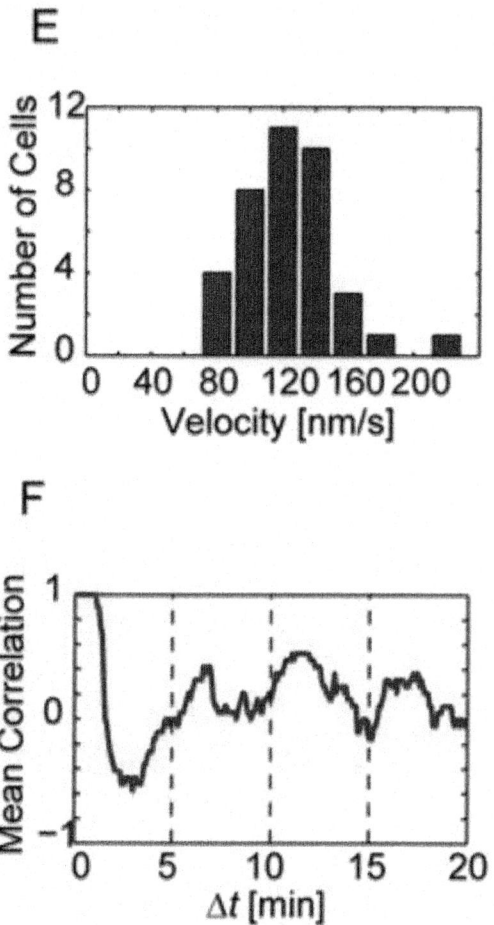

Figure 6. Space-time correlations of CDRs on circular paths. (*A* and *B*) Using micro-contact printing, cells can be patterned into well-defined morphologies. (*B*) On disk-like cells CDRs propagate in lateral direction between the cell nucleus and the cell edge. (*C*) Circular kymograph sampled at the red circle in (*B*). Waves propagating in lateral direction show up as dark stripes. "<"-shaped objects correspond to wave initiation (the green arrow highlights one example) and ">"-shaped objects to wave annihilation. The solid red arrow shows an example of mutual annihilation while the hollow red arrow marks an event in which one pulse survives the collision. (*D*) The apparent high regularity in slopes and frequency of occurrence in C is emphasized in an autocorrelation function $c(\Delta s, \Delta t)$. In this specific example we find propagation velocities of 0.10 μm/s and a typical period of 6 min between two CDR events at the same position (see the cut $c(\Delta s = 0, \Delta t)$ and *F* for the sample average). The cut $c(\Delta s, \Delta t = 0)$ emphasizes the dominant number of four DCRs at the same time on this cell. (*E*) A histogram of velocity data obtained from an autocorrelation analysis of 38 cells.

The mean velocity is 0.12 (± 0.03) μm/s (± SD). (*F*) A cut through the average correlation function of the same 38 cells at constant position. The mean period between two CDR events at the same position is approximately 6 min. The scale bars in *A* and *B* correspond to 50 μm and 25 μm respectively.

We segmented kymographs and calculated autocorrelation functions for a systematic quantitative analysis (Fig 6D). Velocities were extracted from autocorrelation functions using an approach based on the Radon transformation and from these velocity histograms were calculated. We found a mean propagation velocity of 0.12 (±0.03) μm/s (± SD, Fig. 6E). From averaged autocorrelation functions we calculated the temporal correlations at fixed positions. The resulting cut $c(\Delta s = 0, \Delta t)$ shows a pronounced time interval with negative correlations, i.e., a quiescent time before emergence or passage of the next CDR. The mean period between two successive wave events at one position was 6 min (Fig. 6F). CDRs tended to appear in equidistant positions around the nucleus forming patterns in lateral direction. The cut $c(\Delta s, \Delta t = 0)$ in Fig. 6F reflects the four-fold rotational symmetry visible in Fig. 6B.

DISCUSSION AND CONCLUSIONS

CDR Growth Originates from Points

Our results regarding typical CDR sizes and lifetimes largely agree with previously reported values [6–8]. However, so far little emphasis has been put on a characterization of the initiation of CDRs. Rather, the early stages of CDRs have been described as "immature" flat and open structures [7]. Early CDRs are reported to consist of punctual short actin filaments that only later coalesce into circular arrays of bundled actin filaments which are then termed "mature" CDRs [8]. While we did not study the details of the actin structure of CDRs in this work, we showed that CDRs can grow as closed mature rings right from the time point of their initiation. Furthermore, we find that the velocity of CDRs is maximal right after initiation and decays with CDR expansion. Our description is therefore in line with the study by Zeng et al, in which they investigated the dynamics of CDR life courses based on cells that were stimulated with PDGF and then fixed [16]. They describe early CDRs as rings with small radii. In their work the time between stimulation and fixation was quantized with a minimal step width of 2.5 min. However, our results show that the dynamics of CDRs varies on a much shorter time scale than minutes and thus requires live cell imaging with frame rates on the order of seconds (Fig. 4).

CDRs on Cells of Uncontrolled Morphology Lack Typical Velocities

Peleg et al reported CDR propagation velocities of 2.3 (±0.6) µm/s for closing CDRs [17]. In contrast, the experimentally determined radii as a function of time as published by Zeng et al allow to read off velocities on the order of 0.1 µm/s for opening CDRs [16]. However, it is clear from the results of Zeng et al and especially our detailed analysis that the CDR velocity is a dynamic quantity and we thus cannot state a representative value. Instead, we found that the current velocity of a CDR depends on the ratio of its current size relative to its maximal size. Nevertheless, we found CDRs to exhibit reproducible and temporally constant velocity values on microcontact-printed cells where CDRs are forced into fixed sizes by the boundaries set by cell nucleus and cell edge. On this system we found a mean propagation velocity of 0.12 (±0.03) µm/s (± SD). As shown in Fig 4B, the CDR propagation velocities of partially unstable wavefronts can indeed be much higher than those of stable wavefronts. This explains why the velocities reported by Peleg et al exceed the velocities of the study by Zeng et al and our values by one order of magnitude.

CDRs can be Described as Waves in an Active Medium

The aim of our experimental survey was to systematically investigate CDRs with respect to characteristics of theoretical models of actin waves. Several of the novel phenomena we found for CDRs, such as oscillations, annihilations upon collisions, and spiral waves are well known to occur in active media [19, 20]. The idea to describe CDRs as waves in an active medium was first proposed by Zeng et al [16]. If stimulated locally, their system gave rise to concentric propagating wavefronts. The model explains the reversal of CDR propagation direction as a consequence of a slow kinetics of the inhibitory species. The resulting dynamics is highly asymmetric with respect to the time point of maximal CDR size; in the model, the opening process is much slower than the CDR closing. Our results do not support these outcomes of the model by Zeng et al. In cases where CDR wavefronts did neither reach the cell nucleus nor the cell edge or became unstable, we typically found symmetric profiles of the mean radius with respect to the time point of CDR reversal. The model by Zeng et al predicts the concentration of one of the model's constituents to govern the maximal CDR size. In contrast we observed that maximal CDR sizes scale with the cell size and are highly variable.

The model by Peleg et al implicates a correlation between local membrane curvature and wavefront velocity, even though the explicitly considered curvature is that in the xz- and yz-plane [17]. More generally, reaction-diffusion systems often allow for the formulation of an eikonal equation that relates the curvature of the wavefront to its propagation velocity. In contrast, our results show no correlation between local wavefront curvature and propagation velocity. This outcome is puzzling, because we do observe CDRs to propagate as spirals in some cases, and for a persistent spiral pattern there must be a functional relationship between wavefront curvature and velocity [17]. We also found the overall mean curvature not to correlate with the propagation velocity, although we do find the fastest CDR velocity following initiation (smallest radius), and a slowing down as it expands. The overall CDR area seems to govern CDR dynamics.

CDRs Exhibit Phenomena Resembling Characteristics of Actin Waves in Other Cell Types

While our findings of spiral patterns, oscillations and collision annihilation are novel for CDRs, these phenomena have been described in the scope of actin waves in different cell types previously. Numerous theoretical models have been suggested that capture these behaviors [4,5, 20–23]. Prominent examples of actin waves organizing into spirals have been reported for *D. discoideum*. The actin waves in this unicellular slime mold act to establish a leading front in *D. discoideum*'s migration strategy [24–25]. In the process of recovery from actin depolymerization, actin waves in *D. discoideum* form closed, ring-like structures that resemble CDRs in shape and also the propagation velocities are in the same range as the velocities we found for CDRs [25]. Actin waves propagating with similar velocities are also found in neutrophils [4]. In contrast to CDRs, waves in *D. discoideum* and neutrophils are not associated with vertical projections at the dorsal cell side. Waves of phosphatidylinositol(3,4,5)-trisphosphate, that play a major role in the organization of actin waves in *D. discoideum*, explicitly form at the basal side of these cells [26]. In the wave mechanism underlying CDRs, the involvement of a protein species that selectively aggregates at the dorsal cell side is an attractive assumption. In the model by Peleg et al, the selectivity is provided by the deformation of the cell surface and the subsequent accumulation of curvature-sensitive proteins.

CDR Dynamics is Governed by the Limited Availability of Some Reactive Species

We propose that there must be two contributions to the dynamics of wavefront velocities of CDRs. One stems from a reaction-diffusion system that creates waves and makes them propagate. In most cases, however, the actually visible dynamics is dominated by another contribution due to limited resources, which masks the inherent dynamics of the reaction-diffusion system. One could think of several possible effectors, among them membrane tension or limited availability of some reactive protein species. This picture is consistent with our finding of uniform propagation velocities of CDRs on micro-patterned cells. On these cells the CDR size is limited to the space available between cell nucleus and cell edge. Edgar and Bennett observed that CDRs avoid the cell nucleus before and pointed to a possible role of intermediate filaments in this phenomenon [27]. Regardless of the underlying mechanism, our experimental design ensures that CDRs do not change their size upon propagation. This implies that for propagating CDRs on micro-patterned cells the membrane area remains constant. Also the amount of proteins incorporated in a CDR should be constant. With this spatial confinement we also limit the curvature of wavefronts to a specific value. Given the homogeneity of the medium for propagation in lateral direction on disk-like cells, the exhibited constant velocities might follow as a natural consequence. We therefore found a way to allow investigations of the wave propagation without disturbing effects due to limited species or membrane area.

As discussed in Allard and Mogilner's work, travelling waves provide an economic way to support cell motility in situations of limited resources of, e.g., the actin regulating protein VASP or actin itself [2]. Since we do observe correlation between CDR stalling and drastic actin depletion (Fig. 3A), the limiting species could indeed be actin.

Holmes et al investigated the effects of limited resources of regulatory proteins on the propagation of actin waves in a theoretical study [28]. Their reaction-diffusion system produces waves that stall close to the boundary of the system, a phenomenon, which is studied in detail in a succeeding work and called "wave pinning" [29]. Interestingly, the pinned waves in their study exhibit local oscillations around their average position much like the stalled CDRs close to cell edges do in our study. In a different scenario, their model produces waves that slow down and reverse their propagation direction upon approach of boundaries. The effects due to limited resources are thus not only able to explain the slowing down of propagating waves, but also their reversal.

In this picture, the closing of CDRs does not require a contractile element as proposed by Hoon et al [7]. Rather, just like the closed circular shape is a natural consequence of a wave in an active medium that was stimulated in one point, closing back of the wave to one point is a natural consequence of the reflection of these waves from boundaries such as the cell edge or the cell nucleus. This idea is also in accord with our finding that CDR geometries tend to mimic the cell shape.

Mutual CDR Annihilation Suggests the Existence of an Invisible Protein Field Preceding Actin Waves

In their current review, Allard and Mogilner call the mutual annihilation of colliding wavefronts a "signature of excitation waves" [2]. Zeng et al hypothesized before that colliding CDRs would mutually annihilate [16]. Their idea was thereby supported by the fact that they never found intersecting CDRs on fixed cells. Our results partially agree with this picture. We indeed observed mutual annihilation of wavefronts when CDRs happened to collide on random-shaped cells. However, on cells on microcontact printed substrates we also found collisions in which one of the two wavefronts survived. Even though this behavior might seem to contradict an active medium description at first glance, theoretical work by Argentina et al has shown that the FitzHugh-Nagumo model supports crossing of wave pulses [30]. The FitzHugh-Nagumo model is a prototype model of an active medium that supports local oscillations and wave propagation including expanding rings exhibiting instabilities [31]. The model however fails to reproduce our observation of waves that annihilate before they actually make contact, regardless of whether the visible wave corresponds to the activating or the inhibiting species. However, this observation strongly implies that a wave of an actin regulator precedes the visible actin wavefronts. The molecular identification of this wave and a detailed investigation of CDR collisions are future challenges and will add fundamentally to our understanding of CDRs.

CDR Oscillations Support an Active Medium Description

Contrary to statements by Buccione et al [6] and Itoh and Hasegawa [8] that CDRs would only form once, our results indicate that CDRs have a strong tendency to reappear periodically with typical intervals of 5–6 minutes. Especially for cells on microcontact printed substrates, repeated formation of CDRs was the rule rather than the exception. However, the procedure for CDR stimulation that is normally followed in the literature relies on addition of growth factors, such as PDGF, to the cell medium. For cells, this means a sudden disturbance of their biochemical state. In contrast, it was our focus to study cells over long times under constant biochemical conditions. We thus refrained from growth factor stimulation. Instead, we relied on the growth factors contained in FBS, which is a standard constituent of cell culture medium. This limits our knowledge about the details of the biochemical trigger of wave formation but enabled long-term experiments.

If we assume that under our regular experimental conditions cells were stimulated by growth factors contained in the FBS, we can formulate two hypotheses that account for the observed oscillatory reappearances of CDRs. The first is based on the fact that CDRs internalize the receptors that lead to their stimulation via endocytosis [32]. In this picture, growth factor stimulation would lead to CDR formation, which goes along with receptor internalization. As soon as the internalized receptors were renewed, the permanent exposure to growth factors in the cell medium again would lead to CDR formation and so on. Under this assumption, the observed periods between two succeeding CDRs would correspond to the time scale of receptor renewal. In this picture, however, cells stimulated with PDGF should also continuously form re-appearing CDRs, which is not reported in the literature [6, 8, 10].

The second hypothesis is based on characteristics of active media. From a theoretical perspective, wave propagation and oscillatory states are closely related phenomena and several models support both behaviors. Classical systems like the FitzHugh-Nagumo model respond with oscillatory behavior to stimuli, if a limit cycle trajectory in phase space is reached upon excitation. In this case, only a single stimulation would be needed to trigger oscillatory CDR formation. The period between two successive CDRs would correspond to the recovery time of the active medium.

We cannot make definite statements about whether repeated CDR formations were due to repeated stimulations or due to a biochemical state of cells that does not need continuous stimulation to repeatedly trigger waves. Indeed, experiments under FBS-free conditions indicated that cells are even capable of spontaneous CDR formation in the absence of growth factors. Therefore, one might speculate that CDRs cannot only form due to growth factor stimulation but also spontaneously. Spontaneous formation of actin waves has indeed been described in *D. discoideum* and theoretically investigated by Whitelam et al [21]. Recent work by Wu et al shows that coupling of two oscillators can account for spatio-dynamic actin patterns in mast cells, providing an attractive line of thought for the interpretation of patterns we observe on disk-like cells (see $c(\Delta s, \Delta t = 0)$ in Fig. 6D) and future experiments [5].

CONCLUDING REMARKS AND OUTLOOK

We showed in this work that CDRs are a phenomenon that is in agreement with recent theoretical models of actin waves. The apparently complex dynamics of CDRs on cells of uncontrolled morphology can be understood if we assume that a limited resource influences their dynamics. The impact of this limiting resource scales with CDR size. Consequently, CDRs of fixed sizes propagated with constant velocities and regular patterns of reappearance.

Most cell types attain irregular shapes when adhering to substrates. We showed in this work that actin-membrane waves strongly depend on boundary geometry, which complicates quantitative studies of these waves. Furthermore, the propagation velocities of actin waves are of the same order of magnitude as the velocities of changes of cell shape, e.g., the velocities of lamellipodia protrusion [33–35]. This raises questions for the appropriate frame of reference for wave propagation on the time-dependent cell morphology, even though there has been progress along these lines recently [20]. On cells adhering to symmetrically shaped protein patches controlling their morphology, waves propagate in a medium that is homogeneous in lateral direction. Especially, there are no boundaries in this direction. We thus propose that microcontact printing provides ideal means for the study of protein waves in cells. This holds especially true, because it allows for a direct comparability with theoretical studies, in which concepts like shapes of reduced complexity and periodic boundary conditions are often used [21, 23, 36].

One of the most prominent features of CDRs is their ability to strongly deform the cell surface [6, 11, 12]. For a full understanding, their three-dimensional nature must thus be taken into account. Peleg et al took the first steps towards such a three-dimensional modeling [17]. It will be a future task to characterize protein dynamics in three dimensions also experimentally. For this, the quick dynamics of CDRs and their large extension into vertical direction are the major challenges.

The three-dimensional nature of CDRs is also the main feature differentiating them from other kinds of actin waves like, e.g., the spiraling waves on the basal side of *D. discoideum* [37] or lateral membrane waves along cells spreading on planar substrates [38]. However, all of these waves are observed on cells of quasi-circular morphology, which highlights the relevance that symmetry has for the formation of well-observable actin waves of unperturbed propagation.

MATERIALS AND METHODS

Cell Culture

We used NIH 3T3 (ATCC CRL1658) fibroblasts for experiments involving microcontact printing. The NIH 3T3 cell culture was a gift from Louis Lim (Institute of Molecular and Cell Biology, ASTAR Singapore). For long-term experiments we employed a genetically modified variant termed NIH 3T3 X2 [39]. Compared with standard 3T3 cells, an enlarged fraction of X2 cells exhibited CDR formation. Cells were grown under standard conditions in Dulbecco's MEM containing 3.7 g/L $NaHCO_3$, 4.5 g/L D-Glucose

(Biochrom), 100 µg/ml Penicillin/Streptomycin (PAA), and 10% Fetal Bovine Serum (Biochrom). Cells were split at 80% confluency using Trypsin/EDTA (Biochrom). Cells were mycoplasma free.

Microcontact Printing

We followed the protocol by Théry et al for preparation of microcontact-printed substrates with minor modifications [40]. We created a silicon master containing disk structures using deep reactive ion etching. From this master, polydimethylsiloxan (PDMS, Sylgard 184 silicone elastomere) negatives were cast that served as stamps for microcontact printing.

We used a Kinpen 11 (Neoplas Control) for activation of glass substrates under sterile working conditions. The glass substrates were then stamped with human fibronectin (Roche) that was allowed to adhere to PDMS stamps before. After this the substrates were treated with PLL(20)-g-[3, 5]-PEG(2) (SuSoS Surface Technology) to prevent cell adhesion outside of fibronectin-printed areas. We used disks with an area of 3000 $(\mu m)^2$.

Sample Preparation and Imaging

For imaging of random-shaped cells, we plated cells at intermediate confluency on plasma-treated glass bottom dishes. After cells were fully spread, we washed the samples with Phosphate Buffer Saline (PBS) and added fresh DMEM. Imaging was then started and typically lasted several hours. For experiments on cells on fibronectin patterns we plated cells on microcontact-printed substrates and allowed them to adhere for 20 minutes. Samples were then thoroughly rinsed with PBS and supplied with fresh DMEM. Imaging was started thereafter and lasted typically one hour. For fluorescence microscopy cells were transfected with the actin binding peptide pLifeAct–TagGFP2. We used human-platelet derived growth factor BB (hPDGF-BB, Cell Signaling Technology) at a concentration of 30 ng/ml in serum-free DMEM in experiments with growth factor stimulation on naturally shaped cells. For experiments with cells on disk-like protein patches much lower concentrations were needed (1 ng/ml).

Live cell imaging was performed using a Zeiss Axio Oberver.Z1 equipped with a Zeiss incubation system consisting of Heating Unit XL S, Temp Module S, a Pecon Heating Insert P S1 and a CO_2 Module S. All experiments were carried out at a temperature of 37°C and 5% CO_2. For low-resolution and low-magnification imaging, a Zeiss Achro Plan 10x with a numerical aperture of 0.25 was used with a 1.6x optovar lens. For intermediate resolution and magnification we utilized a Zeiss Plan Apochromat 40x with a numerical aperture of 0.95. A Zeiss AxioCam MRm was employed for image acquisition in conjunction with Zeiss AxioVision Software. For fluorescence imaging, samples were illuminated with a Zeiss HXP120 mercury lamp, a 488 nm BrightLine HC filter (AHF Analysetechnik) and a Zeiss 76 HE reflector filter set.

Image Processing and Contour Analysis

All image and data processing was carried out in MATLAB2012b (The Mathworks) and FIJI [41] using custom-written routines.

For visualization of spiral wave patterns (Fig. 3C), we segmented time-lapse sequences via visually determined gray value thresholds and subsequent manual removal of artifacts. The resulting binary time-lapse sequences were then displayed using a 3-d viewer included in FIJI, following [25]. In 3-d views the background color black was set to transparent and the white pixels were displayed as volumetric objects. The surfaces of these objects were then rendered in x-y-t-space.

We used active contours to track the wavefronts of CDRs. Comparing images of fluorescence microscopy and images of phase contrast microscopy, we found that maximal actin concentrations corresponded to minima in phase contrast. We thus used phase contrast imaging to track CDRs and could therefore avoid issues such as photobleaching and phototoxicity. For the implementation of the active contour algorithms we followed Xu and Prince [42] and created a graphical user interface in MATLAB allowing user control and interaction. A rough initial guess of the position of the contour was based on a user input in form of a drawing. The program then let the contour converge towards the minimum of image intensity, which was the position of the CDR wavefront. For all following frames of a time-lapse sequence the program used the final state of the last iteration as the initial guess for the next frame etc. To exclude the bending energy of active contours to affect measurements of contour curvature, we performed one final relaxation step. We implemented this relaxation in the following way. For each contour point a computer program sampled an image intensity profile in normal direction of the contour. The profile was then locally approximated with a cubic function. The position of the contour point was then changed towards the global minimum of this cubic function. The data we obtained from our ruffle tracking routines were sets of contour points. The distance between adjacent contour points was 0.8 μm, which matches the resolution limit of the 10x objective. We performed a final smoothing of contours with a Sobel kernel with a width of the order of our resolution limit.

The individual contour points were the basis for calculation of contour velocities. For each contour point we calculated rays in normal direction with respect to the local contour shape. We then determined the intersection point of this ray with the contour of the next frame. We thereby interpolated linearly between the contour points of the contour of the next frame. The distance between contour point and intersection point divided by the time interval between two frames corresponded to the local velocity.

ACKNOWLEDGMENTS

We thank Arik Yochelis for fruitful discussions and comments, Eike Brauns and Melanie Kirsch from the IMSAS at the University of Bremen for the microfabrication of the waver that served for the production of PDMS stamps for microcontact printing, and Anja Bammann and Alexander Seupt for experimental assistance. EB and HGD are grateful for the hospitality of the Mechanobiology Institute Singapore where this project was initiated.

AUTHOR CONTRIBUTIONS

Conceived and designed the experiments: EB HGD NG. Performed the experiments: EB. Analyzed the data: EB HGD NG. Contributed reagents/materials/analysis tools: EB CGK. Wrote the paper: EB HGD. Contributed a cell line essential for experiments: CGK.

REFERENCES

1. Pollard TD, Borisy GG (2003) Cellular motility driven by assembly and disassembly of actin filaments. Cell 112: 453–465. doi: 10.1016/S0092-8674(03)00120-X. pmid:12600310

2. Allard J, Mogilner A (2013) Traveling waves in actin dynamics and cell motility. Curr Opin Cell Biol 25: 107–115. doi: 10.1016/j.ceb.2012.08.012. pmid:22985541

3. Bretschneider T, Anderson K, Ecke M, Muller-Taubenberger A, Schroth-Diez B, et al. (2009) The three-dimensional dynamics of actin waves, a model of cytoskeletal self-organization. Biophys J 96: 2888–2900. doi: 10.1016/j.bpj.2008.12.3942. pmid:19348770

4. Weiner OD, Marganski WA, Wu LF, Altschuler SJ, Kirschner MW (2007) An Actin-Based Wave Generator Organizes Cell Motility. PLoS Biol 5: e221. doi: 10.1371/journal.pbio.0050221. pmid:17696648

5. Wu M, Wu X, Camilli PD (2013) Calcium oscillations-coupled conversion of actin travelling waves to standing oscillations. PNAS 110: 1339–1344.

doi: 10.1073/pnas.1221538110. pmid:23297209

6. Buccione R, Orth JD, McNiven MA (2004) Foot and mouth: podosomes, invadopodia and circular dorsal ruffles. Nat Rev Mol Cell Biol 5: 647–657. doi: 10.1038/nrm1436. pmid:15366708

7. Hoon JL, Wong WK, Koh CG (2012) Functions and regulation of circular dorsal ruffles. Mol Cell Biol 32: 4246–4257. doi: 10.1128/MCB.00551-12. pmid:22927640

8. Itoh T, Hasegawa J (2012) Mechanistic insights into the regulation of circular dorsal ruffle formation. J Biochem 153: 21–29. doi: 10.1093/jb/mvs138. pmid:23175656

9. Mellström K, Höglund A-S, Nistér M, Heldin C-H, Westermark B, et al. (1983) The effect of platelet-derived growth factor on morphology and motility of human glial cells. J Muscle Res Cell Motil 4: 589–609. doi: 10.1007/BF00712117.

10. Payne LJ, Eves RL, Jia L, Mak AS (2014) p53 Down Regulates PDGF-Induced Formation of Circular Dorsal Ruffles in Rat Aortic Smooth Muscle Cells. PLoS One 9: e108257. doi: 10.1371/journal.pone.0108257. pmid:25247424

11. Dowrick P, Kenworthy P, McCann B, Warn R (1993) Circular ruffle formation and closure lead to macropinocytosis in hepatocyte growth factor/scatter factor-treated cells. Eur J Cell Biol 61: 44–53. pmid:8223707

12. Mercer J, Helenius A (2009) Virus entry by macropinocytosis. Nat Cell Biol 11: 510–520. doi: 10.1038/ncb0509-510. pmid:19404330

13. Julicher F, Kruse K, Prost J, Joanny J (2007) Active behavior of the Cytoskeleton. Phys Rep 449: 3–28. doi: 10.1016/j.physrep.2007.02.018.

14. Marchetti MC, Joanny JF, Ramaswamy S, Liverpool TB, Prost J, et al. (2013) Hydrodynamics of soft active matter. Rev Mod Phys 85: 1143–1189. doi: 10.1103/RevModPhys.85.1143.

15. Kruse K, Julicher F (2005) Oscillations in cell biology. Curr Opin Cell Biol 17: 20–26. doi: 10.1016/j.ceb.2004.12.007. pmid:15661515

16. Zeng Y, Lai T, Koh CG, LeDuc PR, Chiam K-H (2011) Investigating Circular Dorsal Ruffles through Varying Substrate Stiffness and Mathematical Modeling. Biophys J 101: 2122–2130. doi: 10.1016/j.bpj.2011.09.047. pmid:22067149

17. Peleg B, Disanza A, Scita G, Gov N (2011) Propagating cell-membrane waves driven by curved activators of actin polymerization. PLoS One 6: e18635. doi: 10.1371/journal.pone.0018635. pmid:21533032

18. Keener JP (1986) A Geometrical Theory for Spiral Waves in Excitable

Media. Siam J Appl Math 46: 1039–1056. doi: 10.1137/0146062.

19. Cross MC, Hohenberg PC (1993) Pattern formation outside of equilibrium. Rev Mod Phys 65: 851–1112. doi: 10.1103/RevModPhys.65.851.

20. Dreher A, Aranson IS, Kruse K (2014) Spiral actin-polymerization waves can generate amoeboidal cell crawling. New J Phys 16: 055007. doi: 10.1088/1367-2630/16/5/055007.

21. Whitelam S, Bretschneider T, Burroughs NJ (2009) Transformation from Spots to Waves in a Model of Actin Pattern Formation. Phys Rev Lett 102: 198103–198104. doi: 10.1103/PhysRevLett.102.198103. pmid:19519000

22. Doubrovinski K, Kruse K (2011) Cell motility resulting from spontaneous polymerization waves. Phys Rev Lett 107: 258103. doi: 10.1103/PhysRevLett.107.258103. pmid:22243118

23. Gholami A, Enculescu M, Falcke M (2012) Membrane waves driven by forces from actin filaments. New J Phys 14: 115002. doi: 10.1088/1367-2630/14/11/115002.

24. Gerisch G, Bretschneider T, Muller-Taubenberger A, Simmeth E, Ecke M, et al. (2004) Mobile actin clusters and traveling waves in cells recovering from actin depolymerization. Biophys J 87: 3493–3503. doi: 10.1529/biophysj.104.047589. pmid:15347592

25. Schroth-Diez B, Gerwig S, Ecke M, Hegerl R, Diez S, et al. (2009) Propagating waves separate two states of actin organization in living cells. HFSP J 3: 412–427. doi: 10.2976/1.3239407. pmid:20514132

26. Gerisch G, Schroth-Diez B, Muller-Taubenberger A, Ecke M (2012) PIP3 waves and PTEN dynamics in the emergence of cell polarity. Biophys J 103: 1170–1178. doi: 10.1016/j.bpj.2012.08.004. pmid:22995489

27. Edgar AJ, Bennett JP (1997) Circular ruffle formation in rat basophilic leukemia cells in response to antigen stimulation. Eur J Cell Biol 73: 132–140. pmid:9208226

28. Holmes WR, Carlsson AE, Edelstein-Keshet L (2012) Regimes of wave type patterning driven by refractory actin feedback: transition from static polarization to dynamic wave behaviour. Phys Biol 9: 046005. doi: 10.1088/1478-3975/9/4/046005. pmid:22785332

29. Mata MA, Dutot M, Edelstein-Keshet L, Holmes WR (2013) A model for intracellular actin waves explored by nonlinear local perturbation analysis. J Theor Biol 334: 149–161. doi: 10.1016/j.jtbi.2013.06.020. pmid:23831272

30. Argentina M, Coullet P, Krinsky V (2000) Head-on collisions of waves in an excitable FitzHugh-Nagumo system: a transition from wave annihilation to classical wave behavior. J Theor Biol 205: 47–52. doi: 10.1006/jtbi.2000.2044. pmid:10860699

31. Vasiev BN (2004) Classification of patterns in excitable systems with lateral inhibition. Phys Lett A 323: 194–203. doi: 10.1016/j. physleta.2004.01.068.

32. Orth JD, McNiven MA (2006) Get Off My Back! Rapid Receptor Internalization through Circular Dorsal Ruffles. Cancer Res 66: 11094–11096. doi: 10.1158/0008-5472.CAN-06-3397. pmid:17145849

33. Dubin-Thaler BJ, Giannone G, Dobereiner HG, Sheetz MP (2004) Nanometer analysis of cell spreading on matrix-coated surfaces reveals two distinct cell states and STEPs. Biophys J 86: 1794–1806. doi: 10.1016/S0006-3495(04)74246-0. pmid:14990505

34. Giannone G, Dubin-Thaler BJ, Rossier O, Cai Y, Chaga O, et al. (2007) Lamellipodial actin mechanically links myosin activity with adhesion-site formation. Cell 128: 561–575. doi: 10.1016/j.cell.2006.12.039. pmid:17289574

35. Ryan GL, Petroccia HM, Watanabe N, Vavylonis D (2012) Excitable actin dynamics in lamellipodial protrusion and retraction. Biophys J 102: 1493–1502. doi: 10.1016/j.bpj.2012.03.005. pmid:22500749

36. Doubrovinski K, Kruse K (2008) Cytoskeletal waves in the absence of molecular motors. Eur Phys Lett 83: 18003. doi: 10.1209/0295-5075/83/18003.

37. Taniguchi D, Ishihara S, Oonuki T, Honda-Kitahara M, Kaneko K, et al. (2013) Phase geometries of two-dimensional excitable waves govern self-organized morphodynamics of amoeboid cells. PNAS 110: 5016–5021. doi: 10.1073/pnas.1218025110. pmid:23479620

38. Döbereiner HG, Dubin-Thaler BJ, Hofman JM, Xenias HS, Sims TN, et al. (2006) Lateral membrane waves constitute a universal dynamic pattern of motile cells. Phys Rev Lett 97: 038102. doi: 10.1103/PhysRevLett.97.038102. pmid:16907546

39. Singh P, Gan CS, Guo T, Phang H-Q, Sze SK, et al. (2011) Investigation of POPX2 phosphatase functions by comparative phosphoproteomic analysis. Proteomics 11: 2891–2900. doi: 10.1002/pmic.201100044. pmid:21656682

40. Théry M, Piel M (2009) Adhesive Micropatterns for Cells: A Microcontact Printing Protocol. Cold Spring Harb Protoc 4. doi: 10.1101/pdb.prot5255

41.   Schindelin J, Arganda-Carreras I, Frise E, Kaynig V, Longair M, et al. (2012) Fiji: an open-source platform for biological-image analysis. Nat Methods 9: 676–682. doi: 10.1038/nmeth.2019. pmid:22743772

42.   Xu C, Prince JL (1998) Snakes, shapes, and gradient vector flow. IEEE Trans Image Process 7: 359–369. doi: 10.1109/83.661186. pmid:18276256

Chapter 4

QUANTITATIVE ANALYSIS OF LIGAND-EGFR INTERACTIONS: A PLATFORM FOR SCREENING TARGETING MOLECULES

Wei-Ting Kuo[1], Wen-Chun Lin[1], Kai-Chun Chang[2], Jian-Yuan Huang[1], Ko-Chung Yen[1], InChi Young[1], Yu-Jun Sun[1], Feng-Huei Lin[1,3]

[1] Institute of Biomedical Engineering, National Taiwan University, Taipei, Taiwan

[2] Graduate Institute of Clinical Dentistry, National Taiwan University, Taipei, Taiwan

[3] Institute of Biomedical Engineering and Nanomedicine, National Health Research Institutes, Miaoli, Taiwan

ABSTRACT

Epidermal growth factor receptor (EGFR) is often constitutively stimulated in many cancers owing to the binding of ligands such as epidermal growth factor (EGF). Therefore, it is necessary to investigate the interaction between EGFR and its targeting biomolecules. The main aim of this study was to estimate the binding affinity and adhesion force of two targeting molecules, anti-EGFR monoclonal antibody (mAb LA1) and the peptide GE11 (YHWYGYTPQNVI), with respect to EGFR and to compare these values with those obtained for the ligand, EGF. Surface plasmon resonance (SPR) was used to determine the equilibrium dissociation constant (K_D) for evaluating the binding affinity. Atomic force microscopy (AFM) was performed to estimate the adhesion force. In the case of EGFR, the K_D of EGF, GE11, and mAb LA1 were 1.77×10^{-7}, 4.59×10^{-4} and 2.07×10^{-9}, respectively, indicating that the binding affinity of mAb LA1 to EGFR was higher than that of EGF, while the binding affinity of GE11 to EGFR was the lowest among the three molecules. The adhesion force between EGFR and mAb LA1 was 210.99 pN, which is higher than that observed for EGF (209.41 pN), while the adhesion force between GE11 and EGFR was the lowest (59.51 pN). These results suggest that mAb LA1 binds to EGFR with higher binding affinity than EGF and GE11. Moreover, the adhesion force between mAb LA1 and EGFR was greater than that observed for EGF and GE11. The SPR and AFM experiments confirmed the interaction between the receptor and targeting molecules. The results of this study might aid the screening of ligands for receptor targeting and drug delivery.

INTRODUCTION

Epidermal growth factor receptor (EGFR) and its ligand, epidermal growth factor (EGF), are overexpressed in many malignancies, including cancers of the head and neck, breast, kidney, lung, colon, ovary, prostate, brain and spine, pancreas, and bladder [1]. EGFR is activated when a ligand such as EGF binds to its extracellular domain and induces conformational changes. These changes cause the EGFR to form homodimers or heterodimers with other receptors [2]. This leads to the activation of the tyrosine kinase domain of EGFR and the auto-phosphorylation of its C-terminal tyrosine residues. These events subsequently lead to the activation of the downstream signaling pathway [3]. Overexpression of EGFR is associated with several hallmarks of cancer, including inhibition of apoptosis, sustained angiogenesis, proliferation and survival, and tissue invasion and metastasis [4]. Several EGFR inhibitors that block ligand binding have been developed. These inhibitors have been shown to arrest cell growth and induce apoptosis in cancer cells [5]. In recent years, researchers have focused on identifying targeting biomolecules for more efficient drug delivery. Identifying the ligand-receptor interactions will aid in the design of optimal targeting molecules and drugs for cancer therapy.

Surface plasmon resonance (SPR) and atomic force microscopy (AFM) are powerful techniques used to analyze biomolecular interactions [6]. The SPR biosensor technology is used to measure reaction kinetics and to calculate the affinity constants of biomolecular interactions [7, 8]. In this method, the receptor is immobilized on the activated surface of a sensor chip, and a solution containing the ligand is then flowed over the surface of the chip. Binding of the interacting ligand to the surface-immobilized receptors alters the mass of the surface layer. The corresponding change in the refractive index and the shift of the resonant angle of reflected light is then detected. These changes can be monitored in real time by plotting the resonance signal as a function of time [9, 10]. The AFM is a high resolution scanning machine [11], and it is also a useful tool for the measurement of the adhesion forces between ligands and receptors [12, 13]. In this method, the AFM tip is coated with the ligand while the receptor is immobilized on the substrate. The tip is then brought in contact with the surface of substrate, enabling the formation of the ligand-receptor complex. Subsequently, the tip is retracted from the surface and the rupture force required for the dissociation of the ligand from the receptor is determined by estimating the extent of deflection of the cantilever, detected by a laser beam aimed at the cantilever and reflected onto a photodiode detector [14].

In this study, we determined the binding affinity and adhesion force of two targeting biomolecules—anti-EGFR monoclonal antibody (mAb LA1) and peptide GE11 (YHWYGYTPQNVI)—to EGFR, and compared these values with those obtained for EGF, which is the main *in vivo* competitor for the receptor during clinical application. If the binding affinity and adhesion force of the targeting molecules to EGFR are higher than that of EGF, they could compete with EGF for EGFR binding and block the subsequent activation of cellular signaling pathways. Therefore, the SPR and AFM techniques can be used to screen molecules to discover new EGFR-binding molecules for efficient drug delivery.

MATERIALS AND METHODS

Materials and Reagents

Sensor CM5 chips, HBS-EP buffer (10 mM Hepes, 150 mM NaCl, 3mM EDTA, and 0.005% Tween-20), N-Hydroxysuccinimide (NHS), 1-Ethyl-3-(3-dimethylaminopropyl) carbodiimide (EDC), and ethanolamine hydrochloride (EA) were obtained from Biacore Life Science (Uppsala, Sweden). EGFR was purchased from Sigma-Aldrich (Missouri, USA). EGF was purchased from PeproTech (New Jersey, USA). Anti-EGFR monoclonal antibody (mAb LA1) was purchased from Millipore (Massachusetts, USA). Peptide GE11 with sequence YHWYGYTPQNVI was custom synthesized by Millipore (Massachusetts, USA). AFM tips were purchased from Asylum Research (California, USA).

Preparation of EGFR on Sensor CM5 Chips

There are dextran matrix covered with carboxyl groups on CM5 chip. The chips were equilibrated with HBS-EP buffer, followed by addition of 0.1 M NHS and 0.4 M EDC mixture to activate dextran matrix to create succinimide esters. EGFR was passed over at a concentration of 10 μg/ml in 10 mM sodium acetate (pH 4.5). The esters reacted spontaneously with amino groups of EGFR. And EA was added to block the residual N-hydroxysuccinimide ester.

Binding Affinity Analysis

The binding affinity of EGF, peptide GE11 and anti-EGFR antibody to EGFR was measured by using a Biacore X SPR system (Biacore Life Science, Uppsala, Sweden) at room temperature. The chip without EGFR on the other flow channel was used as control. PBS was used as the running buffer and 50 mM NaOH was used for regeneration of the chip surface. For each concentration of EGF, peptide GE11 and anti-EGFR antibody, the flow rate of association and dissociation was 10 µl/min and regeneration was 100 µl/min. The concentrations were 0.25, 0.5, 1, 2, 4 µM for EGF, 10, 20, 40, 80, 160 µM for peptide GE11, and 0.75, 1.5, 3 µM for anti-EGFR antibody. The equilibrium dissociation constant (K_D) was obtained to evaluate the binding affinity by using the BIAEvaluation software (Biacore Life Science, Uppsala, Sweden).

Modification of AFM Tips

EGF, peptide GE11 and anti-EGFR antibody would be individually immobilized on the succinimide-modified silicon nitride cantilever tips and coating was performed only on the extreming end of the cantilever to avoid any influence on the cantilever spring constant. The process was briefly described as followings. The tips were immersed in the solution of EGF, peptide GE11 and anti-EGFR antibody with concentration of 1 mg/ml, respectively, for 24 hours at 4°C and then washed phosphate buffer solution (PBS). Finally, the residual succinimide ester was deactivated by EA for 30 minutes at 4°C. The modified tips were stored in PBS at 4°C for later use.

Adhesion Force Measurements

The topography and microstructure of the chip was obtained by MFP-3D-BIO AFM (Asylum Research, California, USA) with tapping mode. The binding force between the test sample and EGFR was measured on AFM by contact mode in PBS at room temperature. The location of immobilized EGFR on the chip was identified to allow tip moving on for later measurement. Stress-strain curve (force-distance curve) was obtained by moving the surface-modified tip to the EGFR-immobilized location, holding it on for several seconds to allow binding to occur and then retracting. All the measurements were executed at the same loading rate. The spring constant of cantilever was determined in air (measured values from 0.09 to 0.18 N/m). Curves showing significant non-specific interactions as well as those showing a zero interaction were not analyzed.

RESULTS

Immobilization of EGFR on Sensor CM5 Chips

EGFR was immobilized on Sensor CM5 chips using the amine coupling method (Fig. 1A). We used SPR to monitor the change in resonance units

(RUs), i.e. change in the refractive index of the chip, at every second. First, ethyl(dimethylaminopropyl)carbodiimide/*N*-hydroxysuccinimide   (EDC/ NHS) were reacted with the carboxyl groups on the dextran layer of the chip. Then, EGFR was immobilized on the chip via the free primary amine groups. Finally, any residual free amine groups were blocked with ethanolamine (EA). Immobilization was considered complete when a value of 12,000 RU of EGFR was achieved on an experimental flow channel (Fig. 1B).

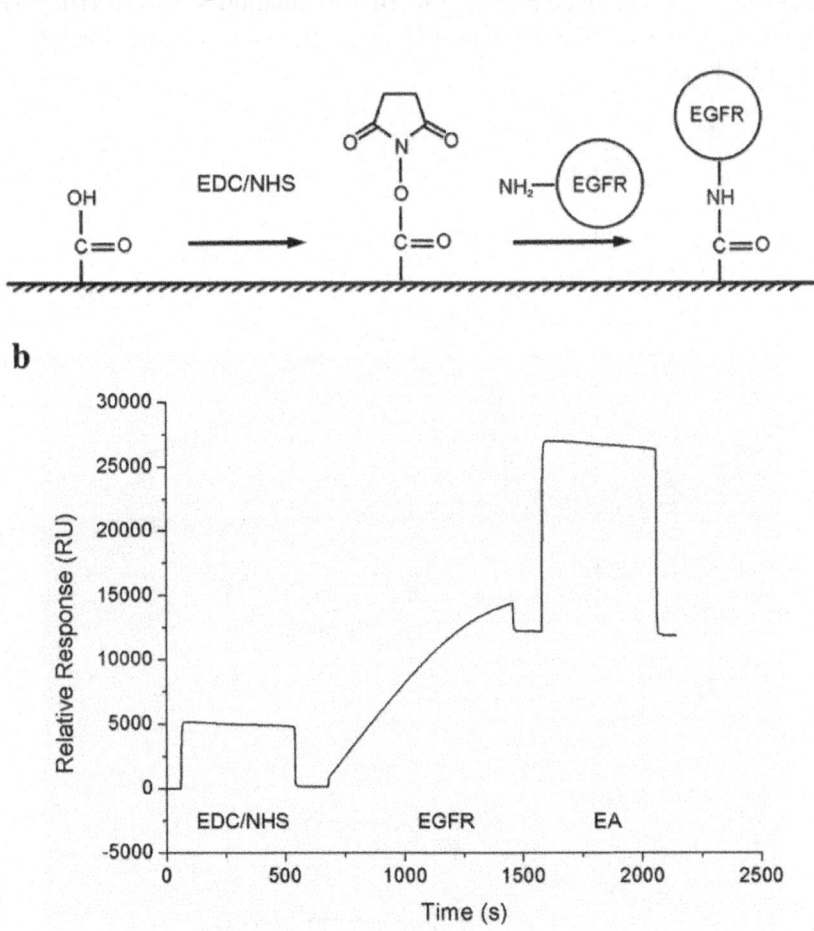

Figure 1. Immobilization of EGFR to the Sensor CM5 chips. (a) EGFR was immobilized to the surface of chip by amine coupling method. (b) SPR Sensorgram of the immobilization of EGFR.

Estimation of Binding Affinity by SPR

The binding affinity between EGF and EGFR was estimated by flowing the ligand (EGF) over the EGFR-immobilized Sensor CM5 chip. Different concentrations of EGF were injected into the flow-channel and then passed over the EGFR-immobilized Sensor M5 chip. As shown in Fig. 2A, the binding affinity between EGF and EGFR was estimated in terms of the dissociation constant (K_D), calculated using the BIA evaluation software (Biacore Life Science). The K_D value between EGF and EGFR was approximately 1.77×10^{-7} (Table 1).

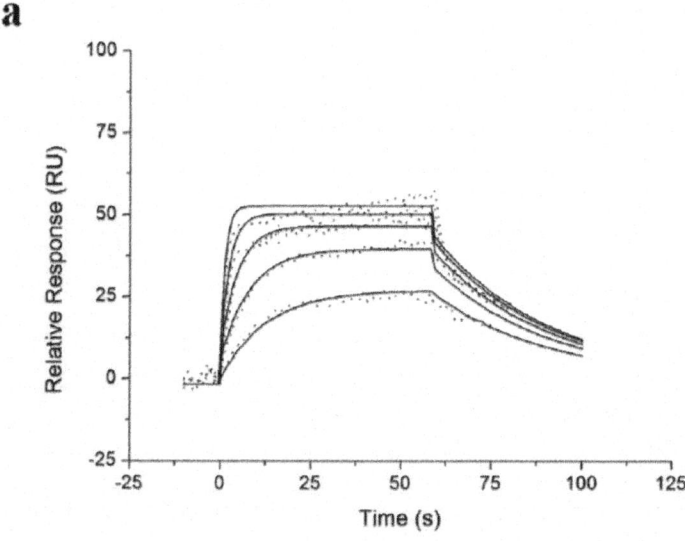

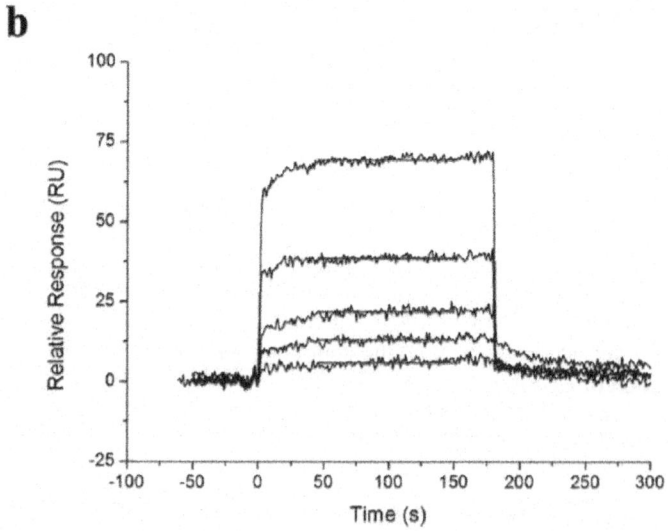

c

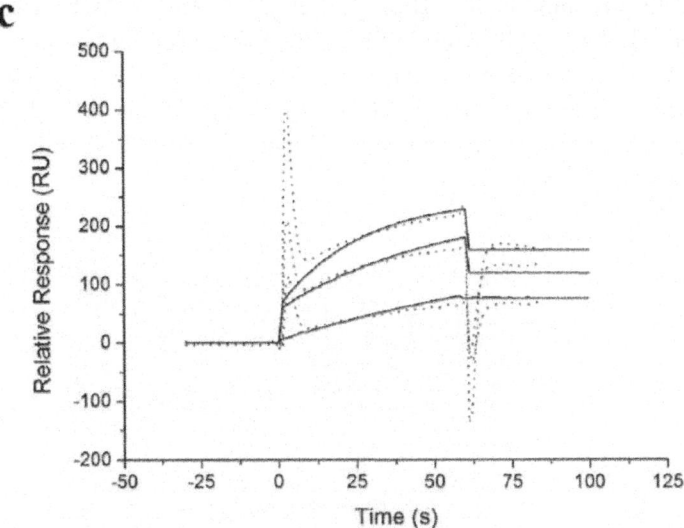

Figure 2. SPR sensorgrams of ligands to EGFR immobilized surface. (a) EGF to EGFR. (b) GE11 to EGFR. (c) mAb to EGFR.

Table 1. K_D of ligands to EGFR based on Fig. 2

Ligands	EGF	GE11	mAb
K_D (M)	1.77×10^{-7}	4.59×10^{-4}	2.07×10^{-9}

K_D: equilibrium dissociation constant

doi:10.1371/journal.pone.0116610.t001

The binding affinities of GE11 and mAb LA1 to EGFR were also estimated as described above (Fig. 2B and 2C). The K_D value between GE11 and EGFR was 4.59×10^{-4} and that between GE11 and EGFR was 2.07×10^{-9} (Table 1).

Estimation of Adhesion Force by AFM

The tip of the AFM (coated with immobilized EGF) was mounted on the cantilever of the AFM and scanned on Sensor CM5 chips to trace the location of the immobilized EGFR. AFM images of the chip surface with and without immobilized EGFR are shown in Fig. 3. The chip without immobilized EGFR had a relatively smooth surface (Fig. 3A), while the chip coated with immobilized EGFR appeared rough (Fig. 3B). The AFM images were processed using the IGOR Pro MFP-3D software (Asylum Research). Clear pictures could be obtained after image processing for both the control (Fig.

3C) and EGFR-immobilized (Fig. 3D) chips. Spikes as high as 25 nm could be observed on the EGFR-immobilized chip (Fig. 3D), indicating the location of the EGFRs. The adhesion force between EGF and EGFR was calculated, sorted into a histogram, and fitted to a single Gaussian curve (Fig. 4A). The Gaussian peak of the histogram was located at 209.41 pN (Table 2).

a

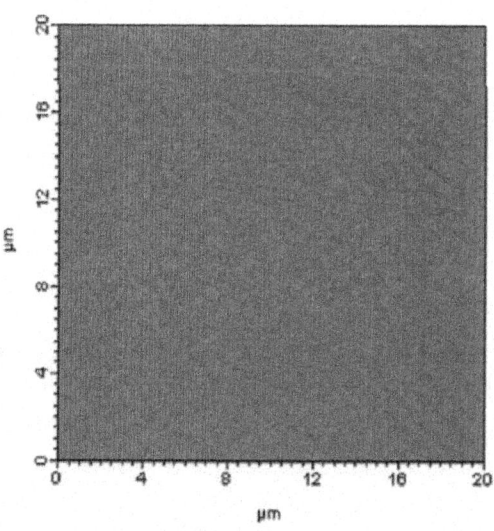

b

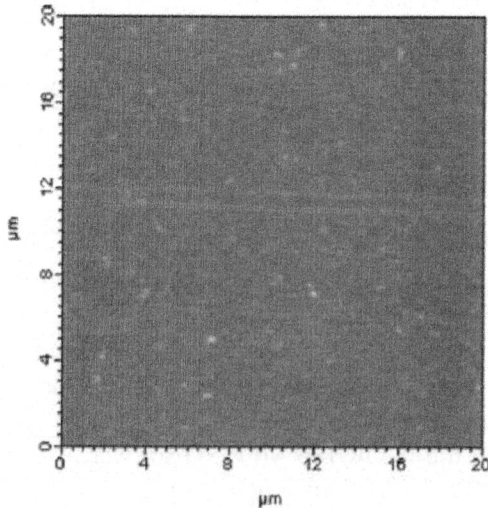

c

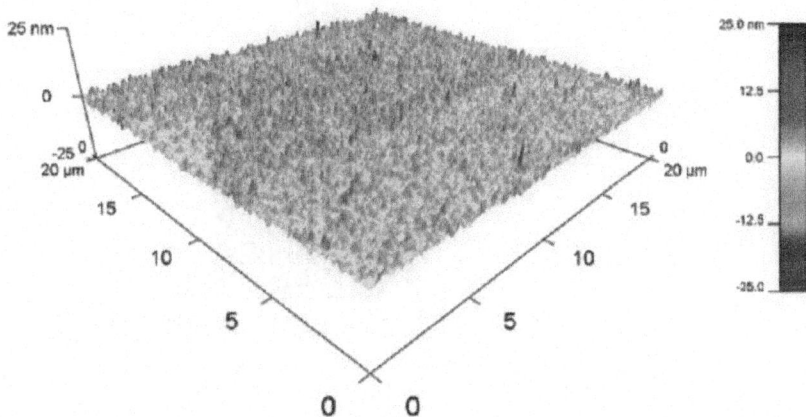

d

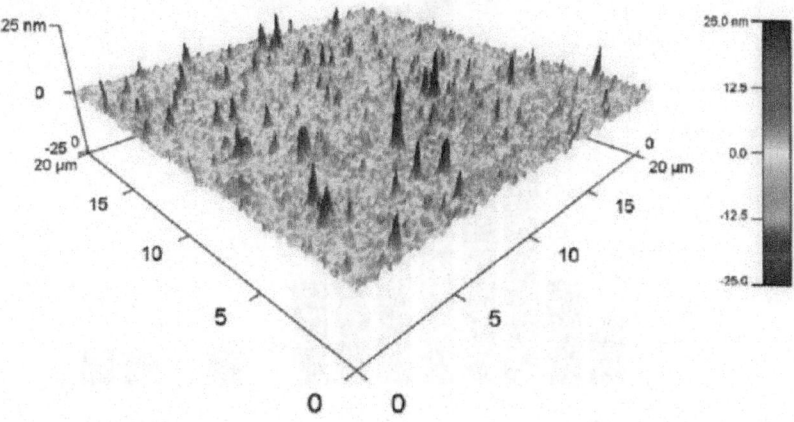

Figure 3. AFM pictures of Sensor CM5 chip surface and surface immobilized with EGFR. (a) The height image of chip. (b) The height image of EGFR immobilized chip. (c)(d) The three-dimensional view of (a) and (b) with a pseudocolor scale ranging from low (purple) to high (red).

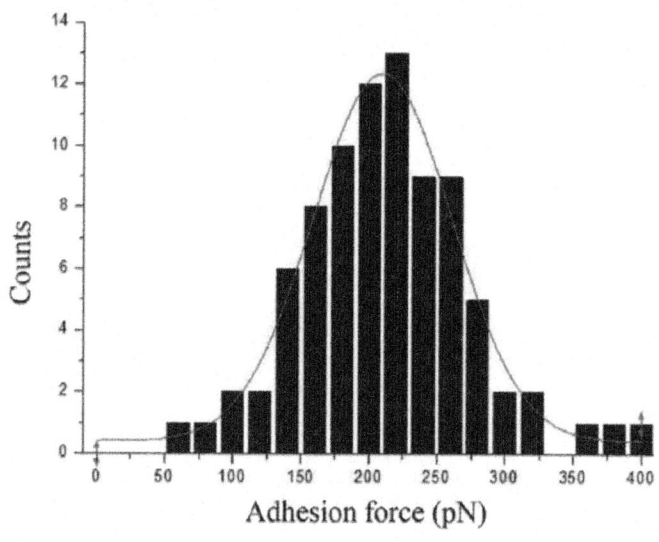

(a)

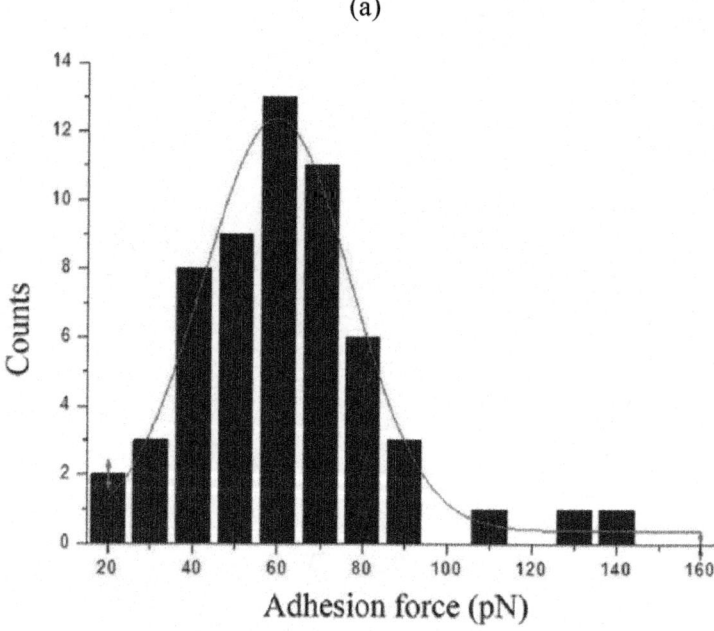

(b)

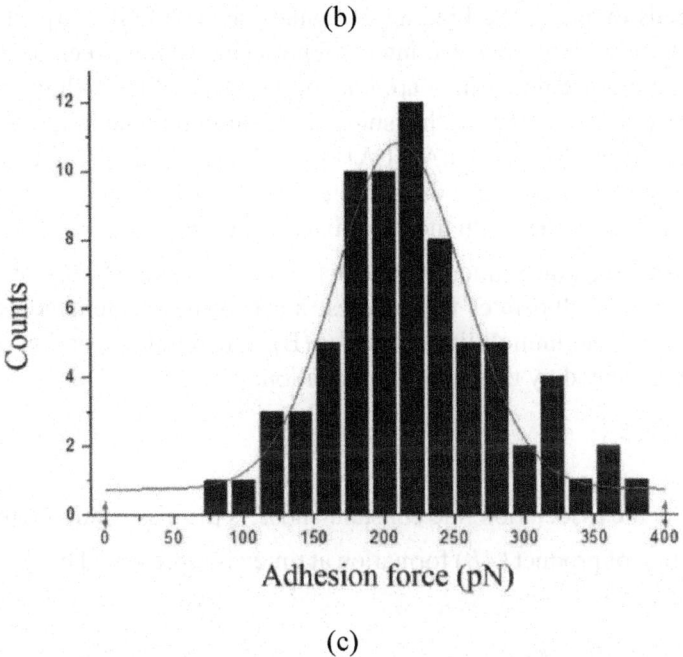

Adhesion force (pN)

(c)

Figure 4. Adhesion force histograms with fitting Gaussian (red line) of ligands to EGFR immobilized surface. (a) EGF to EGFR. (b) GE11 to EGFR. (c) mAb to EGFR.

Table 2. Adhesion force of ligands to EGFR based on Fig. 4

Ligands	EGF	GE11	mAb
Adhesion force (pN)	209.41	59.51	210.99

doi:10.1371/journal.pone.0116610.t002

The same procedure was used to calculate the adhesion force between the targeting molecules, GE11 and mAb LA1, and EGFR. As shown in Fig. 4B and 4C, the Gaussian peak of the histogram was located at 59.51 pN for GE11, and at 210.99 pN for mAb (Table 2).

DISCUSSION

Recently, increased understanding of the molecular mechanism of tumor biology has led to the development of EGFR-targeting biomolecules, which exhibit improved target selectivity toward cancer cells [15]. Nevertheless, not all EGFR targeting biomolecules are equally effective, although all of them were designed to target EGFR. EGFR is often constitutively stimulated in cancer cells owing to the binding of ligands such as EGF [16]. Therefore, it is necessary to investigate the interaction mechanism between EGFR and its targeting biomolecules, which are known bind to EGFRs with higher affinity than their ligands (EGF). In this study, we determined the binding affinity of two targeting molecules—mAb LA1 and GE11—to EGFR and determined the adhesion force between these molecules and the receptor. Moreover, we compared these values with those obtained for EGF.

We used the kinetic method to calculate the binding affinity by SPR [17–19]. This model illustrates the simplest mechanism of interaction between a ligand (A) and an immobilized receptor (B). The binding reaction of A and B can be represented by the following equation:

$$A + B \overset{k_a}{\underset{k_d}{\longleftrightarrow}} AB \tag{1}$$

where k_a is the association rate constant and k_d is the dissociation rate constant.

The rate of product (AB) formation at time t is represented by the following equation:

$$\frac{d[AB]}{dt} = k_a[A][B] - k_d[AB] \tag{2}$$

After reaction time t has elapsed, the equation may be represented as follows:

$$[B] = [B]_0 - [AB] \tag{3}$$

where $[B]_0$ is the concentration of B at $t = 0$.

On combining Equations (2) and (3), the following equation is obtained:

$$\frac{d[AB]}{dt} = k_a[A]([B]_0 - [AB]) - k_d[AB] \tag{4}$$

The number of AB complexes formed at the surface is proportional to the signal, R. Therefore, Equation (4) may be represented as follows:

$$\frac{dR}{dt} = k_a C(R_{max} - R) - k_d R \tag{5}$$

where dR/dt is the rate of formation of surface-associated complexes, C is the concentration of A, and R_{max} is the capacity of A bound to B at the surface. The integrated form of the rate equation may be represented as follows:

$$R_t = \frac{Ck_a R_{max}}{Ck_a + k_d}\left[1 - e^{-\{(Ck_a + k_d)t\}}\right] \tag{6}$$

This integrated rate equation describes the association phase of the binding curve.

The rate of dissociation of the complex (AB) is represented by the following equation:

$$\frac{dR}{dt} = -k_d R$$

(7)

The integrated form of the rate equation is as follows:

$$R_t = R_0 e^{-k_d t}$$

(8)

where R_t is the response at time t and R_0 is the amplitude of the response.

We used Equations (6) and (8) to independently fit the data obtained for the association and dissociation phases, respectively. These equations predict the binding kinetic parameters. We flowed different concentrations of EGF, GE11, and mAb LA1 over the immobilized EGFRs, and then calculated the equilibrium dissociation constant (K_D) using the BIAEvaluation software (Fig. 2). Our results revealed that the K_D of EGF-EGFR binding was higher than that of mAb-EGFR binding, and lower than that of GE11-EGFR binding (Table 1). There is an inverse relationship between K_D and affinity. Therefore, the binding affinity of mAb to EGFR was higher than that of EGF to EGFR. The binding affinity of GE11 to EGFR was the lowest among the three molecules.

Next, we used AFM to measure the adhesion force between the ligands and EGFR by recording a force curve, which is a plot of cantilever deflection (d), converted from a position-sensitive photo-diode (PSPD). This deflection distance, as a function of sample position along the z-axis, is then converted into the force (F) acting on the spring constant (k) of the cantilever tip, and can be represented by the Hooke's law [20] as follows:

$$F = k \times d$$

(9)

The spring constant is determined from the individual frequency resonances and shape factors, and is expressed as follows:

$$k = 2w(\pi f L)\sqrt{\frac{\rho^3}{E}}$$

(10)

where w is the width of the cantilever, f is the measured resonant frequency, L is the length of cantilever, ρ is the density of the cantilever material, and E is the elastic modulus (Young's modulus) of the cantilever material [21]. In this force curve, the adhesion force is characterized as the maximum force required to facilitate the separation of the ligand-receptor partners after contact. Therefore, the adhesion force of the interaction between the ligand and receptor can be estimated by measuring the increase in the volume of the force [22]. In this study, we individually coated EGF, GE11, and mAb LA1 on the tip of the AFM and immobilized EGFR on a chip to determine the adhesion force between the ligand and the receptor (Fig. 3). The results revealed that the adhesion force of the mAb-EGFR interaction was higher than that of the EGF-EGFR interaction, while the adhesion force of the GE11-EGFR interaction was the lowest among the three molecules (Fig. 4; Table 2).

GE11 was synthesized by the random peptide phage display method as an EGFR-targeting ligand that could not activate the receptor [23]. It has been conjugated to many biomaterials that interact with EGFR efficiently and with high specificity for imaging purposes and for drug delivery to EGFR-overexpressing tumors [24–26]. Nevertheless, Abourbeh et al. demonstrated that the EGFR-binding affinity/inhibitory potency of EGF are several orders of magnitude higher than those of GE11 by in vitro radioactive binding studies [27]. This was in contradiction to the results obtained by Li et al., who first reported the synthesis the GE11 peptide [23]. In this study, we used two non-radioactive methods, SPR and AFM, to confirm that the binding affinity and adhesion force of GE11 to EGFR was lower than that of EGF to EGFR.

CONCLUSIONS

The results of our study revealed that mAb LA1 had higher adhesion force and binding affinity to EGFR compared with EGF and GE11. SPR and AFM analyses confirmed the interaction between the receptor and the targeting ligands. The results of this study might aid in the screening of ligands for receptor targeting and drug delivery.

AUTHOR CONTRIBUTIONS

Conceived and designed the experiments: WTK WCL FHL. Performed the experiments: WTK WCL KCC JYH KCY ICY YJS. Analyzed the data: WTK WCL FHL. Contributed reagents/materials/analysis tools: KCC JYH KCY ICY YJS. Wrote the paper: WTK FHL.

REFERENCES

1.   Grandis JR, Sok JC (2004) Signaling through the epidermal growth factor receptor during the development of malignancy. Pharmacol Ther 102: 37–46. pmid:15056497 doi: 10.1016/j.pharmthera.2004.01.002

2.   Yarden Y (2001) The EGFR family and its ligands in human cancer. signalling mechanisms and therapeutic opportunities. Eur J Cancer 37 Suppl 4: S3–S8. pmid:11597398

3. Nyati MK, Morgan MA, Feng FY, Lawrence TS (2006) Integration of EGFR inhibitors with radiochemotherapy. Nat Rev Cancer 6: 876–885. pmid:17036041 doi: 10.1038/nrc1953

4. Hanahan D, Weinberg RA (2000) The hallmarks of cancer. Cell 100: 57–70. pmid:10647931 doi: 10.1016/s0092-8674(00)81683-9

5. Capdevila J, Elez E, Macarulla T, Ramos FJ, Ruiz-Echarri M, et al. (2009) Anti-epidermal growth factor receptor monoclonal antibodies in cancer treatment. Cancer Treat Rev 35: 354–363. doi: 10.1016/j.ctrv.2009.02.001. pmid:19269105

6. Piehler J (2005) New methodologies for measuring protein interactions in vivo and in vitro. Curr Opin Struct Biol 15: 4–14. pmid:15718127 doi: 10.1016/j.sbi.2005.01.008

7. Fägerstam LG, Frostell-Karlsson A, Karlsson R, Persson B, Rönnberg I (1992) Biospecific interaction analysis using surface plasmon resonance detection applied to kinetic, binding site and concentration analysis. J Chromatogr 597: 397–410. pmid:1517343 doi: 10.1016/0021-9673(92)80137-j

8. Myszka DG, Jonsen MD, Graves BJ (1998) Equilibrium analysis of high affinity interactions using BIACORE. Anal Biochem 265: 326–330. pmid:9882410 doi: 10.1006/abio.1998.2937

9. Wilson WD (2002) Analyzing biomolecular interactions. Science 295: 2103–2105. pmid:11896282 doi: 10.1126/science.295.5562.2103

10. Cooper MA (2002) Optical biosensors in drug discovery. Nat Rev Drug Discov 1: 515–528. pmid:12120258 doi: 10.1038/nrd838

11. Engel A, Müller DJ (2000) Observing single biomolecules at work with the atomic force microscope. Nat Struct Mol Biol 7: 715–718.

12. Florin EL, Moy VT, Gaub HE (1994) Adhesion forces between individual ligand-receptor pairs. Science 264: 415–417. pmid:8153628 doi: 10.1126/science.8153628

13. Hinterdorfer P, Baumgartner W, Gruber HJ, Schilcher K, Schindler H (1996) Detection and localization of individual antibody-antigen recognition events by atomic force microscopy. Proc Natl Acad Sci U S A 93: 3477–3481. pmid:8622961 doi: 10.1073/pnas.93.8.3477

14. Müller DJ, Dufrêne YF (2008) Atomic force microscopy as a multifunctional molecular toolbox in nanobiotechnology. Nat Nanotechnol 3: 261–269. doi: 10.1038/nnano.2008.100. pmid:18654521

15. Yarden Y, Sliwkowski MX (2001) Untangling the ErbB signalling network. Nat Rev Mol Cell Biol 2: 127–137. pmid:11252954 doi: 10.1038/35052073

16. Mendelsohn J, Baselga J (2000) The EGF receptor family as targets for cancer therapy. Oncogene 19: 6550–6565. pmid:11426640 doi: 10.1038/sj.onc.1204082

17. O'Shannessy DJ, Brigham-Burke M, Soneson KK, Hensley P, Brooks I (1993) Determination of rate and equilibrium binding constants for macromolecular interactions using surface plasmon resonance: use of nonlinear least squares analysis methods. Anal Biochem 212: 457–468. pmid:8214588 doi: 10.1006/abio.1993.1355

18. O'Shannessy DJ (1994) Determination of kinetic rate and equilibrium binding constants for macromolecular interactions: a critique of the surface plasmon resonance literature. Curr Opin Biotechnol 5: 65–71. pmid:7764646 doi: 10.1016/s0958-1669(05)80072-2

19. MacKenzie CR, Hirama T, Deng SJ, Bundle DR, Narang SA, et al. (1996) Analysis by surface plasmon resonance of the influence of valence on the ligand binding affinity and kinetics of an anti-carbohydrate antibody. J Biol Chem 271: 1527–1533. pmid:8576148 doi: 10.1074/jbc.271.3.1527

20. Allen S, Chen X, Davies J, Davies MC, Dawkes AC, et al. (1997) Detection of antigen-antibody binding events with the atomic force microscope. Biochemistry 36: 7457–7463. pmid:9200694 doi: 10.1021/bi962531z

21. Cleveland JP, Manne S 1, Bocek D, Hansma PK 1 (1993) A nondestructive method for determining the spring constant of cantilevers for scanning force microscopy. Rev Sci Instrum 64: 403–405. doi: 10.1063/1.1144209

22. Lee CK, Wang YM, Huang LS, Lin S (2007) Atomic force microscopy: determination of unbinding force, off rate and energy barrier for protein-ligand interaction. Micron 38: 446–461. pmid:17015017 doi: 10.1016/j.micron.2006.06.014

23. Li Z, Zhao R, Wu X, Sun Y, Yao M, et al. (2005) Identification and characterization of a novel peptide ligand of epidermal growth factor receptor for targeted delivery of therapeutics. FASEB J 19: 1978–1985. pmid:16319141 doi: 10.1096/fj.05-4058com

24. Dejesus OT (2012) Synthesis of [64Cu]Cu-NOTA-Bn-GE11 for PET imaging of EGFR-rich tumors. Curr Radiopharm 5: 15–18. pmid:21864245 doi: 10.2174/1874471011205010015

25. Master AM, Qi Y, Oleinick NL, Gupta AS (2012) EGFR-mediated intracellular delivery of Pc 4 nanoformulation for targeted photodynamic

therapy of cancer: in vitro studies. Nanomedicine 8: 655–664. doi: 10.1016/j.nano.2011.09.012. pmid:22024195

26. Tang H, Chen X, Rui M, Sun W, Chen J, et al. (2014) Effects of Surface Displayed Targeting Ligand GE11 on Liposome Distribution and Extravasation in Tumor. Mol Pharm.

27. Abourbeh G, Shir A, Mishani E, Ogris M, Rödl W, et al. (2012) PolyIC GE11 polyplex inhibits EGFR-overexpressing tumors. IUBMB Life 64: 324–330. doi: 10.1002/iub.1002. pmid:22362419

Chapter 5

QUANTITATIVE PHENOTYPING-BASED IN VIVO CHEMICAL SCREENING IN A ZEBRAFISH MODEL OF LEUKEMIA STEM CELL XENOTRANSPLANTATION

Beibei Zhang[1], Yasuhito Shimada[1,2,3,4,5], Junya Kuroyanagi[1], Noriko Umemoto[1,5], Yuhei Nishimura[1,2,3,4,5], Toshio Tanaka[1,2,3,4,5]

[1] Department of Molecular and Cellular Pharmacology, Pharmacogenomics and Pharmacoinformatics, Mie University Graduate School of Medicine, Edobashi, Tsu, Mie, Japan

[2] Mie University Medical Zebrafish Research Center, Edobashi, Tsu, Mie, Japan

[3] Department of Bioinformatics, Mie University Life Science Research Center, Edobashi, Tsu, Mie, Japan

[4] Department of Omics Medicine, Mie University Industrial Technology Innovation, Edobashi, Tsu, Mie, Japan

[5] Department of Systems Pharmacology, Mie University Graduate School of Medicine, Edobashi, Tsu, Mie, Japan

ABSTRACT

Zebrafish-based chemical screening has recently emerged as a rapid and efficient method to identify important compounds that modulate specific biological processes and to test the therapeutic efficacy in disease models, including cancer. In leukemia, the ablation of leukemia stem cells (LSCs) is necessary to permanently eradicate the leukemia cell population. However, because of the very small number of LSCs in leukemia cell populations, their use in xenotransplantation studies (*in vivo*) and the difficulties in functionally and pathophysiologically replicating clinical conditions in cell culture experiments (*in vitro*), the progress of drug discovery for LSC inhibitors has been painfully slow. In this study, we developed a novel phenotype-based *in vivo* screening method using LSCs xenotransplanted into zebrafish. Aldehyde dehydrogenase-positive (ALDH+) cells were purified from chronic myelogenous leukemia K562 cells tagged with a fluorescent protein (Kusabira-orange) and then implanted in young zebrafish at 48 hours post-fertilization. Twenty-four hours after transplantation, the animals were treated with one of eight different therapeutic agents (imatinib, dasatinib, parthenolide, TDZD-8, arsenic trioxide, niclosamide, salinomycin, and thioridazine). Cancer cell proliferation, and cell migration were determined by high-content imaging. Of the eight compounds that were tested, all except imatinib and dasatinib selectively inhibited ALDH+ cell proliferation in zebrafish. In addition, these anti-LSC agents suppressed tumor cell migration in LSC-xenotransplants. Our approach offers a simple, rapid, and reliable *in vivo* screening system that facilitates the phenotype-driven discovery of drugs effective in suppressing LSCs.

INTRODUCTION

Leukemia stem cells (LSCs) comprise a population of cancer stem cells (CSCs) in hematological malignancies. They possess characteristics similar to those of normal stem cells, specifically, the ability to serve as progenitor cells, but in this case they give rise to all cancer cell types, including chronic myelogenous leukemia (CML), rather than the cells of normal hematopoiesis [1]–[4]. LSCs represent a malignant reservoir of disease that is believed to drive relapse and resistance to chemotherapy [4]. Imatinib mesylate, a BCR-ABL tyrosine kinase inhibitor, has revolutionized the treatment of CML and as such is a model for targeted therapy in other cancers. However, in recent years, the efficacy of imatinib in disease eradication has been challenged [5] because of the resistance of LSCs [6], [7]. Moreover, resistance to the newer tyrosine kinase inhibitors, such as dasatinib and nilotinib, has also been documented [8],[9]. Therapeutic failure in the permanent eradication of leukemia by anti-cancer drugs such as imatinib has stimulated interest in LSC-targeted drug discovery as a rational cancer therapeutic strategy. Although the pathophysiological functions of LSCs cannot be demonstrated under culture conditions, compounds that inhibit their growth have recently been identified by *in vitro* screening [10]. Nonetheless, preclinical evaluation of their therapeutic potential is relatively slow mainly because of the very small population of LSCs available for testing in animal models[11]–[13].

Over the last few decades, a zebrafish-based screening method has emerged as a high-throughput and cost-effective alternative to other animal models and as such has been used to assess the efficacy and toxicity of several chemical compounds [14], [15]. Young zebrafish can be easily raised in 96-well plates and the maintenance cost is less than 1% of that of mice [16]. In addition, the transparent body wall of the fish enables phenotype-based screening of functional internal organs, which can be imaged using fluorescent and/or luminescent probes[17], [18]. As a cancer model, the immaturity of the young zebrafish immune system allows the xenotransplantation of human cancer cells into the fish as early as 48 h post-fertilization (hpf)[19]. The advantages of zebrafish xenotransplantation have been demonstrated in several studies in which *in vivo* fluorescent imaging was used to evaluate tumorigenesis, tumor angiogenesis, and metastatic phenotype [20]–[22]. However, despite the advantages of this method, image acquisition and quantification are labor-intensive and thus not conducive for high-throughput chemical screening. Here, we describe a rapid and phenotype-based zebrafish xenotransplant assay that is compatible with automated high-content imaging in 96-well plates. The method was tested by evaluating the efficacy of imatinib, dasatinib, parthenolide, TDZD-8, arsenic trioxide, niclosamide, salinomycin, and thioridazine in preventing LSC proliferation, tumor cell migration *in vivo*.

RESULTS

Zebrafish Xenotransplantation Assay for Screening LSC Inhibitors

A schematic representation of the experimental design is provided in Fig. 1. Cultured K562 (K562-KOr) cells stably expressing Kusabira-orange (KOr) fluorescent protein were subjected to fluorescence-activated cell sorting (FACS) to obtain aldehyde dehydrogenase-positive (ALDH+) cells and ALDH- cells, which were subsequently transplanted into the yolk sac of 48 hpf zebrafish. The xenotransplantation procedures are depicted in the Supporting Information (Fig. S1). Twenty-four hours post-injection (hpi), cancer-positive fish with similar tumor mass were selected. The variation in Kusabira-orange integrated fluorescence intensity between the recipients at 72 hpf is shown in Fig. S2A. Xenotransplantation was successful in about 73% of the zebrafish. The average survival rate post transplantation at 72 hpf before treatment was above 84% throughout our study (Fig. S2B). The larvae were transferred to a 96-well plate and imaged under anesthesia using a high-content imager. The eight therapeutic test compounds were then added to the 96-well plate using an automated pipetting workstation. Forty-eight hours later, the larvae were imaged again and cancer progression, including tumor size and cell migration, was analyzed.

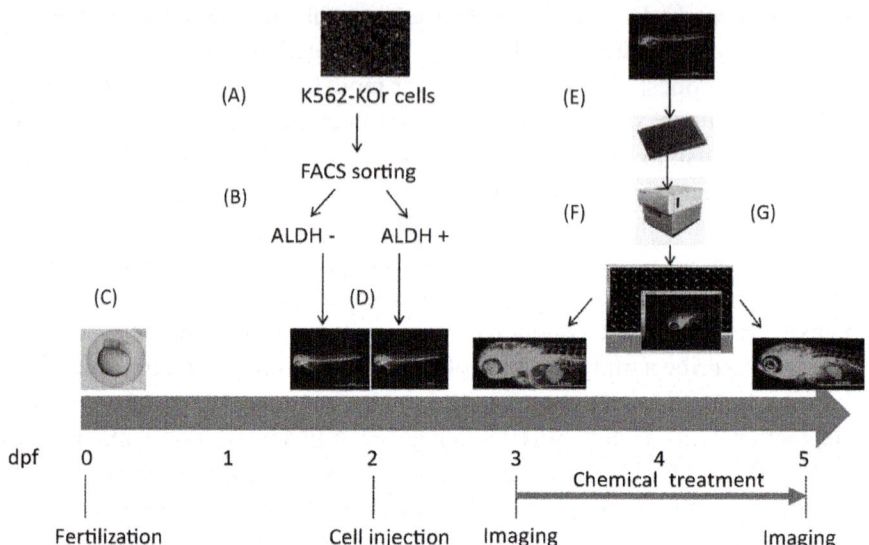

Figure 1. Schematic representation of the experimental design for LSC inhibitor screening in zebrafish. (A) Cultured K562 cells expressing a Kusabira-orange (KOr) fluorescent protein. Scale bar: 50 μm. (B) Sorting of ALDH- and ALDH+ cell populations. (C) A preparation of zebrafish embryos. Scale bar: 500 μm. (D) Xenotransplantation of ALDH- and ALDH+ cells into zebrafish at 48 hpf. Scale bar: 1.0 mm. (E, F) Xenografted zebrafish were transferred into 96-well plates at 72 hpf, imaged using a high-content imaging system, and then treated with the test compounds. Scale bar: 1.0 mm. (G) At 120 h (48 h after treatment), the xenografted zebrafish were imaged again to evaluate the effects of the chemicals. Scale bar: 1.0 mm.

ALDH+ K562-KOr Cells have LSC Properties

K562 cells with high ALDH activity (ALDH+) comprised ~3.8% of the total cell population (Fig. 2A), similar to the yield in a previous study [23]. The LSC properties of these ALDH+ cells were determined by qPCR analysis of CD133 mRNA, a biomarker of CSCs [24]. CD133 expression was higher in ALDH+ than in ALDH- cells ($P<0.01$, Fig. 2B). FACS analyses of the two subpopulations for CD34, another CSC biomarker, showed that it was much more highly expressed in the ALDH+ (61.4%) than in the ALDH- (2.1%) population (Fig. 2C). CD38 and Lineage (Lin) were also negative in ALDH+ cells but not in ALDH- cells, as seen by immunofluorescent staining (Fig. 2D). In addition, the *in vitro* proliferation of ALDH+ cells was greater than that of ALDH- cells at 72 h ($P<0.01$, Fig. 2E). Six days after xenotransplantation (a relatively long duration for zebrafish), ALDH+ cells were more tumorigenic than ALDH- cells (Fig. 3A and B) and exhibited greater distal migration to the tail region (Fig. 3C).

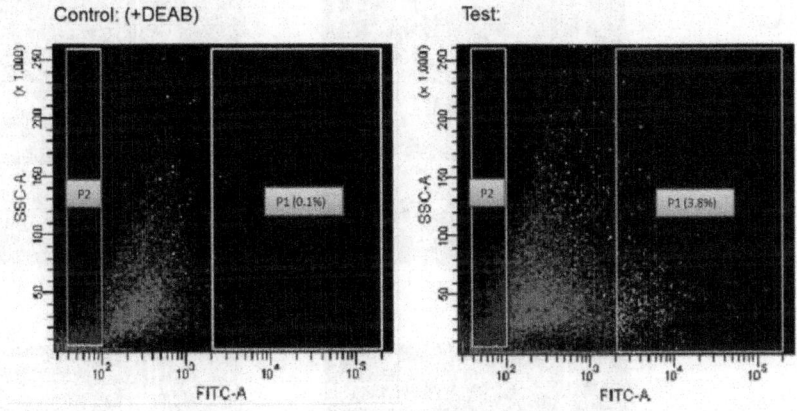

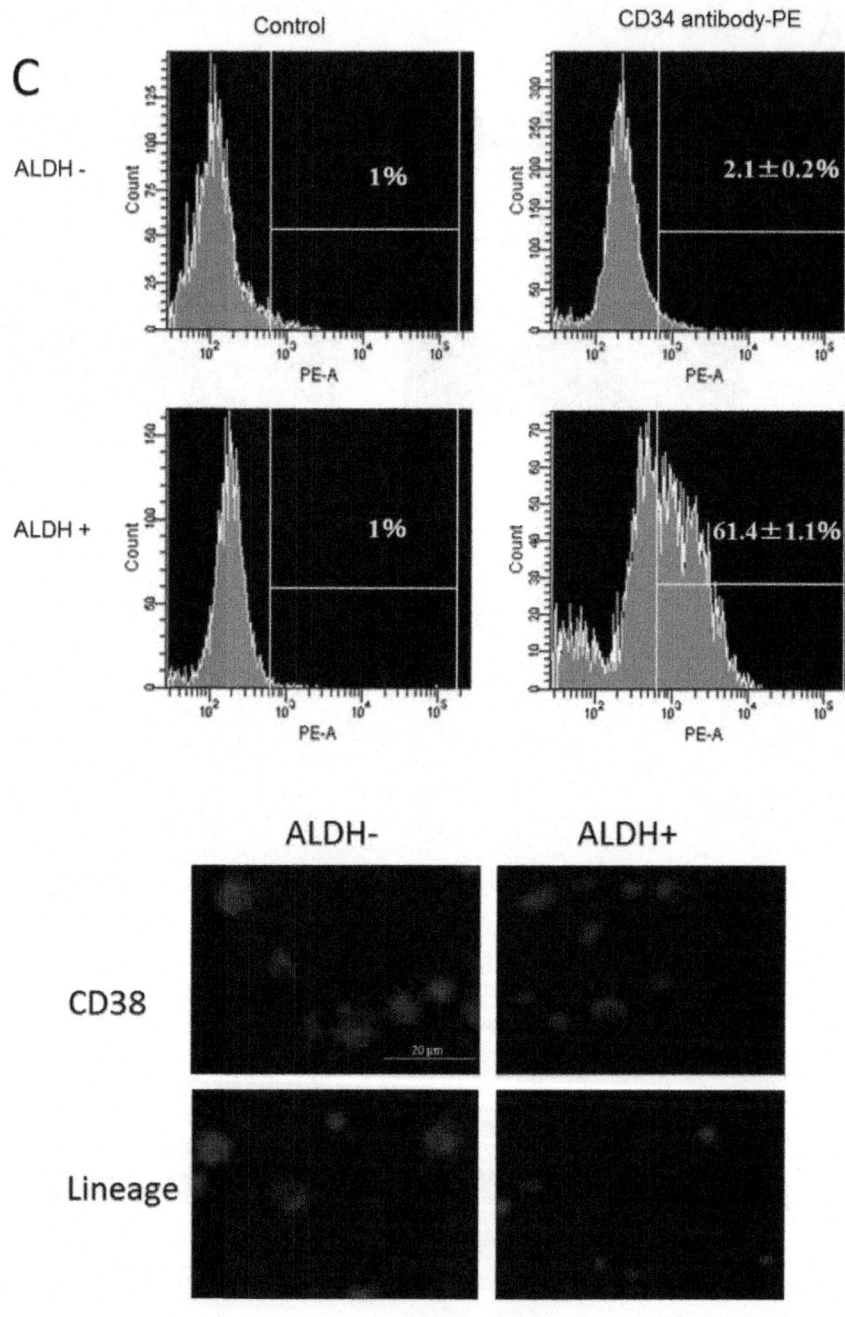

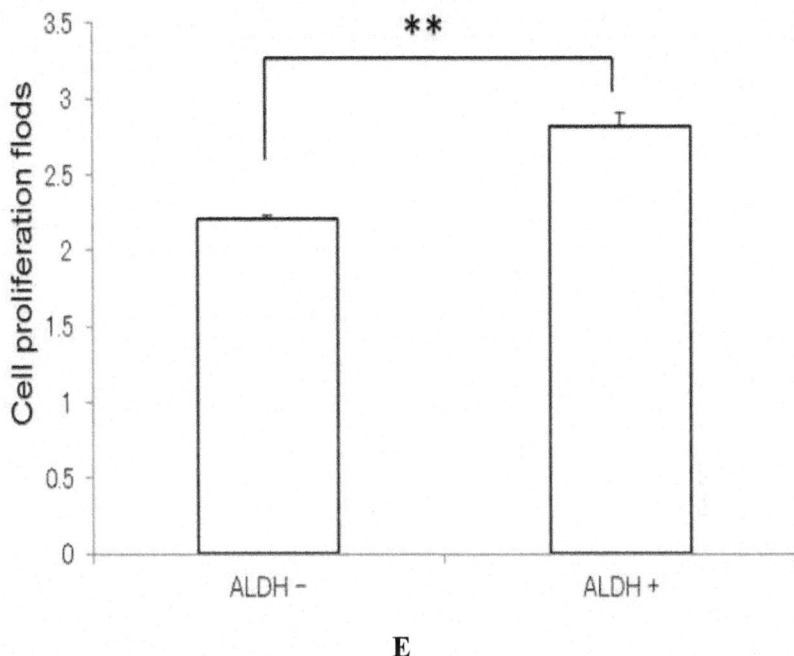

E

Figure 2. ALDH+ K562-KOr cells showed LSC properties. (A) FACS analysis for ALDH+ cells (P1 and P2 indicate the ALDH+ and ALDH- subpopulations, respectively). (B) qPCR for the CSC marker CD133 in ALDH+ cells (C) CD34-positive cells in the ALDH+ cell population (n=3), *$P<0.05$. (D) The negative of CD38 and Lin in ALDH+ cells determined by immunofluorescent staining. Blue, nucleus; red, CD38 or Lin. Scale bar: 20 μm. (E) Cell proliferation at 72 h differed in ALDH- and ALDH+ cells (n=3), **$P<0.01$.

A

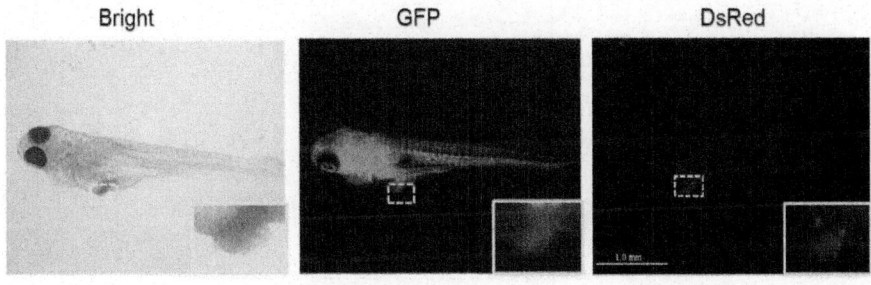

B
*
percentage (%) with obvious tumor mass
70
60
50
40
30
20
10
0
ALDH-
ALDH +
C
*
percentage (%) with tail metastasis
50
45
40
35
30
25
20
15
10
5
0
ALDH-
ALDH +

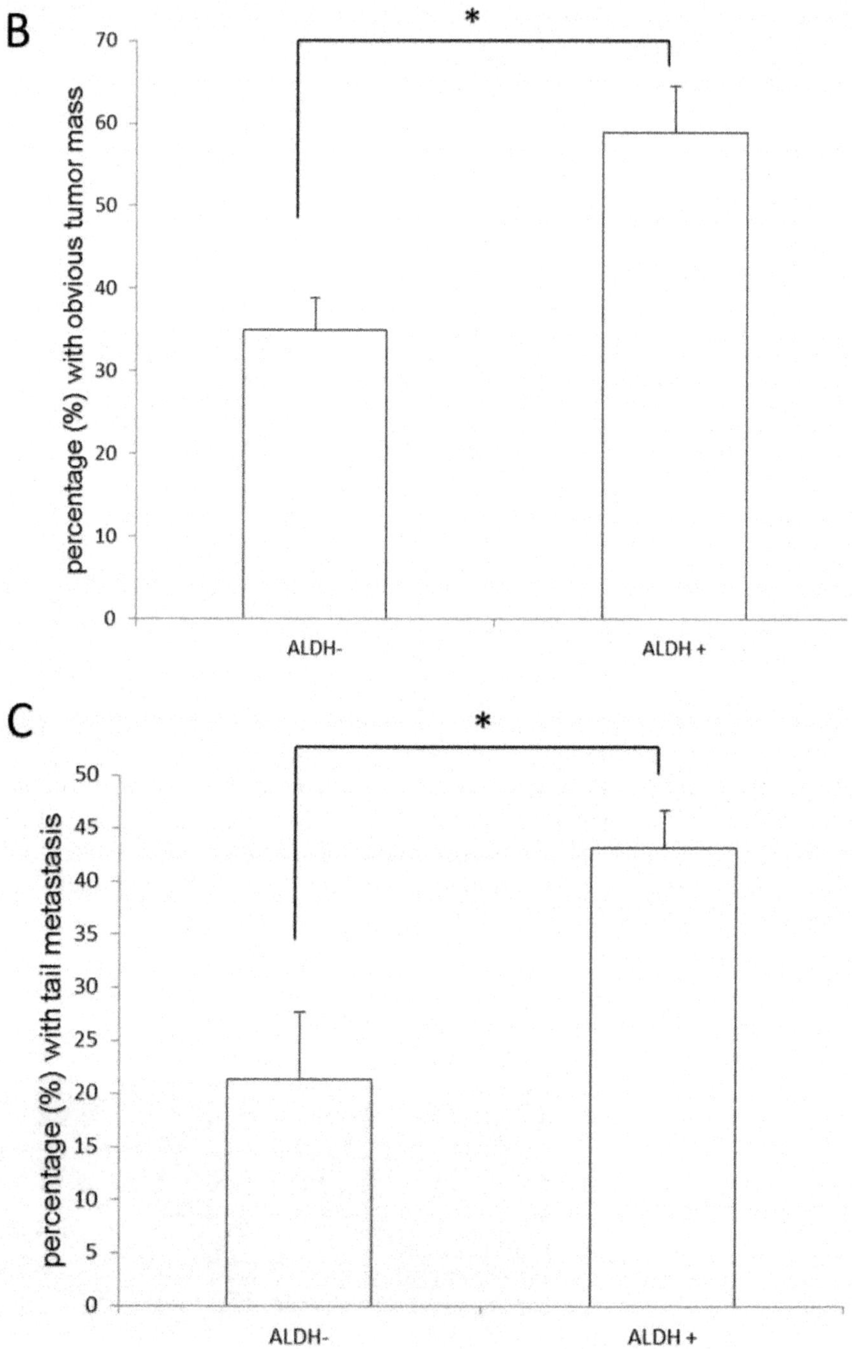

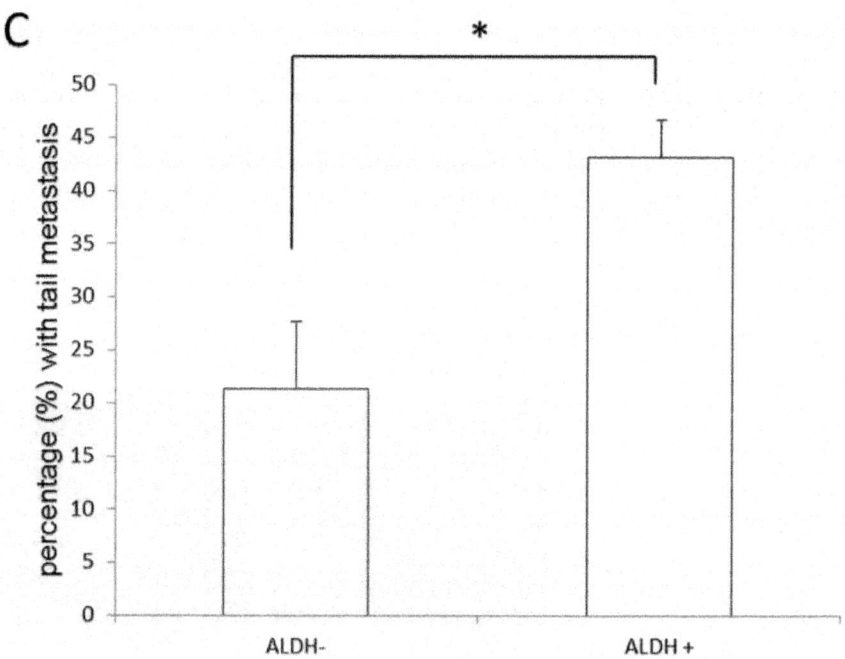

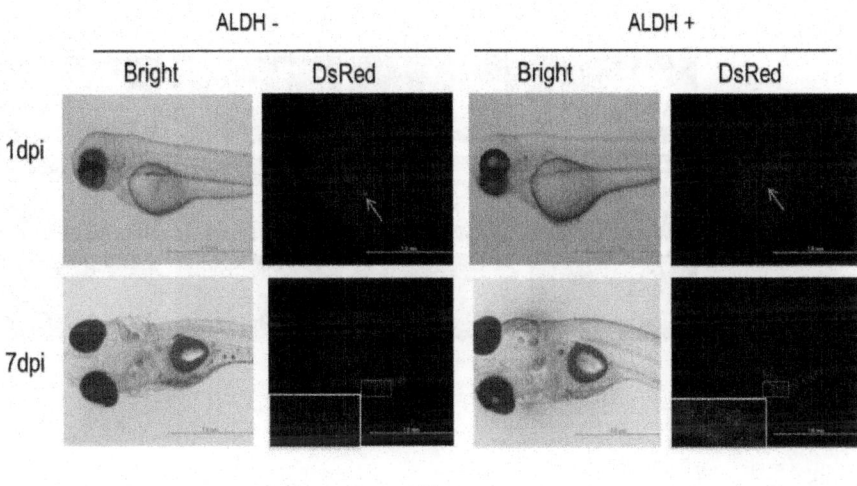

D

Figure 3. LSC ability in zebrafish xenograft. (A) Typical images of cancer xenografts at 6 days post-injection (dpi). Scale bar: 1.0 mm. The implanted tumor is framed in yellow (the magnification shows the outlined area). (B) In 6-dpi xenografts, ALDH+ cells exhibited greater tumorigenicity than ALDH- cells. *P<0.05, based on 3 independent experiments. (C) The distal (tail region) migration of ALDH+ cells was greater than that of ALDH- cells. *P<0.05, based on 3 independent experiments. (D) In single cell xenotransplants, ALDH+ cells proliferated at 7 dpi whereas ALDH- cells were no longer detectable. Scale bar: 1.0 mm. The implanted tumor is framed in yellow (the magnification shows the outlined area).

To validate these LSC properties in the zebrafish xenografts, we conducted an *in vivo* limiting dilution assay. Transplanted zebrafish with single cancer cell in transplant site were collected and the two cell populations (ALDH- and ALDH+) were analyzed. ALDH+ cells were observed to proliferate after 7 days while ALDH- cells were no longer detectable (Fig. 3D). Consistent with these findings, tumorigenesis capacity in zebrafish xenotransplanted with the ALDH+ population was also much higher than in fish xenotransplanted with ALDH- cells for 72 hpi (P<0.05, Fig. 4A and B). These results showed that the ALDH+ population of K562-KOr cells contained putative LSCs.

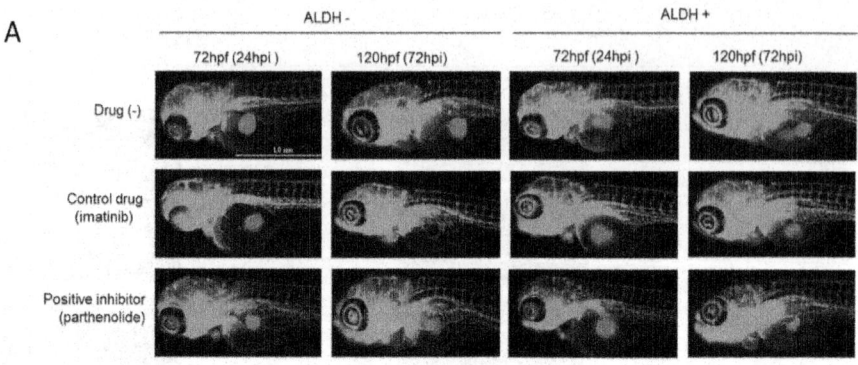

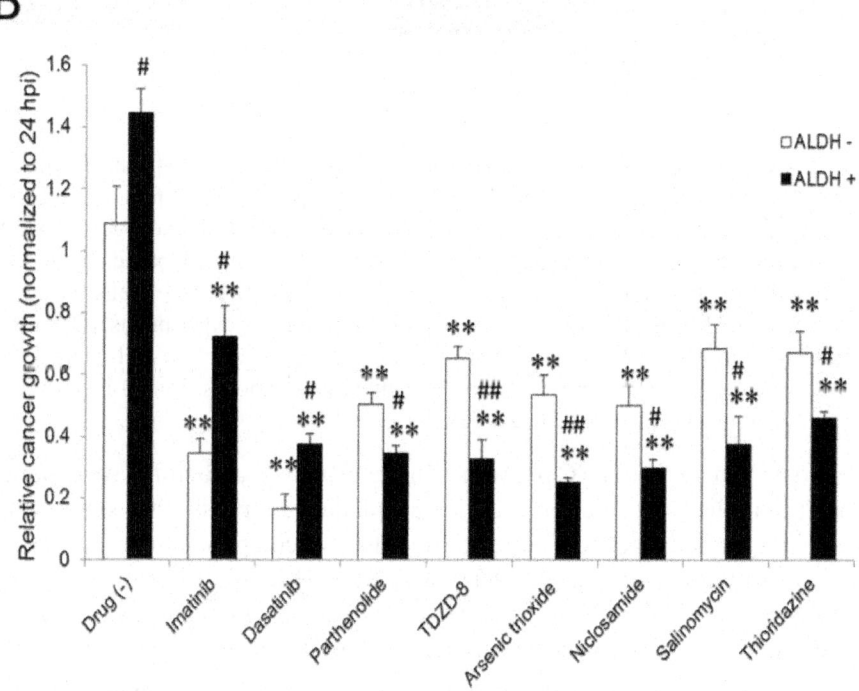

C

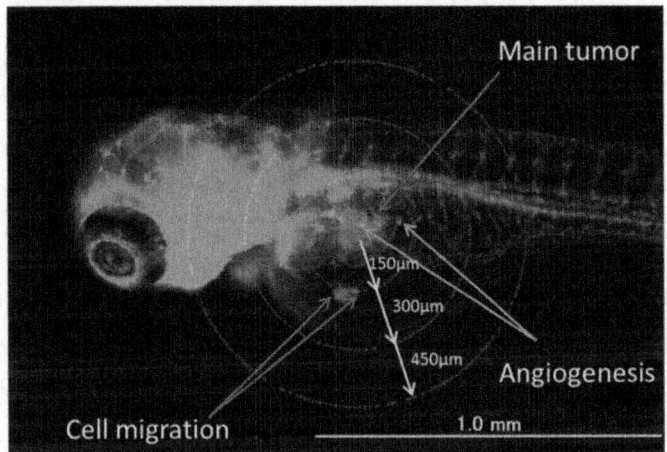

D

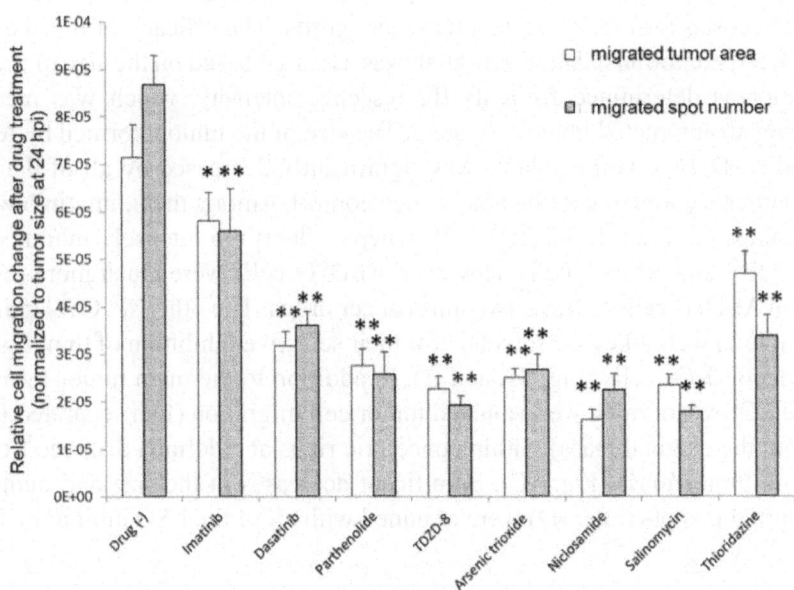

Figure 4. The effects of the test compounds on tumor inhibition in LSC-xenografted zebrafish. (A) Typical images of xenotransplanted zebrafish either not treated or treated with the control drug (imatinib) or an LSC inhibitor (parthenolide). Scale bar: 1.0

mm. (B) Tumor proliferation in xenotransplanted zebrafish. The test compounds (0.5 µM) were administered at 72–96 hpf. The ratio of the fluorescence intensity at 120 and 72 hpf was calculated as an indicator of the increase in tumor size, i.e. tumor proliferation (n=15–25). ∗∗P<0.01 vs. no drug; #P<0.05 and ##P<0.01 vs. ALDH- cells. (C) Representative image of the main tumor, angiogenesis, and cell migration. Scale bar: 1.0 mm. (D) Quantitative analysis of cell migration. The number of migrated tumors and their sizes (area) were measured within concentric rings at a defined distance from the main tumor and grouped accordingly. The LSC inhibitors decreased the size and the number of migrated tumors in LSC xenotransplants (n=15–25). ∗P<0.05 ∗∗ P<0.01 vs. no drug.

Treatment of Zebrafish Xenotransplants with LSC Inhibitors

The survival ratios of normal 3 days post-fertilization (dpf) zebrafish treated with the eight different therapeutic test compounds for 48 h are shown in Table 1. These data show that imatinib, dasatinib, TDZD-8, and arsenic trioxide had the fewest side effects, based on median lethal doses (LD_{50}) >10 µM, followed by parthenolide and thioridazine (LD_{50}: 1 µM–10 µM) and niclosamide and salinomycin (LD_{50}: 0.5 µM~1 µM). Because at a concentration of 0.5 µM none of the compounds caused obvious morphological changes, this concentration was chosen for use in the zebrafish xenografts. The efficacy of the chemicals in K562-xenotransplanted zebrafish was assessed based on the size of the main tumor as determined from its fluorescence intensity, which was measured using an automated imaging system. The size of the tumors formed by ALDH- and ALDH+ xenotransplants was significantly decreased by all of the tested chemicals compared to the non-treated control. Among them, imatinib was the weakest inhibitor of ALDH+ cells whereas dasatinib intensely inhibited both ALDH+ and ALDH- cells. However, ALDH+ cells were much more resistant than ALDH- cells to these two anti-cancer drugs. The other six CSC inhibitors (Table 2) were likewise screened for their selective inhibition of tumorigenesis in K562-KOr cells (Fig. 4A and B). In addition to the main tumor formed by the LCS xenografts, we evaluated tumor cell migration (foci separated further from the original mass) within concentric rings at a defined distance from the main tumor mass (Fig. 4C). Significant decreases in the size and number of migration spots (Fig. 4D) were obtained with all of the LSC inhibitors.

Table 1. Survival rate (%) of normal zebrafish treated with the test chemicals

Chemicals	Concentrations(μM)						
	0.125	0.25	0.5	1	2.5	5	10
No drug	100±0	100±0	100±0	100±0	100±0	100±0	100±0
Imatinib	100±0	100±0	100±0	100±0	100±0	100±0	100±0
Dasatinib	100±0	100±0	100±0	100±0	100±0	100±0	100±0
Parthenolide	100±0	100±0	100±0	100±0	100±0	100±0	0±0
TDZD-8	100±0	100±0	100±0	100±0	100±0	100±0	100±0
Arsenic trioxide	100±0	100±0	100±0	100±0	100±0	100±0	100±0
Niclosamide	100±0	100±0	100±0	0±0	0±0	0±0	0±0
Salinomycin	100±0	100±0	100±0	0±0	0±0	0±0	0±0
Thioridazine	100±0	100±0	100±0	100±0	100±0	0±0	0±0

Zebrafish were exposed to different concentrations of the test compounds for 48 h from 72 hpf. Three independent experiments were performed (n = 10 under each condition per test).
doi:10.1371/journal.pone.0085439.t001

Table 2. Characteristics of the test chemicals

Compound	Leukemia type	Inhibited subpopulation	Main mechanism	Model	Reference
Imatinib	CML	Normal leukemia cells	Tyrosine kinase inhibitor	Mouse, human	[8,9]
Dasatinib	CML	CD34+CD38-	Inhibits CrKL phosphorylation	Mouse, human	[8,9]
Parthenolide	AML,CML	CD34+CD38-	Inhibits NF-κB, activates p53, stimulates ROS production	Mouse	[30]
TDZD-8	AML,CML,ALL	CD34+CD38-	Inhibits NF-κB, oxidative stress	Mouse	[31]
Arsenic trioxide	APL,CML	CD34+CD38-	Reduces PML, blocks NF-κB, stimulates ROS production	Mouse	[32,35]
Niclosamide	AML	CD34+CD38-	Inactivates NF-κB, stimulates ROS production	Mouse	[36]
Salinomycin	CLL	CD44+	Inhibits the Wnt pathway and NF-κB, induces oxidative stress	Mouse	[37,38]
Thioridazine	AML	CD45+CD33+	Inhibits DR signaling, antioxidant activity	Mouse	[39,40]

CML, chronic myelogenous leukemia; AML, acute myelogenous leukemia; ALL, acute lymphoblastic leukemia; APL, acute promyelocytic leukemia; CLL, chronic lymphocytic leukemia; ROS, reactive oxygen species; PML, promyelocytic leukemia protein; DR, dopamine receptor.
doi:10.1371/journal.pone.0085439.t002

To explore the relationship of ROS and the anti-LSC effect, we cultured ALDH+ cells with 10 μM of the six LSC inhibitors except imatinib and dasatinib for 24 h. After treatment, the ALDH+ cell survival ratio decreased significantly ($P<0.01$, Fig. 5A). A determination of ROS status after chemical treatments showed that all six compounds induced the overproduction of ROS in ALDH+ K562 cells ($P<0.01$, Fig. 5B and C).

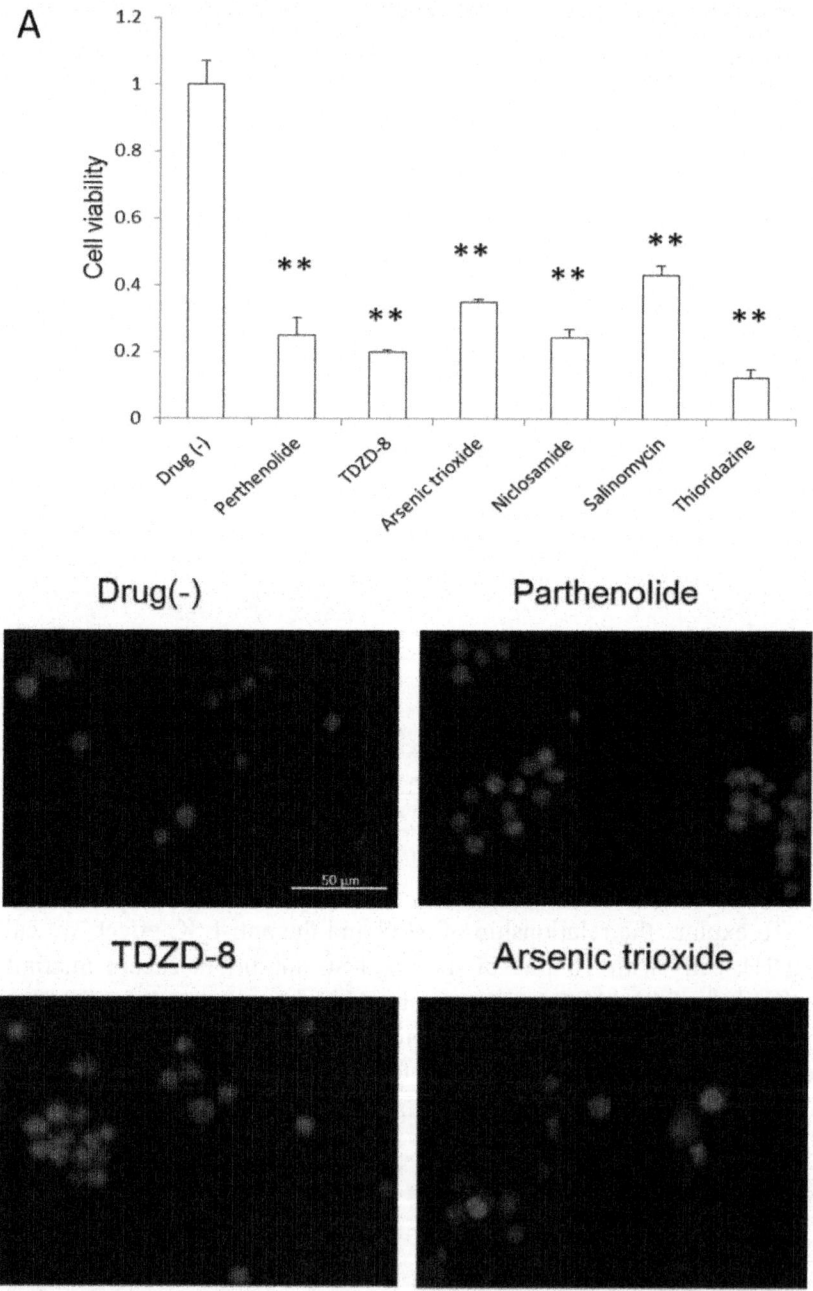

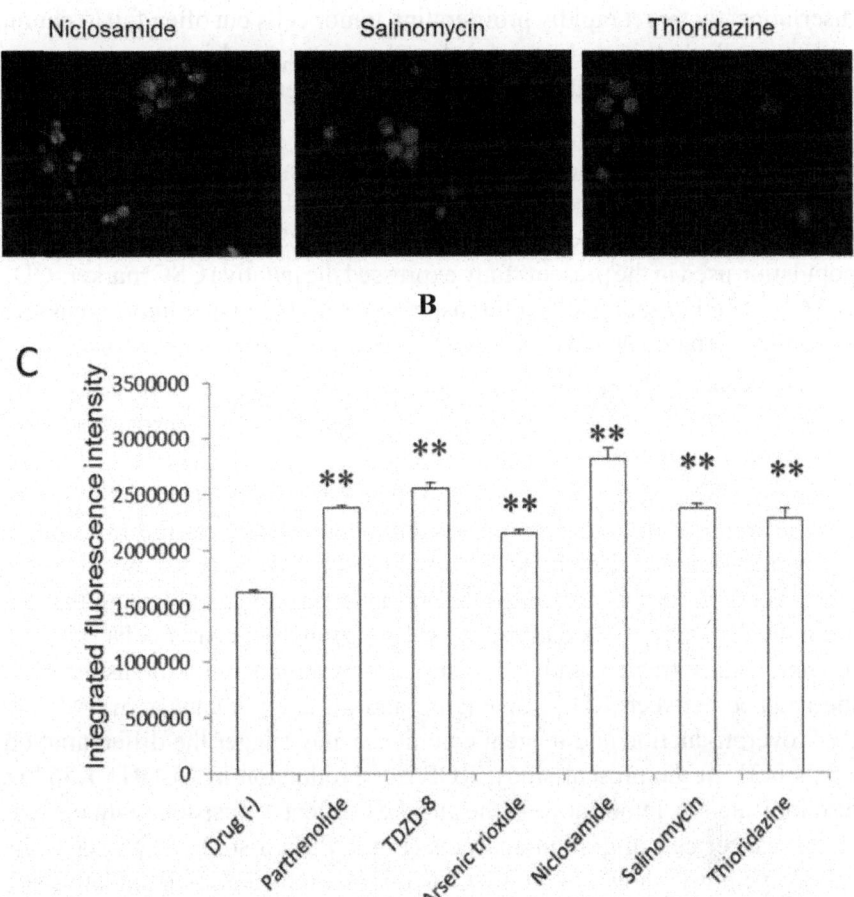

Figure 5. ROS overproduction by the LSC inhibitors. (A) LSC inhibitors (10 μM) significantly inhibited the *in vitro* proliferation of ALDH+ cells 24 h after treatment (n=3), **$P<0.01$. (B) Typical images showing ROS production in ALDH+ cells. LSC inhibitors (10 μM) significantly induced *in vitro* ROS production in cells treated for 24 h. Blue, nucleus; red, ROS. Scale bar: 50 μm. (C) Quantitative analysis of ROS in the cells (n=4), **$P<0.01$.

DISCUSSION

The advantages of therapeutically targeting the self-renewing LSC or CSC cell populations include less toxicity and fewer side effects. Moreover, this approach is more potent than standard chemotherapeutic agents, which nondiscriminately target rapidly proliferating tumor cells but often fail to eliminate resistant cells. In this study, we developed a phenotype-driven *in vivo* screening method in which a zebrafish model is used to determine LSC inhibition.

In previous studies, cell populations with high ALDH activity were shown in serial or secondary transplantation assays to exhibit CSC properties [25], [26] and high ALDH activity has successfully been used to identify LSCs from clinical samples [27], [28]. The ALDH+ K562 cell population used in the present study expressed the putative CSC markers CD133 and CD34 and in zebrafish xenotransplants exhibited higher tumorigenesis and imatinib resistance than ALDH- cells, consistent with other reports [27], [29].

The LSC inhibition results support the findings of previous studies examining the efficacies and mechanisms of action of the tested compounds. The six that inhibited LSCs in our zebrafish model system are also potent inhibitors of CSCs, both *in vitro* and in rodent models [8], [11],[30]–[41]. With the exception of thioridazine, these compounds were reported to cause ROS overproduction and inactivate NF-κB [30]–[41]. In CSCs, low ROS levels protect cells from DNA damage during tumor seeding, suggesting that LSCs are more sensitive to oxidative stress than normal leukemia cells [42]. Thus, in tumor cells treated with ROS-stimulating compounds, a disturbance of the balance between ROS scavenging and production causes overwhelming ROS overproduction and in stem cell niches may trigger the differentiation of CSCs [43]. In the present study, ROS overproduction in ALDH+ K562 cells also indicated its importance in the anti-LSC effect. Consistent with the results of our cancer cell migration analyses (Fig 4D), in a study of prostate cancer salinomycin was shown to inhibit CSC migration by inducing oxidative stress, as demonstrated in a wound healing assay [37].

Cross-talk between ROS and NF-κB is well-established [44] and has been implicated in the mechanism of action of the NF-κB pathway inhibitor parthenolide, which preferentially inhibits the stem cell population of breast cancer cells [12]. The ability of thioridazine to block dopamine receptors may explain the reduced growth of CSC malignancies achieved with this drug [40],[45] and suggests a relationship between NF-κB and dopamine receptor signaling [46], [47].

A number of recent zebrafish leukemia xenograft studies indicated that zebrafish are a useful animal model in cancer research and chemotherapeutic drug screening [48]–[50]. Several methods have been developed for the *in vitro* evaluation of cell proliferation (MTT assay), metastasis (under-agarose migration assay), and angiogenesis (endothelial tube formation assay). Our zebrafish model provides an ideal platform for the simultaneous evaluation *in vivo* of LSC proliferation, angiogenesis, and metastasis (migration) as well as drug-related side effects. The small number of cells (100–200 cells/injection) required for the assay and the high-throughput screening (in 96-well-plate format) overcome the bottlenecks that arise because of the limited number of CSCs (~0.1% of the total cancer cell population). In addition, with our zebrafish-based method multiple novel candidate anti-LSC agents can be tested in small amounts (nM or µM concentrations, 200 µl volume per animal). Thus, imaging-based LSC xenotransplant screening in zebrafish offers distinct advantages over other animal models and can greatly accelerate the phenotype-driven discovery of anti-LSC agents.

MATERIALS AND METHODS

Ethical Approval

All animal experiments were conducted according to the Animal Welfare and Management Act (Ministry of Environment of Japan) and complied with international guidelines. Ethical approval from the local Institutional Animal Care and Use Committee was not sought, since this law does not mandate the protection of fish. After the experiments, the fish were sacrificed at 5 (or 8/9) dpf by an overdose of anesthesia.

Chemicals

All of the test compounds (imatinib, dasatinib, parthenolide, TDZD-8, arsenic trioxide, niclosamide, salinomycin, thioridazine) were purchased from Sigma-Aldrich (St. Louis, MO). Stock solutions (10 mM) were dissolved in dimethyl sulfoxide (DMSO; Sigma-Aldrich). For anesthesia, 100 ppm 2-phenoxyethanol (2-PE; Wako Pure Chemical Industries, Osaka, Japan) was diluted in E3 medium (5 mM NaCl, 0.17 mM KCl, 0.4 mM $CaCl_2$, and 0.16 mM $MgSO_4$).

Zebrafish

The care and breeding of the zebrafish followed previously described protocols [51]. Because of the greater transparency of their bodies, which facilitates *in vivo* monitoring of tumor angiogenesis, nacre/rose/fli1:egfp zebrafish, obtained by cross-breeding nacre/rose mutants and fli1:egfp transgenic zebrafish, were used in the experiments [52]. Three days before xenotransplantation, individual female zebrafish were placed in mating tanks with males. The next morning, mating was initiated by light stimuli and the resulting fertilized eggs were collected. These eggs were incubated in E3 medium at 28°C, removing the dead eggs and replenishing the medium every day until the experiments were conducted.

Preparation of K562-KOr Cells

K562 cells were obtained from the RIKEN Cell Bank (Tokyo, Japan) and pre-cultured in RPMI1640 medium (Life Technologies, Carlsbad, CA) supplemented with 10% heat inactivated fetal bovine serum (Life Technologies), 100 U penicillin G/ml and 100 µg streptomycin (Sigma-Aldrich)/ml at 37°C in 5% CO_2. The cells were transfected with the Kusabira-orange (KOr) fluorescent protein expression vector phKO1-MN1 (Amalgaam, Tokyo, Japan) using LipofectAMINE 2000 (Life Technologies) according to the manufacturer's instructions. Twenty-four hours after transfection, cells stably expressing KOr (K562-KOr cells) were selected in medium containing 800 µg geneticin/ml (Roche Diagnostics, Mannheim, Germany). After one week of culture, the KOr-expressing cells were purified by FACSAria flow cytometry (BD Biosciences, San Jose, CA) and further cultured.

LSCs from K562-KOr Cells

K562-KOr cells positive or negative for ALDH were sorted in an ALDEFLUOR assay (StemCell Technologies, Vancouver, Canada) followed by FACSAria flow cytometry (BD Biosciences), according to the manufacturer's instructions. To confirm the LSC character of the ALDH+ cells, CD34 expression was assayed by incubating the cells with anti-CD34-phycoerythrin (PE)-conjugated antibody (Beckman Coulter, Krefeld, Germany) followed by FACSAria flow cytometry according to the manufacturer's instructions. CD38 and Lineage (Lin) expression was confirmed using immunofluorescent staining with APC anti-human CD38 antibody (BioLegend, San Diego, CA, USA) and APC anti-human Lineage cocktail (BioLegend) according to the manufacturer's instructions.

Total RNA Extraction, cDNA Synthesis, and qPCR

Total RNA was purified from the cells using the RNeasy mini kit (Qiagen, Hilden, Germany) according to the manufacturer's instructions. The first-strand cDNA was synthesized from 200 ng of total RNA using the SuperScript III cDNA synthesis kit (Life Technologies) with random primers (Life Technologies). RT-PCR was performed using Power SYBR Green Master Mix (Applied Biosystems, Foster City, CA) and a 7300 real-time PCR system (Applied Biosystems) as recommended by the manufacturer. The target gene was amplified using the primers CD133 (5′-ATC TGC AGT GGA TCG AGT TCT CT -3′ and 5′-ACA CAG AAA GAC ATC AAC AGC AGT AT-3′). The data were normalized with respect to the human housekeeping gene β-actin (ACTB), amplified with the primers 5′- TGT GCT ATC CCT GTA CGC CTC -3′ and 5′- GTA GAT GGG CAC AGT GTG GGT GA -3′.

Cell Proliferation Assay

Sorted ALDH+ and ALDH- cells were cultured in 96-well plates at a density of 3000 cells per well in RPMI1640 medium supplemented with 1% heat inactivated fetal bovine serum, 100 U penicillin G/ml and 100 µg streptomycin ml at 37°C in 5% CO_2. After treatment of the cells with the test compounds for 24 h, cell proliferation was measured using the CellTiter-Glo luminescent cell viability assay (Promega, Madison, WI, USA). Luminescence signals were measured in a Victor2 fluorescent plate reader (PerkinElmer Boston, MA, USA).

Intracellular ROS Quantification

The cells were fluorescently stained for intracellular ROS status by incubating them in 5 µM CellROX Deep Red detection reagent (Life Technologies) for 30 min, followed by three washes in PBS. After nuclear staining with 40 µg Hoechst 33342 dye (Dojin, Kumamoto, Japan)/ml for 5 min, images were captured to detect the intracellular ROS signal using the ImageXpress MICRO high content screening system (Molecular Devices, Sunnyvale, CA, USA). Cell fluorescence was quantified using the accompanying software.

Leukemia Cell Xenotransplantation

Just before xenotransplantation, 48-hpf zebrafish were dechorionized using 2 mg pronase (Roche Diagnostics)/ml as described previously [53], anesthetized, and arrayed on a holding sheet. ALDH+ and ALDH- cells (1×10^6 cells each) were separately suspended in 50 µl of Hanks' balanced salt solution (Life Technologies). The glass needles used to inject the cells were made from a GD-1 glass capillary (Narishige, Tokyo, Japan) using a PP-830 gravity puller (Narishige) and fine-polished with an EG-44 microforge (Narishige). The number of injected cells was counted microscopically by transferring the same volume of injected cells on glass slides using the same glass capillary tubes and injection pressure (FemtoJet, Eppendorf, Hamburg, Germany) in each experiment, as described in a previous study [54]. The avascular region of the yolk sac was then injected with a volume of the above-described suspension containing 100–200 cells using the glass needles and the FemtoJet injection system (Eppendorf, Hamburg, Germany). The xenotransplanted zebrafish were subsequently maintained at 32°C.

High-Content Imaging

Twenty-four hours after xenotransplantation (72 hpf), the successfully xeno-transplanted zebrafish were transferred in 50 µl of anesthetic solution into a

96-well imaging plate (353219; BD Biosciences). After gentle centrifugation (300 G, 30 s), the zebrafish were imaged live in an ImageXpressMICRO (Molecular Devices, Sunnyvale, CA) using the image acquisition program to automatically detect and image the fish in each well as follows: The overall well was imaged by prescanning, with 4-view (9-views for 48 wells) photographs obtained using a FITC filter (Semrock, Rochester, NY) and a 2-power lens (Plan Apo; Nikon, Tokyo, Japan). The zebrafish body was recognized based on the GFP intensity in the image and the stage was then moved such that the center of brightness was the center of view. Five images were taken continuously using a FITC filter and a 2-power lens, moving the system in the z direction by 40 μm each time. A composite image was then created from the best-focused images. For dual-wavelength (EGFP and KOr) imaging, serial radiography was performed using the same tetramethylrhodamine isothiocyanate (TRITC) method used for KOr and the best-focused composite image was created. A 4-power lens (S Fluor; Nikon) was used to obtain ten images, continuously moving the lens in the z direction by 20 μm each time. The Cool SNAP HQ (Roper Scientific, Tucson, AZ) CCD camera was used, with camera binning and gain both set to 1.

Chemical Treatment

After initial imaging of the fish, a JANUS automated workstation (Perkin Elmer, USA) was used to replace the anesthetic solution with 100 μl of the test compound in E3 medium. The 96-well plate was shaken for 30 s on an MTS2 shaker (IKA Labortcchnik, Staufen, Germany) and then incubated at 32°C as described above. After 24 h, the medium was replaced with fresh chemical-containing medium, again using a JANUS automated workstation. Forty-eight hours after treatment (120 hpf), the zebrafish were imaged again as described above.

Image Analysis

Tumor size, and cell migration were analyzed using an imaging-based method and MetaXpress software (Molecular Devices). From the transplanted tumor clusters, the main tumor was identified using a multi-wavelength cell scoring application module based on the KOr (TRITC filter) images, calculating the area, total luminance value, and average radius of the tumors. Blood vessels were distinguished based on the detection of cell bodies, as described in the tumor angiogenesis image analysis program. Metastatic tumors were identified using the Transfluor application module. Concentric circles with radii of 150 μm, 300 μm, and 450 μm were drawn from the center of brightness and the number and size (area) of the metastases (migration) were calculated with respect to their distance from the center (0–150 μm, 150–300 μm, 300–450 μm, and >450 μm).

Statistical Analysis

Data are expressed as the mean ± SEM. Differences between two groups were compared using Student's t-test. For multiple comparisons, a one-way ANOVA followed by Dunnett's test for multiple comparisons was used. $P<0.05$ was considered statistically significant.

SUPPORTING INFORMATION

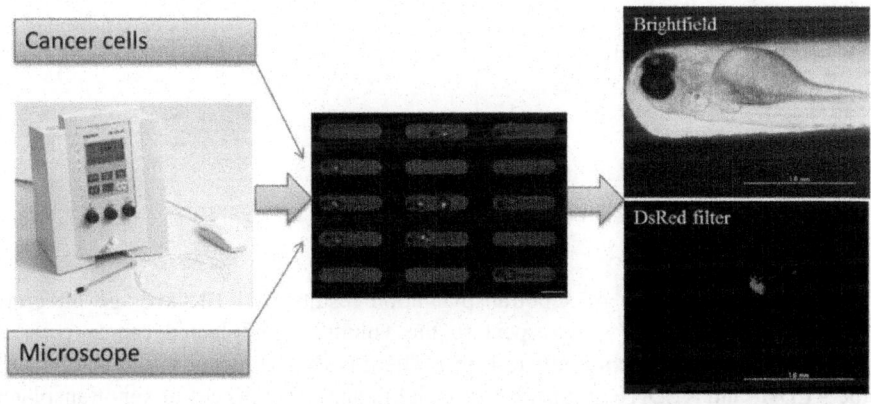

Figure S1. Xenotransplantation procedures.

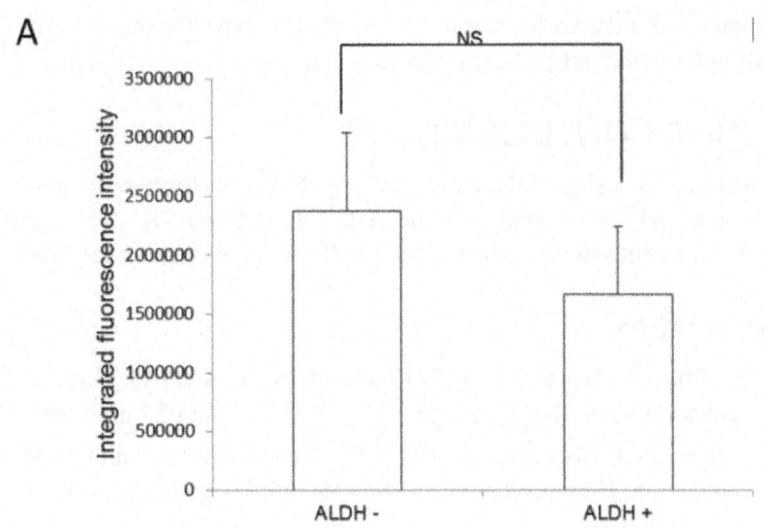

B

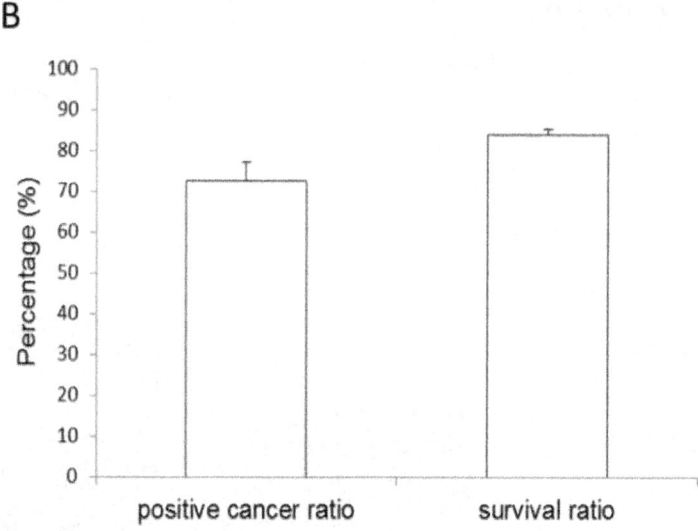

Figure S2. Description of xenotransplantation results. (A) The average integrated fluorescence intensity (with respect to the volume of implanted cancer cells) of ALDH- and ALDH+ xenografts (24 hpi). There is no significant difference between the ALDH- and ALDH+ groups. NS, not significant. (B) Successful xenotransplantation and survival ratio.

ACKNOWLEDGMENTS

We thank S. Ichikawa for her assistance in the experiments, M. Ariyoshi for breeding the fish, and R. Ikeyama and Y. Tamura for secretarial assistance.

AUTHOR CONTRIBUTIONS

Conceived and designed the experiments: YS TT. Performed the experiments: BZ YS JK NU YN. Analyzed the data: BZ YS JK NU YN. Contributed reagents/materials/analysis tools: BZ YS JK NU YN. Wrote the paper: BZ YS.

REFERENCES

1.   Savona M, Talpaz M (2008) Getting to the stem of chronic myeloid leukaemia. Nat Rev Cancer 8: 341–350. doi: 10.1038/nrc2368

2.   Heaney NB, Holyoake TL (2007) Therapeutic targets in chronic myeloid leukaemia. Hematol Oncol 25: 66–75. doi: 10.1002/hon.813

3.   Yong AS, Keyvanfar K, Eniafe R, Savani BN, Rezvani K, et al. (2008) Hematopoietic stem cells and progenitors of chronic myeloid leukemia

express leukemia-associated antigens: implications for the graft-versus-leukemia effect and peptide vaccine-based immunotherapy. Leukemia 22: 1721–1727. doi: 10.1038/leu.2008.161

4.   Stuart SA, Minami Y, Wang JY (2009) The CML stem cell: evolution of the progenitor. Cell Cycle 8: 1338–1343. doi: 10.4161/cc.8.9.8209

5.   Rice KN, Jamieson CH (2010) Molecular pathways to CML stem cells. Int J Hematol 91: 748–752. doi: 10.1007/s12185-010-0615-8

6.   Graham SM, Jorgensen HG, Allan E, Pearson C, Alcorn MJ, et al. (2002) Primitive, quiescent, Philadelphia-positive stem cells from patients with chronic myeloid leukemia are insensitive to STI571 in vitro. Blood 99: 319–325. doi: 10.1182/blood.v99.1.319

7.   Angstreich GR, Matsui W, Huff CA, Vala MS, Barber J, et al. (2005) Effects of imatinib and interferon on primitive chronic myeloid leukaemia progenitors. Br J Haematol 130: 373–381. doi: 10.1111/j.1365-2141.2005.05606.x

8.   Copland M, Hamilton A, Elrick LJ, Baird JW, Allan EK, et al. (2006) Dasatinib (BMS-354825) targets an earlier progenitor population than imatinib in primary CML but does not eliminate the quiescent fraction. Blood 107: 4532–4539. doi: 10.1182/blood-2005-07-2947

9.   Thomas X (2012) Philadelphia chromosome-positive leukemia stem cells in acute lymphoblastic leukemia and tyrosine kinase inhibitor therapy. World J Stem Cells 4: 44–52. doi: 10.4252/wjsc.v4.i6.44

10. Pei S, Jordan CT (2012) How close are we to targeting the leukemia stem cell? Best Pract Res Clin Haematol 25: 415–418. doi: 10.1016/j.beha.2012.10.003

11. Jordan CT (2007) The leukemic stem cell. Best Pract Res Clin Haematol 20: 13–18. doi: 10.1016/j.beha.2006.10.005

12. Zhou J, Zhang H, Gu P, Bai J, Margolick JB, et al. (2008) NF-kappaB pathway inhibitors preferentially inhibit breast cancer stem-like cells. Breast Cancer Res Treat 111: 419–427. doi: 10.1007/s10549-007-9798-y

13. Gupta PB, Onder TT, Jiang G, Tao K, Kuperwasser C, et al. (2009) Identification of selective inhibitors of cancer stem cells by high-throughput screening. Cell 138: 645–659. doi: 10.1016/j.cell.2009.06.034

14. Zon LI, Peterson RT (2005) In vivo drug discovery in the zebrafish. Nat Rev Drug Discov 4: 35–44. doi: 10.1038/nrd1606

15. Jing L, Zon LI (2011) Zebrafish as a model for normal and malignant hematopoiesis. Dis Model Mech 4: 433–438.

16. Pichler FB, Laurenson S, Williams LC, Dodd A, Copp BR, et al. (2003)

Chemical discovery and global gene expression analysis in zebrafish. Nat Biotechnol 21: 879–883. doi: 10.1038/nbt852

17. Delvecchio C, TiefenbachH J, Krause M (2011) The zebrafish: a powerful platform for in vivo, HTS drug discovery. Assay Drug Dev Technol 9: 354–361. doi: 10.1089/adt.2010.0346

18. Snaar-Jagalska BE (2009) ZF-CANCER: developing high-throughput bioassays for human cancers in zebrafish. Zebrafish 6: 441–443. doi: 10.1089/zeb.2009.0614

19. Konantz M, Balci TB, Hartwig UF, Dellaire G, Andre MC, et al. (2012) Zebrafish xenografts as a tool for in vivo studies on human cancer. Ann NY Acad Sci 1266: 124–137. doi: 10.1111/j.1749-6632.2012.06575.x

20. Peal DS, Peterson RT, Milan D (2010) Small molecule screening in zebrafish. J Cardiovasc Transl Res 3: 454–460. doi: 10.1007/s12265-010-9212-8

21. Nicoli S, Ribatti D, Cotelli F, Presta M (2007) Mammalian tumor xenografts induce neovascularization in zebrafish embryos. Cancer Res 67: 2927–2931. doi: 10.1158/0008-5472.can-06-4268

22. Marques IJ, Weiss FU, Vlecken DH, Nitsche C, Bakkers J, et al. (2009) Metastatic behaviour of primary human tumours in a zebrafish xenotransplantation model. BMC Cancer 9: 128. doi: 10.1186/1471-2407-9-128

23. Nakamura S, Yokota D, Tan L, Nagata Y, Takemura T, et al. (2012) Down-regulation of Thanatos-associated protein 11 by BCR-ABL promotes CML cell proliferation through c-Myc expression. Int J Cancer 130: 1046–1059. doi: 10.1002/ijc.26065

24. Koyama-Nasu R, Takahashi R, Yanagida S, Nasu-Nishimura Y, Oyama M, et al. (2013) The Cancer Stem Cell Marker CD133 Interacts with Plakoglobin and Controls Desmoglein-2 Protein Levels. PLOS One 8: e53710. doi: 10.1371/journal.pone.0053710

25. Szabo AZ, Fong S, Yue L, Zhang K, Strachan LR, et al. (2012) The CD44(+) ALDH(+) Population of Human Keratinocytes is Enriched for Epidermal Stem Cells with Long Term Repopulating Ability. Stem Cells 31: 786–799. doi: 10.1002/stem.1329

26. Hess DA, Wirthlin L, Craft TP, Herrbrich PE, Hohm SA, et al. (2006) Selection based on CD133 and high aldehyde dehydrogenase activity isolates long-term reconstituting human hematopoietic stem cells. Blood 107: 2162–2169. doi: 10.1182/blood-2005-06-2284

27. Fleischman AG (2012) ALDH marks leukemia stem cell. Blood 119: 3376–3377. doi: 10.1182/blood-2012-02-406751

28. Gerber JM, Qin L, Kowalski J, Smith BD, Griffin CA, et al. (2011) Characterization of chronic myeloid leukemia stem cells. Am J Hematol 86: 31–37. doi: 10.1002/ajh.21915

29. Moreb JS, Ucar D, Han S, Amory JK, Goldstein AS, et al. (2012) The enzymatic activity of human aldehyde dehydrogenases 1A2 and 2 (ALDH1A2 and ALDH2) is detected by aldefluor, inhibited by diethylaminobenzaldehyde and has significant effects on cell proliferation and drug resistance. Chem Biol Interact 195: 52–60. doi: 10.1016/j.cbi.2011.10.007

30. Guzman ML, Rossi RM, Karnischky L, Li X, Peterson DR, et al. (2005) The sesquiterpene lactone parthenolide induces apoptosis of human acute myelogenous leukemia stem and progenitor cells. Blood 105: 4163–4169. doi: 10.1182/blood-2004-10-4135

31. Guzman ML, Li X, Corbett CA, Rossi RM, Bushnell T, et al. (2007) Rapid and selective death of leukemia stem and progenitor cells induced by the compound 4-benzyl, 2-methyl, 1,2,4-thiadiazolidine, 3,5 dione (TDZD-8). Blood 110: 4436–4444. doi: 10.1182/blood-2007-05-088815

32. Davison K, Mann KK, Miller WH Jr (2002) Arsenic trioxide: mechanisms of action. Semin Hematol 39 (2 Suppl 1)3–7. doi: 10.1053/shem.2002.33610

33. Fuchs O (2010) Transcription factor NF-kappaB inhibitors as single therapeutic agents or in combination with classical chemotherapeutic agents for the treatment of hematologic malignancies. Curr Mol Pharmacol 3: 98–122. doi: 10.2174/1874467211003030098

34. Ito K, Bernardi R, Morotti A, Matsuoka S, Saglio G, et al. (2008) PML targeting eradicates quiescent leukaemia-initiating cells. Nature 453: 1072–1078. doi: 10.1038/nature07016

35. Miller WH Jr, Schipper HM, Lee JS, Singer J, Waxman S (2002) Mechanisms of action of arsenic trioxide. Cancer Res 62: 3893–3903.

36. Jin Y, Lu Z, Ding K, Li J, Du X, et al. (2010) Antineoplastic mechanisms of niclosamide in acute myelogenous leukemia stem cells: inactivation of the NF-kappaB pathway and generation of reactive oxygen species. Cancer Res 70: 2516–2527. doi: 10.1158/0008-5472.can-09-3950

37. Ketola K, Hilvo M, Hyotylainen T, Vuoristo A, Ruskeepaa AL, et al. (2012) Salinomycin inhibits prostate cancer growth and migration via induction of oxidative stress. Br J Cancer 106: 99–106.

38. Lu D, Choi MY, Yu J, Castro JE, Kipps TJ, et al. (2011) Salinomycin inhibits Wnt signaling and selectively induces apoptosis in chronic lymphocytic leukemia cells. Proc Natl Acad Sci USA 108: 13253–13257. doi: 10.1073/pnas.1110431108

39. Rodrigues T, Santos AC, Pigoso AA, Mingatto FE, Uyemura SA, et al. (2002) Thioridazine interacts with the membrane of mitochondria acquiring antioxidant activity toward apoptosis—potentially implicated mechanisms. Br J Pharmacol 136: 136–142. doi: 10.1038/sj.bjp.0704672

40. Sachlos E, Risueno RM, Laronde S, Shapovalova Z, Lee JH, et al. (2012) Identification of drugs including a dopamine receptor antagonist that selectively target cancer stem cells. Cell 149: 1284–1297. doi: 10.1016/j.cell.2012.03.049

41. Zhelev Z, Ohba H, Bakalova R, Hadjimitova V, Ishikawa M, et al. (2004) Phenothiazines suppress proliferation and induce apoptosis in cultured leukemic cells without any influence on the viability of normal lymphocytes. Cancer Chemother Pharmacol 53: 267–275. doi: 10.1007/s00280-003-0738-1

42. Liu L, Chen R, Huang S, Wu Y, Li G, et al. (2011) Knockdown of SOD1 sensitizes the CD34+ CML cells to imatinib therapy. Med Oncol 28: 835–839. doi: 10.1007/s12032-010-9529-9

43. Abdel-Wahab O, Levine RL (2010) Metabolism and the leukemic stem cell. J Exp Med 207: 677–680. doi: 10.1084/jem.20100523

44. Morgan MJ, Liu ZG (2011) Crosstalk of reactive oxygen species and NF-kappaB signaling. Cell Res 21: 103–115. doi: 10.1038/cr.2010.178

45. Yuan LB, He Q, Guo YM (2007) Mechanism of apoptosis-inducing effects of dopamine on K562 leukemia cells. Journal of Zhejiang University (Medical Science) 36: 191–195.

46. Zhen X, Zhang J, Johnson GP, Friedman E (2001) D(4) dopamine receptor differentially regulates Akt/nuclear factor-kappa b and extracellular signal-regulated kinase pathways in D(4)MN9D cells. Mol Pharmacol 60: 857–864.

47. Takeuchi Y, Fukunaga K (2004) Different effects of five dopamine receptor subtypes on nuclear factor-kappaB activity in NG108-15 cells and mouse brain. J Neurochem 88: 41–50. doi: 10.1046/j.1471-4159.2003.02129.x

48. Corkery DP, Dellaire G, Berman JN (2011) Leukaemia xenotransplantation in zebrafish—chemotherapy response assay in vivo. Br J Haematol 153: 786–789. doi: 10.1111/j.1365-2141.2011.08661.x

49. Pruvot B, Jacquel A, Droin N, Auberger P, Bouscary D, et al. (2011) Leukemic cell xenograft in zebrafish embryo for investigating drug efficacy. Haematologica 96: 612–616. doi: 10.3324/haematol.2010.031401

50. Smithen DA, Forrester AM, Corkery DP, Dellaire G, Colpitts J, et al. (2013) Investigations regarding the utility of prodigiosenes to treat leukemia. Org Biomol Chem 11: 62–68. doi: 10.1039/c2ob26535d

51. Westerfield M (2007) The zebrafish book: A guide for the labortory use of zebrafish danio *(Branchydanio) rerio. Fifth edition: University of Oregon Press.

52. Liu Z, Liu F (2012) Cautious use of fli1a:EGFP transgenic zebrafish in vascular research. Biochem Biophys Res Commun 427: 223–226.

53. Yang IH, Lee D, Lee SH, Kang JY (2008) Characterization of proteolytically digested zebrafish chorion as extracellular matrix. Conf Proc IEEE Eng Med Biol Soc 2008: 1837–1840. doi: 10.1109/iembs.2008.4649537

54. Haldi M, Ton C, Seng WL, McGrath P (2006) Human melanoma cells transplanted into zebrafish proliferate, migrate, produce melanin, form masses and stimulate angiogenesis in zebrafish. Angiogenesis 9: 139–151. doi: 10.1007/s10456-006-9040-2

Chapter 6

QUANTITATIVE ANALYSES OF FORCE-INDUCED AMYLOID FORMATION IN CANDIDA ALBICANS ALS5P: ACTIVATION BY STANDARD LABORATORY PROCEDURES

Cho X. J. Chan[1,2,3], Ivor G. Joseph[1] , Andy Huang[1] , Desmond N. Jackson[1] , Peter N. Lipke[1,2]

[1] Biology Department, Brooklyn College City University of New York, New York, New York, United States of America

[2] The Graduate Center, City University of New York, New York, New York, United States of America

[3] Haskins Laboratories and the Department of Chemistry and Physical Sciences, Pace University, New York, New York, United States of America

ABSTRACT

Candida albicans adhesins have amyloid-forming sequences. In Als5p, these amyloid sequences cluster cell surface adhesins to create high avidity surface adhesion nanodomains. Such nanodomains form after force is applied to the cell surface by atomic force microscopy or laminar flow. Here we report centrifuging and resuspending *S. cerevisiae* cells expressing Als5p led to 1.7-fold increase in initial rate of adhesion to ligand coated beads. Furthermore, mechanical stress from vortex-mixing of Als5p cells or *C. albicans* cells also induced additional formation of amyloid nanodomains and consequent activation of adhesion. Vortex-mixing for 60 seconds increased the initial rate of adhesion 1.6-fold. The effects of vortex-mixing were replicated in heat-killed cells as well. Activation was accompanied by increases in thioflavin T cell surface fluorescence measured by flow cytometry or by confocal microscopy. There was no adhesion activation in cells expressing amyloid-impaired Als5p^{V326N} or in cells incubated with inhibitory concentrations of anti-amyloid dyes. Together these results demonstrated the activation of cell surface amyloid nanodomains in yeast expressing Als adhesins, and further delineate the forces that can activate adhesion *in vivo*. Consequently there is quantitative support for the hypothesis that amyloid forming adhesins act as both force sensors and effectors.

INTRODUCTION

Yeast cell surface adhesins, such as the *Candida albicans* adhesin Als5p and *Saccharomyces cerevisiae* flocculins Flo1p and Flo11p, mediate cell-to-cell aggregation and cell-to-surface adhesion. Within the mid-regions of many adhesins are 6-7-amino acid sequences predicted by TANGO (http://tango.crg.es/) to form amyloids [1–3]. A single site mutation (V326N) in the amyloid region of Als5p decreases cell-to-cell aggregation, cell-to-substrate adhesion, and fluorescence of the amyloid-reporting dye thioflavin T [4]. Similarly, anti-amyloid compounds inhibit activation of the *S. cerevisiae* flocculins [2].

Extension forces cluster the adhesins into amyloid-like surface patches [4,5]. Mechanical extension force applied with the tip of an atomic force microscope (AFM) activates the clustering of Als5p (hereafter designated Als5p^{WT}) molecules into nanodomains and the clusters propagate across the cell surface. This clustering is mediated by the amyloid-forming sequence, because the clustering response is absent from a non-amyloid-forming mutant of the protein Als5p^{V326N} [2,4,5]. We have proposed that the pulling on the surface protein results in exposure of the amyloid regions of the protein, which then interact through amyloid stacking to cluster with neighboring Als5p^{WT} molecules in 100–500nm diameter surface nanodomains [4,6]. These nanodomains are highly fluorescent after staining with thioflavin dyes [4,5]. Als5p^{WT} clusters take minutes to form and propagate slowly around the cell surface at a rate of ~20 nm/min. Similarly, hydrodynamic shear from laminar flow can also activate the yeast surface amyloids to increase surface binding, cell-cell aggregation, and formation of mechanically robust biofilms [7]. These changes are consistent with observations that *C.albicans* biofilms grown under flow are more extensive and include more hyphae [8].

These findings correspond to known properties of amyloids. One relevant observation is that shear force can partially unfold proteins, leading to exposure of amyloid-forming sequences. Subsequently these sequences aggregate into β-sheet rich forms that assemble in a cross-β structure, characteristic of amyloid fibril formation [9–14]. For instance, conformational changes in proteins resulting from partial unfolding from their native state facilitate amyloid formation in transthyretin [15] and lysozyme [16]. Shear flow from a Couette cell produces amylogenic precursors in β-lactoglobulin, and enhances fibril formation as well through the alignment and further unfolding of the protein under shear flow, thus resulting in the formation of amyloid precursors and or their maturation into fibers [17,18].

Secondly, amyloid formation itself may be triggered by shear force. When Aβ-peptide is stirred there is an increase in thioflavin T fluorescence as well as growth of amyloid fibers that are not seen with quiescent peptides [18]. Dunstan et al. hypothesized that a possible mechanism of the effect of shear is the alignment of the aggregates, to facilitate assembly into fibrils. This idea is supported by observations that aggregates of proteins such as β-lactoglobulin align under flow [17,19].

Testing such ideas in the yeast adhesins requires the ability to quantify amyloid formation, something we have not been able to do *in vivo*. Therefore we set out to induce and measure activation of adhesion in populations of cells. The development of these assays has led to our realization that the cell surface adhesins are sensitive to activation during cell preparation. Quantitative assays also have confirmed that amyloid-forming adhesins can both sense and respond to force.

RESULTS

Effect of Vortex-Mixing on Adhesion of Als5p-Expressing *S. cerevisiae* Cells

We looked for increases in cell-to-bead adhesion and cell-to-cell aggregation of Als5p[WT]-expressing cells with ligand-coated beads [20]. Suspensions of cells expressing Als5p[WT] were vortex-mixed for 5 minutes at 2500 rpm. The initial onset of adhesion was determined by monitoring size of aggregates in the first 10–15 minutes of aggregation. (These brief assays minimized induction of nanodomains that occurs during standard 45 min assays [4,20]). Cells that had been vortex-mixed formed bigger initial aggregates than cells that were not vortex-mixed (Fig 1A). To quantify the number of cells bound we suspended the aggregates with NaOH and then determined optical density at 600nm. Vortex-mixing of the cells caused an average 1.6-fold increase in adhesion to beads and aggregation (Fig 1B). There was no aggregation in cells expressing empty vector when vortex-mixed (Fig 1A and 1B).

A

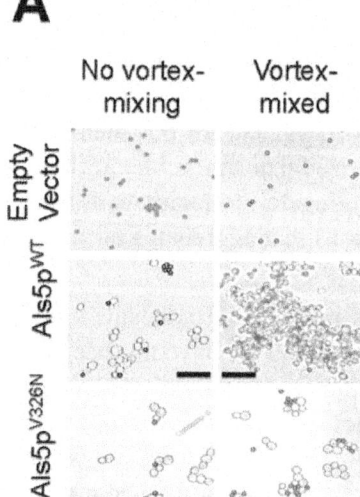

B

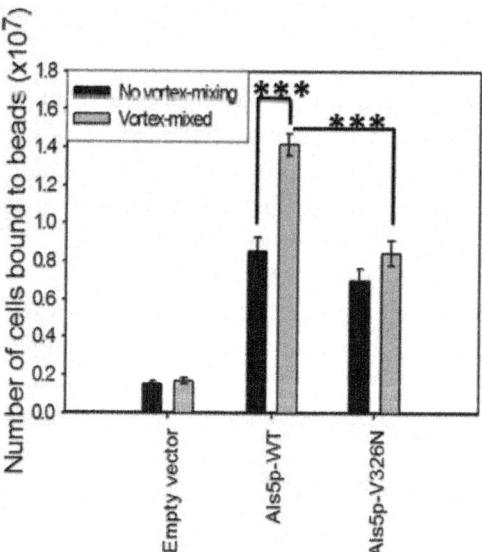

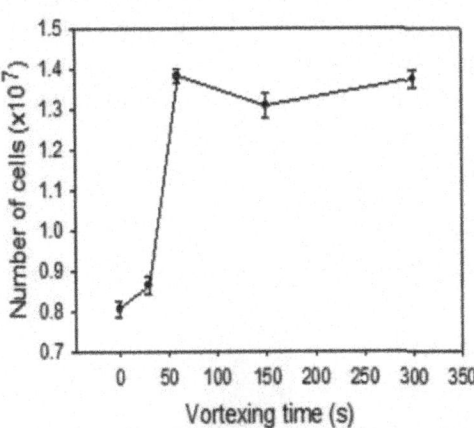

Figure 1. The effects of vortex-mixing on adhesion and aggregation of *S. cerevisiae*. Cells carrying an empty vector or expressing Als5p^{WT} or Als5p^{V326N} were vortex-mixed or not, then aggregated for 10 minutes with heat-denatured BSA-coated magnetic beads: **(A)** Bright-field micrographs of the cells. The dark-colored beads are 1 μm in diameter. Scale bars represent 20μm. **(B)** Quantification of cells adhering to beads. Error bars represent s.d. for n = 4. A student t-test was performed: ******* represents p<0.001. **(C)** Time course for activation of Als5p-expressing cells by vortex mixing at 2500 rpm. Error bars represent s.e.m. for n = 8.

We determined the vortex-mixing time needed to initiate cell adhesion. Mixing for 60 seconds increased the number of cells bound to the beads from (8.1 ±. 4) x 10^6 to (1.4 ±. 03) x 10^7. There was no additional increase with mixing times up to 5 min (Fig 1C). Therefore under these conditions 60 seconds of vortex-mixing was sufficient to increase the adhesion and aggregation of the cells to ligand-coated beads.

Amyloid-Dependence of Vortexed-Induced Adhesion

If force-induced cell adhesion is amyloid-dependent, then vortex-mixing should not activate aggregation on cell expressing the non-amyloid mutant Als5p^{V326N} adhesin. As predicted, there was no increase in the size of the aggregates of the amyloid mutant protein (Fig 1A and 1B). Similarly, amyloid-binding dyes should inhibit this increase. At concentrations above 30 µM, the amyloid-binding dye thioflavin S (ThS) binds to and disrupts amyloids, therefore decreasing adhesion [2,4,21]. This was indeed the case: ThS (0.2 mM) added after vortex-mixing inhibited the binding of Als5p^{WT}-expressing cells to the ligand-coated beads by 6.3-fold (Fig 2A). In the presence of ThS, there were no aggregates formed with Als5p^{V326N}-expressing cells or cells with empty vector, nor was there any effect of vortex-mixing (data not shown).

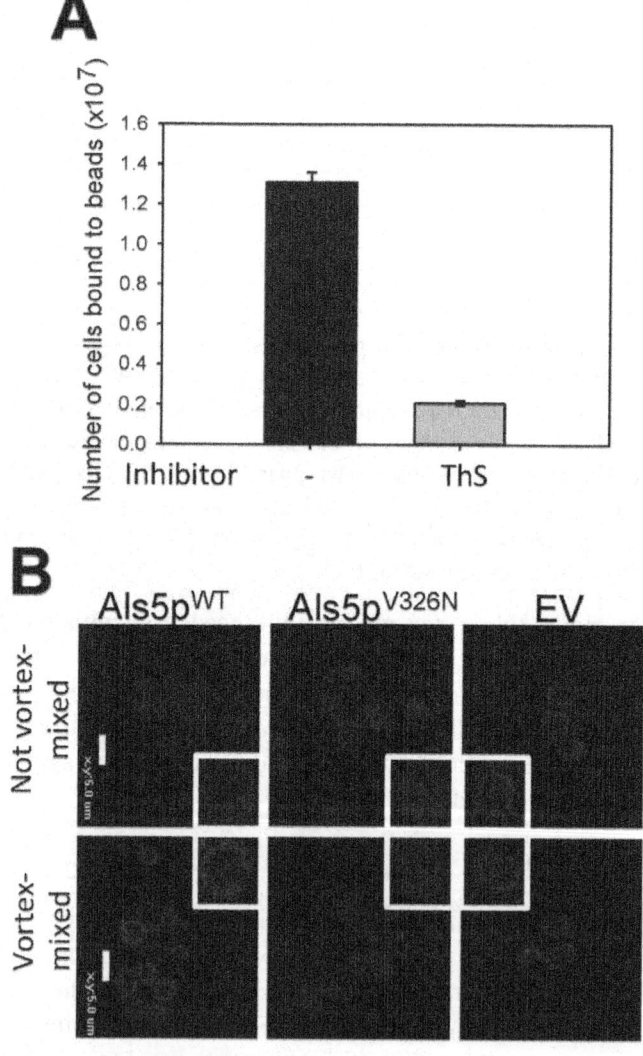

Figure 2. (A) Effects of anti-amyloid dye on vortex-activated adhesion and aggregation of *S. cerevisiae* **cells expressing Als5p^{WT}**. A 10- minute aggregation assay in the

absence or presence of 0.2 mM ThS. (B) Effects of vortex mixing on surface amyloid nanodomains. Confocal micrographs of cells stained with 500 nM ThT without vortex mixing (top row) or after 5 min vortex mixing (bottom row). Scale bars represent 5μm.

Effect of Vortex-Mixing on Cell Surface Thioflavin-T Fluorescence

Amyloid-dependent activation of cell adhesion is mediated by formation of surface amyloid nanodomains that migrate around the cell surface [5]. Sub-inhibitory concentrations of thioflavin T (ThT) or ThS stain amyloids *in vitro* and on yeast cell surfaces [2,4,9,17]. To assay whether the increase in adhesion was accompanied by the formation of amyloids on the surface of cells expressing yeast adhesins, we stained quiescent and vortex-mixed cells with ThT or ThS (1 μM, a non-inhibitory concentration) and analyzed them by flow cytometry. For Als5p^{WT}-expressing cells, vortex-mixing increased the surface fluorescence (S1 and S2 Figs). This increase in surface fluorescence was not seen with cells transformed with EV or cells expressing the non-amyloid Als5p^{V326N}.

The mean cellular fluorescence from flow cytometry correlated with data that vortex-mixing increases the mean cell surface fluorescence (Table 1 and S1 Table). There was a 1.5-fold fluorescence increase due to vortex mixing for cells expressing Als5p^{WT}. Cells with EV or expressing Als5p^{V326N} had little to no increase. Results were similar with ThS (S2 Fig and S1 Table). Therefore, vortex-mixing cells expressing Als5p led to significant increases in surface fluorescence intensity with the amyloid-staining dyes.

Table 1. Effect of vortex-mixing on mean ThT fluorescence of yeast cells

Yeast	Mean ± se		Ratio
	Vortex-mixed	Quiescent	
C. albicans	792 ± 57	541 ± 61	1.51 ± 0.12
S. cerevisiae (Als5p^{WT})	317 ± 25	207 ± 18	1.53 ±. 0.12
S. cerevisiae (Als5p^{V326N})	85 ± 37	99 ± 48	0.85 ± 0.66
S. cerevisiae (EV)	72 ± 24	70 ± 13	1.03 ± 0.38

doi:10.1371/journal.pone.0129152.t001

Nanodomains formed by vortex mixing were also microscopically visible. Cells were vortex-mixed for 5 min. and then stained with ThT (Fig 2B). This increase was not seen with the non-amyloid mutant Als5p^{V326N} or cells with empty vector. These data confirmed that vortex-mixing induced formation of ThT-fluorescent surface nanodomains in cells expressing a yeast adhesin.

Effects of Vortex-Mixing in *C. albicans*

We also assayed increases of aggregation in live *C. albicans* cells. To maximize expression of Als1p and perhaps other adhesins, cells were diluted in fresh YPD media for 45 minutes before the assays [22,23]. When *C. albicans* cells were vortex-mixed at 2500 RPM for 5 min., the initial aggregates were larger than in cells not vortex-mixed (Fig 3A). The mixing time for maximal activation was similar to that for Als5p^{WT}-expressing *S. cerevisiae*, but the initial rate of activation was greater, with a 10% increase within 30 seconds of vortex-mixing and a maximal 2.75-fold increase in initial adhesion rate (Fig 3B and 3D). Activation was accompanied by a slight increase in surface ThT fluorescence (S1 and S2 Figs, Table 1, and S1 Table). Confocal microscopy showed that vortex-mixing induced subtle differences in *C. albicans* surface structure, with some cells having a more uniform and less punctate distribution of surface amyloid (Fig 3C). As with Als5p^{WT}, ThS (200 μM) or CR (500 μM) inhibited the aggregation and adhesion of cells to ligand-coated beads (Fig 3D).

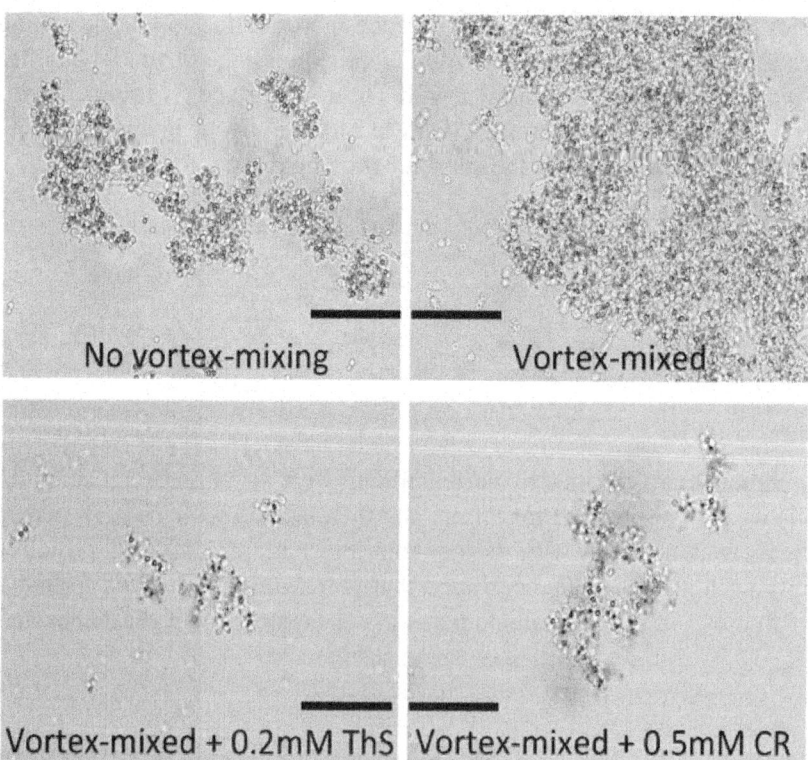

B

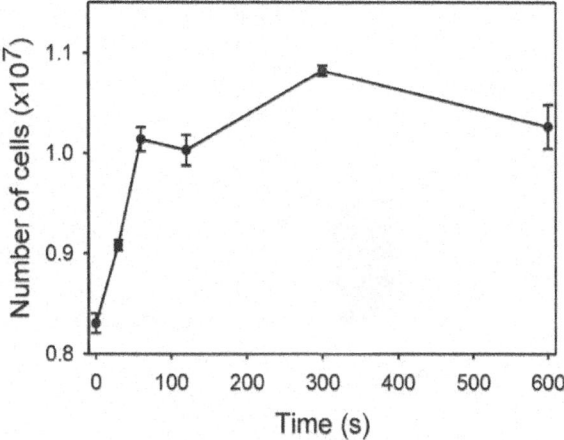

C

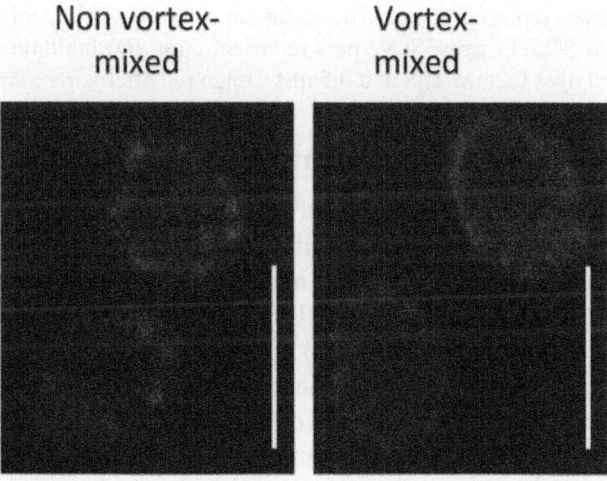

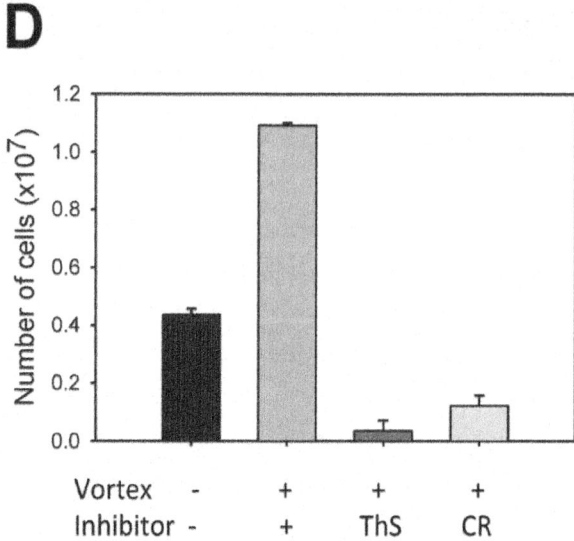

Figure 3. Effect of vortex-mixing on aggregation and surface nanodomains on *C.albicans* **SC5314 cells.** (A) Effects of vortex mixing on initial aggregation. Scale bars represent 100μm. (B)Number of cells bound to beads after vortex mixing for different times. Error bars represent standard deviation for n = 3. (**C**) ThT staining of control and vortex-mixed SC5314 cells. Scale bars represent 5μm. (**D**) Inhibition of aggregation with amyloid dyes 0.2mM ThS and 0.5mM Congo red after vortex-mixing.

Effects of Vortex-Mixing of Heat-Killed Adhesin-Expressing Cells

Heat-killed cells aggregate effectively, and are able to form surface nanodomains [4,5]. Therefore, if vortex-mixing-induced nanodomain formation is independent of cellular metabolism, it should also be apparent in heat-killed cells. Cells expressing Als5p^{WT} or Als5p^{V326N} were heat-killed for 15 minutes at 60°C, then allowed to equilibrate at 25°C for one hour before assay. As expected, there was an increase in bright puncta on the surface of heat-killed Als5p^{WT} cells when vortex-mixed (S3 Fig). In contrast, heat-killed Als5p^{V326N} non-amyloid mutant cells did not show annular staining or nanodomain formation.

Effects of Centrifuging Cells on Aggregation

Because vortex mixing activated amyloid-dependent cell adhesion, it was possible that there might be similar effects from the shear associated with centrifugation and resuspension of the cells. Therefore we grew cultures in medium buffered at pH 5.5 with 50mM MOPS so that aggregation assays could be performed directly on cells. This procedure eliminated the need to centrifuge the cells and resuspend in buffer before assay. Culture aliquots of 1ml were placed directly in test tubes. The samples were then treated in one of three procedures: some tubes were left quiescent, some were centrifuged and the cells resuspended, and some were centrifuged, resuspended, and then vortex-mixed at 2500 rpm for 1 min. Aggregation assays determined the initial rates (Fig 4). Centrifugation increased the initial aggregation values 1.7-fold. Vortex-mixing activated the cells another 1.6-fold. The total fold increase due to centrifugation, resuspension, and vortex mixing was 2.7-fold. Therefore, standard techniques for washing and resuspension of cells can activate formation of amyloid surface nanodomains.

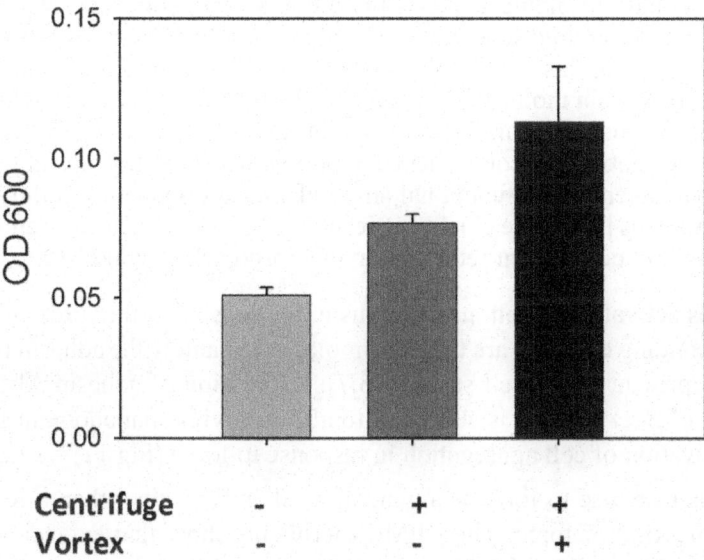

Figure 4. Effect of centrifugation on aggregation. Cells were grown in buffered medium, and aggregated in 10 min. assays. Some of the cells were centrifuged (3200 x g, 3mins) and resuspended before assay, and some of the centrifuged cells were also vortex mixed (2500 rpm, 1 min.) before assay.

DISCUSSION

Our study of activation of aggregation in cell suspensions reinforces the idea that force alone is sufficient to induce surface nanodomains [4,7,24]. This remarkable activity takes place on cell surfaces without need for a signaling or metabolic response in the cell [4,5]. The results also show that shear forces for formation of amyloid nanodomains are similar to those applied to cells in standard laboratory protocols for washing and resuspending yeasts, and vortex mixing. Therefore cell preparation procedures can inadvertently affect aggregation behavior. We have also demonstrated quantification of amyloid formation by flow cytometry and quantification of initial rates of cell adhesion. The techniques and results are valid for *C. albicans* as well as for Als5p displayed on the surface of *S. cerevisiae*.

Fungal Adhesins as Force Sensors

Our results constitute additional support for the idea that Als adhesins themselves sense and respond to force [5,7]. There is a simple model that explains this idea. First, the amyloid sequence-containing T domain in Als5p^{WT} is only marginally stable, and the domain unfolds in response to extension forces [5,7]. Domain unfolding exposes the amyloid core sequence, which in turn interacts with amyloid sequences in nearby Als adhesin molecules to form cell surface nanodomains within minutes [4–6,24–26]. These nanodomains consist of arrayed adhesin molecules, aggregated on the cell surface through amyloid-like interactions of the amino acids in amyloid core sequences [27,28]. Thus, amyloid formation is a consequence of protein conformational change, and depends on presence of a functional amyloid-forming sequence in the T domain of Als proteins [4,5]. The clustering results in very high local concentrations of adhesins, and consequent reduction in macroscopic k_{off} values [6].

This activation is cell-autonomous in the sense that the force sensors and the responding effectors are the same molecules, namely the adhesin molecules already present at the cell surface [5,7]. Expression of adhesin Als5p on the surface of *S. cerevisiae* is sufficient to allow amyloid nanodomain formation and activation of cell aggregation in response to force (Fig 1) [4,29].

This response to force is a general mechanism, rather than a response to a specific kind of force. The AFM experiments show that extension force in one area of the cell can lead to activation of the entire cell as the nanodomains propagate across the surface [6,7]. Similarly, a single area of the cell surface is initially stimulated when a cell binds to a ligand-coated bead; mixing during the adhesion assay generates extension force on the adhesins bound to the bead [21]. Later in the assay, cell-to-cell adhesion may directly stimulate adhesins on parts of cells that are not in contact with a bead. Activation can also follow more global stimulation, as in vortex mixing or laminar flow [7]. Vortex-mixing suspensions at different cell densities did not show differences in activation rates, and this result implied that liquid shear was the activating force, rather than cell-cell collisions. Als protein surface amyloids are present on *C. albicans* abscesses in infected tissue, so the nanodomains must form during the infection process, perhaps due to friction exerted by fungal growth through the host tissue [30,31]. Thus, the data strengthens a generally applicable model of force-activated fungal cell adhesion [4,7,24].

Time and Force for Activation

Nanodomain formation followed after one minute of vortex mixing. This time is significantly shorter than the 25 min interval observed after single molecule stimulation for Als5p^{WT} in AFM experiments [5]. On the other hand, Als-bearing cells start to become globally cell surface activated within 15 minutes in adhesion assays, and 7 minutes under laminar flow [7]. These differences in activation rate are consistent with differences in the frequency of molecular stretching in the three different scenarios. A few molecules are individually stretched in the AFM. In our adhesion assays, cells are gently mixed, usually at 170–200 rpm. This process results in random collisions, adhesions and subsequent stretching of adhesins, so many areas of the cell surface are stimulated in the course of a few minutes. Our unpublished data shows that slower vortex mixing speeds are less effective, and that mixing speed in the adhesion assay itself affects the size of the aggregates, with greater adhesion as the mixing speed becomes faster. Therefore, the speed of mixing both before and during the adhesion assay affect initial rate of cell-to-cell aggregation.

The forces needed to activate cell adhesion are comparable to those the yeast encounter *in vivo* and in the lab. Unfolding of the amyloid-containing T domains of Als5p^{WT} or Als1p in the AFM followed application of forces in the range of 50–100 pN [5,26,29]. This amount of force is similar to that encountered in flowing blood, or under flow in the natural environment [32]. Physiological shear rates *in vivo* are in the range of 100–8000 s^{-1} in blood vessels and the extracellular matrix [32]. Such a shear rate is also similar to that applied by vortex mixing, where a broad range of shear rates, from 200–8000 sec^{-1}, occur depending on sample volume, proximity to air and glass interfaces, vessel geometry, and mixing speeds [33]. The product of shear rate and viscosity (8.9 x 10^{-3} dyne sec cm^{-2}) yields a resulting shear stress of ~10–100 dyne cm^{-2}, or 1–10 pN µm^{-2}. Although this force appears less than the instantaneous force applied in AFM, its application over one minute time would lead to high T domain unfolding probability, relative to the standard AFM contact time of a second or less [29]. Thus, Als proteins show activation under forces such as centrifugation or vortex mixing, as we have demonstrated. Therefore, cell preparation procedures will affect results of cell adhesion assays, and cells will need to be treated gently to achieve baseline aggregation ability (the equivalent of the quiescent cells in our assays).

The consequences of force-induced activation can be easily quantified. Confocal microscopy and flow cytometry with thioflavin T are effective in visualizing and quantifying surface amyloids. Force-activated cells had punctate nanodomains with increased thioflavin T staining, which may also be measured by image analysis (not shown). Collectively, the assays can now be used to quantify amyloids in other fungal adhesins. We have also shown that quantities of adhering cells can be compared by spectrophotometry instead of cell counting after cell dissociation [20]. This procedure has allowed us to measure initial rates of adhesion.

SUMMARY

We have outlined quantitative methods for comparison of cell surface amyloid nanodomains and for cell adhesion in a magnetic bead assay, including showing differences in initial adhesion rate for vortex-stimulated cells. We have also shown that vortex-mixing measurably and reproducibly activated surface amyloid nanodomain formation on populations of cells, allowing us to compare assays on different days and cell cultures. The results support a conclusion that the mechanism and consequences of activation by laminar flow, vortex-mixing, or shaking in aggregations assays are similar. Each of these techniques shows increased surface fluorescence with thioflavin T, sensitivity to anti-amyloid dyes, and comparable kinetics and force requirements. Bioinformatic analyses show that similar amyloid-forming sequences are common in fungal adhesins, as well as some bacterial adhesins [34–37]. Indeed the importance of amyloid interactions has also been demonstrated in several other fungal and bacterial adhesion systems, including assembly of gram negative curlins [38–41], and gram positive adhesins including *Streptococcus mutans* P1 [42], and *Bacillus subtilis* TasA [36,43]. It remains to be seen if any of these other systems also show force-induced clustering and activation.

MATERIALS AND METHODS

Strains and Media

Saccharomyces cerevisiae strain W303-1B *MAT leu2 ura3 ade2 trp1* (Rodney Rothstein, Columbia U.) harboring the empty vector (pJL1-EV) or expressing Als5p^{WT} or Als5p^{V326N} was grown in complete synthetic medium (CSM) lacking tryptophan with galactose as carbon source [4]. Cultures were grown for 48 hours at 24°C at 170 RPM. When desired, cells were heat killed in a water bath at 60°C for 15 minutes and then incubated at room temperature for 1 hour before activation and assay.

Candida albicans strain SC5314 was grown overnight in yeast extract with peptone and 2% glucose (YPD) at 30°C at 170 RPM. Als1p expression was induced by placing an aliquot of cells in fresh YPD medium [23,44] in a 1:10 dilution and shaking at 170 RPM at 30°C for 45 minutes.

Aggregation Assays

Aggregation assays were modified from published procedures [20,21]. Briefly, cells were centrifuged at 4000 RPM for 3 minutes to remove culture media. The cells were then washed gently three times with 10mM tris, 1mM EDTA (TE) pH 7.0 and gently resuspended in the same buffer. The OD_{600nm} of the cell suspension was determined with a Spectronic 21 D+ spectrophotometer, and the suspension was adjusted to 10^8 cells/ml. Aliquots (1 ml) were then placed in test tubes (13x100 mm) either left stationary on the lab bench or vortex-mixed at 2500 RPM for 5 min using a Fisher Scientific multi-tube vortexer. This vortex mixer has an eccentric orbit of 3.6mm. (Standard lab vortex mixers have 5 mm orbits.) Cell suspension (1ml) was mixed with 10^6 BSA-coated magnetic beads. The suspensions were incubated on an orbital shaker for 10 minutes at 170 RPM at 24°C. The assay tubes were placed on a magnet. The unbound cells were gently removed with a pipette and the beads with the cell aggregates were washed once 500 µl of TE buffer. For microscopic viewing, cells were resuspended in 100 µl of TE buffer and 4 µl applied to a glass slide. Microscopic observations were made with an Olympus microscope using a 60X oil objective. For quantification, the aggregates and beads were resuspended in 300 µl 1 M NaOH and shaken gently on an orbital shaker for 20 minutes. The beads were then separated on a magnet, and the OD_{600nm} determined on a 200µl aliquot in a 96-well plate with a Spectronic Genesys plate reader. In this assay OD_{600nm} of 1.0 corresponds to 8.8 x 10^7 cells/ml. Unless otherwise stated, all assays were done on at least two independent cultures, in triplicate for each.

M-280-tosylactivated-magnetic Dynabeads (Invitrogen, Carlsbad, CA) were covalently derivatized with 1mg/ml heat-denatured bovine serum album (BSA) overnight according to the manufacturer's protocol.

Dye Inhibition

Als-expressing *S. cerevisiae* cells or *C. albicans* cells were vortex-mixed or left quiescent for 5 minutes and then ThS or CR) was added. Ligand-coated beads were added, and aggregation assays performed as described above.

Staining Protocols

Stock concentrations of ThS and ThT were made with deionized water and filtered with a 2 µm filter. The concentration was then determined with a spectrophotometer, using Beer's law, using an extinction coefficient of 2.66 x 10^3 L/mol*cm.

Confocal Microscopy

Confocal imaging was done with a Nikon confocal microscope. 10^8 cells were stained with ThT (1 µM) in a final volume of 1ml immediately after vortex-mixing. The cells were vortex-mixed on a low setting with the dye for 5 seconds to resuspend the dye, and then 4 µl of the suspension was placed onto a glass slide for imaging. The stained cells were not washed prior to microscopy. The gain of the microscope was set at 7.75 with the phase at 162. The excitation was at 408nm with an emission detector at 450 ± 35 nm. Pictures were taken at 2048 x 2048 quality. The images were quantified for blue pixel counts using the Image J software with Color Profiler plugin. Six cells per sample were counted together for blue pixels.

Flow Cytometry

Flow cytometry was done with BD Biosciences BD FACS Aria II cell sorter with excitation at 405 nm and an emissions filter of 450 ± 50nm. 10^6 cells were in 12mm x 75 mm tubes with or without vortex-mixing and then brought to a final concentration of 1 µM ThT or 1 µM ThS in a total volume of 1 ml in their respective buffer as mentioned above. The cells were filtered with a 40 µm filter before analysis. A 70-micron nozzle size was used with default sheath pressure, amplitude, and frequency parameters as per manual. 20,000 cells were monitored for each assay.

Centrifugation Aggregation Assay

Cells were grown in Complete Synthetic medium without Trp (Sunrise Science Products), with 40 mg/ml adenine, 2% galactose and 50mM MOPS, pH 5.5, over two nights to an OD 1.0 or more. Aliquots of 1 ml were gently pipetted into test tubes. Some tubes were left quiescent on the bench top, whereas others were centrifuged at 3200 x g for 3 min then gently resuspended in the same medium. Some tubes containing the centrifuged cells were then vortex-mixed at 2500 rpm for 1 minute. Ligand-coated magnetic beads were added to each tube at a 1:10 bead to cell ratio. 10 min aggregation assays were then performed.

SUPPORTING INFORMATION

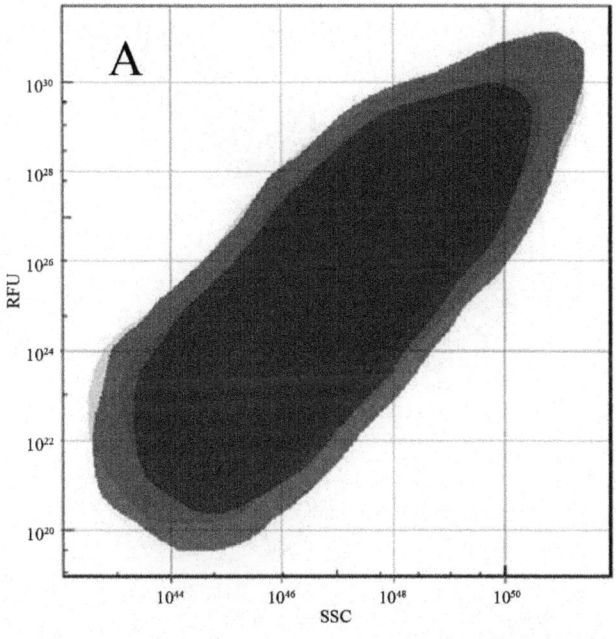

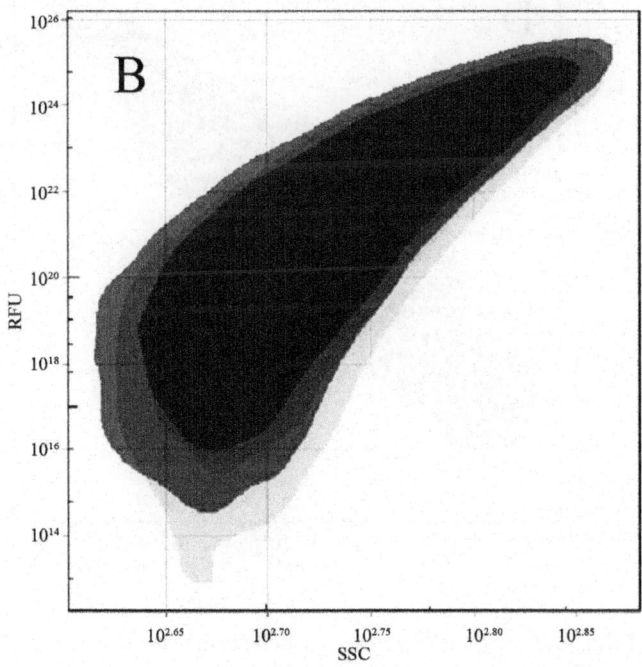

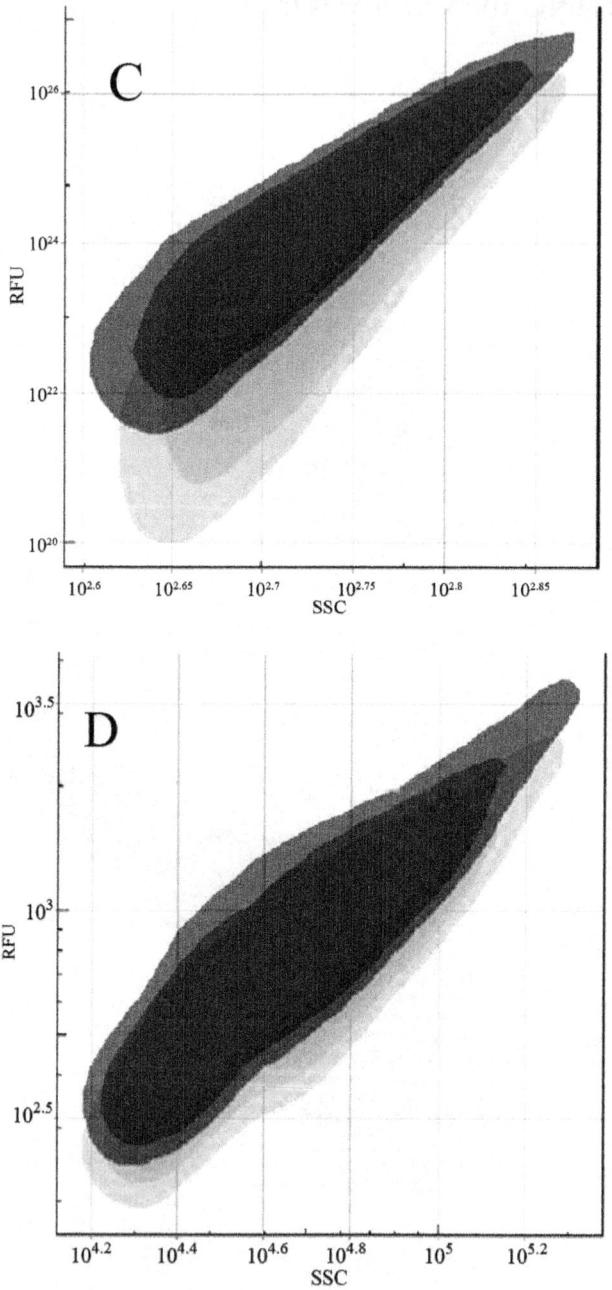

S1 Figure. Effects of vortex-mixing on ThT cell surface fluorescence. FACS analyses of populations of cells stained with ThT 1 μM. Yellow represents unstimulated cells, and blue represents vortex-mixed cells. **(A)** *S. cerevisiae* cells with EV; **(B)** *S. cerevi-*

*siae*cells expressing Als5p^{V326N}; **(C)** *S. cerevisiae* cells expressing Als5^{WT}; **(D)** *C. albicans*SC5314.

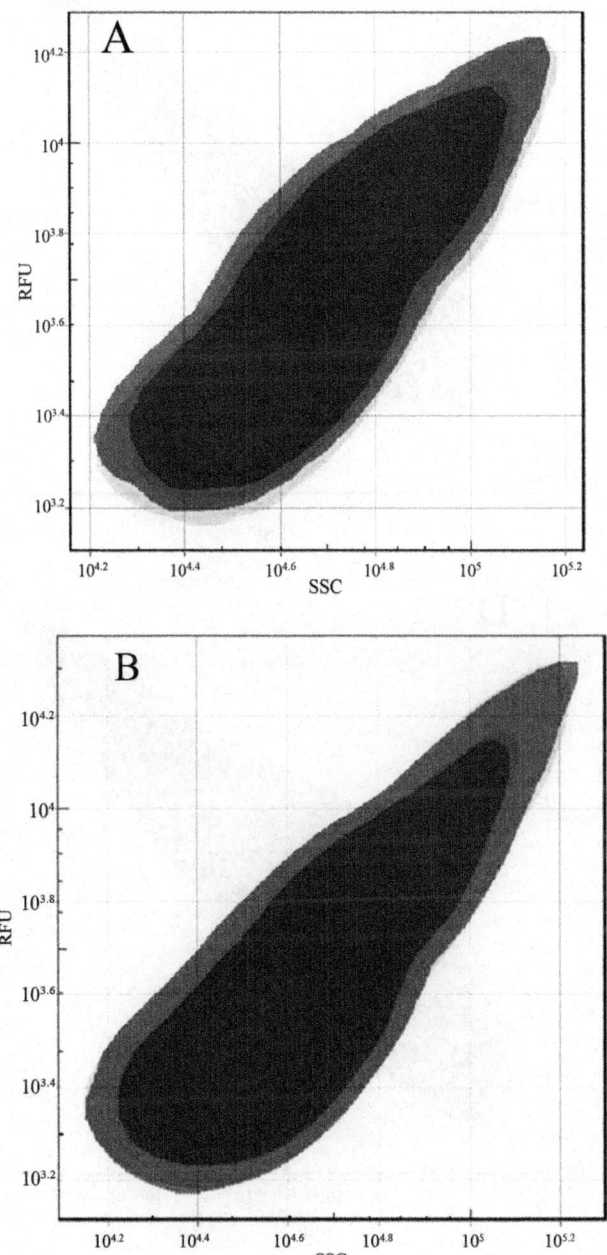

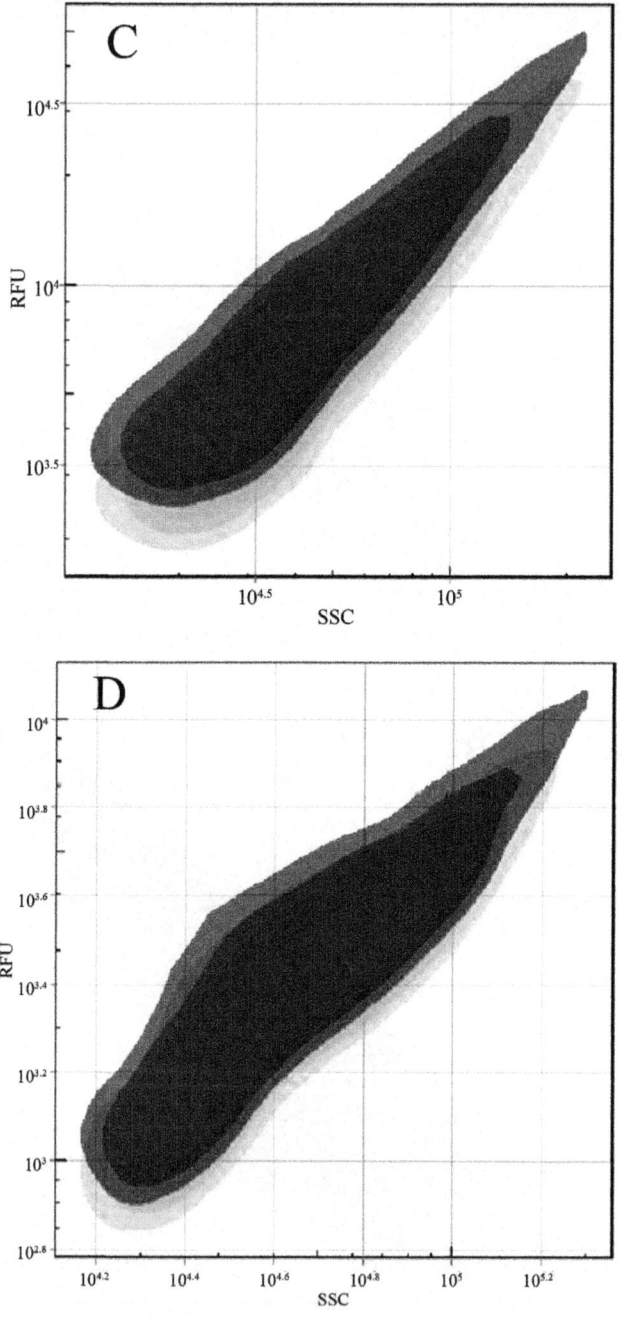

S2 Figure. Effects of vortex-mixing on ThS cell surface fluorescence. FACS analyses of populations of cells stained with ThS 1 μM. Yellow represents unstimulated cells, and blue represents vortex-mixed cells. **(A)** *S. cerevisiae* cells with EV; **(B)** *S. cerevisiae* cells expressing Als5p^{V326N}; **(C)** *S. cerevisiae* cells expressing Als5^{WT}; **(D)** *C. albicans*SC5314.

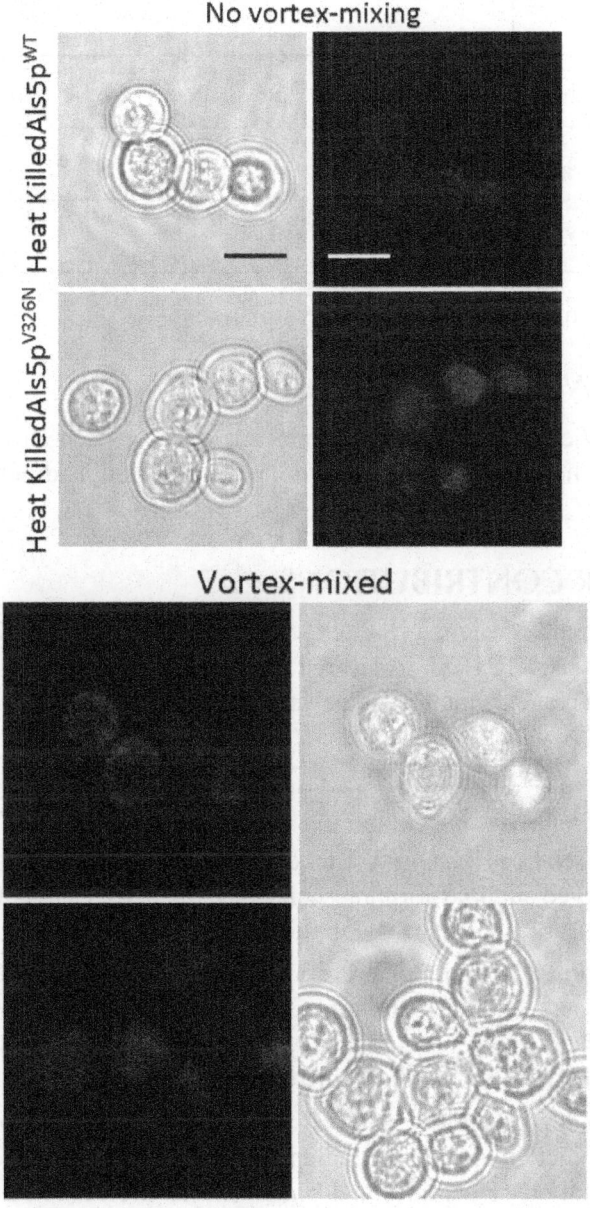

S3 Figure. Effects of vortex-mixing on heat killed Als5p-expressing cells. Cells were vortex-mixed or not then stained with ThT (500 nM). Scale bars represent 5µm.

S1 Table. Effect of vortex-mixing on mean ThS fluorescence of yeast cells

Cells	Vortex-mixed	Fluorescence (mean±se)*	Ratio
C. albicans SC5314	-	2315 ±93	1.3± .133
	+	2890±365	
S. cerevisiae (Als5)	-	6344±391	1.3± .103
	+	7965±656	
S. cerevisiae (V326N)	-	4628±171	1.03±0.072
	+	4761±301	
S. cerevisiae (EV)	-	4442±308	1.01±0.073
	+	4478±100	

*Geometric mean and s.e. for 3 determinations

ACKNOWLEDGMENTS

We thank Victor Gresseau for technical assistance, Brett Branco for helpful insights in shear rates and Juergen Polle for access to and help with the cell sorter.

AUTHOR CONTRIBUTIONS

Conceived and designed the experiments: CXJC PNL. Performed the experiments: CXJC IGJ. Analyzed the data: CXJC IGJ PNL DNJ AH. Contributed reagents/materials/analysis tools: CXJC IGJ PNL DNJ AH. Wrote the paper: CXJC IGJ PNL.

REFERENCES

1.   Otoo HN, Lee KG, Qiu W, Lipke PN. Candida albicans Als adhesins have conserved amyloid-forming sequences. Eukaryot Cell. 2008;7: 776–782. doi: 10.1128/EC.00309-07. pmid:18083824

2.   Ramsook CB, Tan C, Garcia MC, Fung R, Soybelman G, Henry R, et al. Yeast cell adhesion molecules have functional amyloid-forming sequences. Eukaryot Cell. 2010;9: 393–404. doi: 10.1128/EC.00068-09. ; 10.1128/EC.00068-09. pmid:20038605

3.   Fernandez-Escamilla AM, Rousseau F, Schymkowitz J, Serrano L. Prediction of sequence-dependent and mutational effects on the aggregation of peptides and proteins. Nat Biotechnol. 2004;22: 1302–1306. doi: 10.1038/nbt1012. pmid:15361882

4.   Garcia MC, Lee JT, Ramsook CB, Alsteens D, Dufrene YF, Lipke PN.

A role for amyloid in cell aggregation and biofilm formation. PLoS One. 2011;6: e17632. doi: 10.1371/journal.pone.0017632. ; 10.1371/journal. pone.0017632. pmid:21408122

5. Alsteens D, Garcia MC, Lipke PN, Dufrene YF. Force-induced formation and propagation of adhesion nanodomains in living fungal cells. Proc Natl Acad Sci U S A. 2010;107: 20744–20749. doi: 10.1073/ pnas.1013893107. ; 10.1073/pnas.1013893107. pmid:21059927

6. Lipke PN, Garcia MC, Alsteens D, Ramsook CB, Klotz SA, Dufrene YF. Strengthening relationships: amyloids create adhesion nanodomains in yeasts. Trends Microbiol. 2012;20: 59–65. doi: 10.1016/j. tim.2011.10.002. ; 10.1016/j.tim.2011.10.002. pmid:22099004

7. Chan CX, Lipke PN. Role of force-sensitive amyloid-like interactions in fungal catch-bonding and biofilms. Eukaryot Cell. 2014;13: 1136–1142. doi: 10.1128/EC.00068-14. pmid:24681687

8. Dutton LC, Nobbs AH, Jepson K, Jepson MA, Vickerman MM, Aqeel Alawfi S, et al. O-mannosylation in Candida albicans enables development of interkingdom biofilm communities. MBio. 2014;5: e00911–14. doi: 10.1128/mBio.00911-14. pmid:24736223

9. Calamai M, Chiti F, Dobson CM. Amyloid fibril formation can proceed from different conformations of a partially unfolded protein. Biophys J. 2005;89: 4201–4210. doi: 10.1529/biophysj.105.068726. pmid:16169975

10. Harper JD, Lieber CM, Lansbury PT Jr. Atomic force microscopic imaging of seeded fibril formation and fibril branching by the Alzheimer's disease amyloid-beta protein. Chem Biol. 1997;4: 951–959. pmid:9427660 doi: 10.1016/s1074-5521(97)90303-3

11. Sunde M, Blake CC. From the globular to the fibrous state: protein structure and structural conversion in amyloid formation. Q Rev Biophys. 1998;31: 1–39. pmid:9717197 doi: 10.1017/s0033583598003400

12. Serpell LC. Alzheimer's amyloid fibrils: structure and assembly. Biochim Biophys Acta. 2000;1502: 16–30. pmid:10899428 doi: 10.1016/s0925-4439(00)00029-6

13. Dobson CM. Protein misfolding, evolution and disease. Trends Biochem Sci. 1999;24: 329–332. pmid:10470028 doi: 10.1016/s0968-0004(99)01445-0

14. Fandrich M, Zandomeneghi G, Krebs MR, Kittler M, Buder K, Rossner A, et al. Apomyoglobin reveals a random-nucleation mechanism in amyloid protofibril formation. Acta Histochem. 2006;108: 215–219. doi: 10.1016/j.acthis.2006.03.012. pmid:16714052

15. McCutchen SL, Lai Z, Miroy GJ, Kelly JW, Colon W. Comparison of lethal and nonlethal transthyretin variants and their relationship to amyloid disease. Biochemistry. 1995;34: 13527–13536. pmid:7577941 doi: 10.1021/bi00041a032

16. Merlini G, Bellotti V. Lysozyme: a paradigmatic molecule for the investigation of protein structure, function and misfolding. Clin Chim Acta. 2005;357: 168–172. doi: 10.1016/j.cccn.2005.03.022. pmid:15913589

17. Hill EK, Krebs B, Goodall DG, Howlett GJ, Dunstan DE. Shear flow induces amyloid fibril formation. Biomacromolecules. 2006;7: 10–13. doi: 10.1021/bm0505078. pmid:16398490

18. Dunstan DE, Hamilton-Brown P, Asimakis P, Ducker W, Bertolini J. Shear flow promotes amyloid-{beta} fibrilization. Protein Eng Des Sel. 2009;22: 741–746. doi: 10.1093/protein/gzp059. ; 10.1093/protein/gzp059. pmid:19850675

19. Ladner-Keay CL, Griffith BJ, Wishart DS. Shaking alone induces de novo conversion of recombinant prion proteins to beta-sheet rich oligomers and fibrils. PLoS One. 2014;9: e98753. doi: 10.1371/journal.pone.0098753. pmid:24892647

20. Gaur NK, Klotz SA. Expression, cloning, and characterization of a Candida albicans gene, ALA1, that confers adherence properties upon Saccharomyces cerevisiae for extracellular matrix proteins. Infect Immun. 1997;65: 5289–5294. pmid:9393828

21. Rauceo JM, Gaur NK, Lee KG, Edwards JE, Klotz SA, Lipke PN. Global cell surface conformational shift mediated by a Candida albicans adhesin. Infect Immun. 2004;72: 4948–4955. doi: 10.1128/IAI.72.9.4948–4955.2004. pmid:15321986

22. Hoyer LL. The ALS gene family of Candida albicans. Trends Microbiol. 2001;9: 176–180. pmid:11286882 doi: 10.1016/s0966-842x(01)01984-9

23. Hoyer LL, Scherer S, Shatzman AR, Livi GP. Candida albicans ALS1: domains related to a Saccharomyces cerevisiae sexual agglutinin separated by a repeating motif. Mol Microbiol. 1995;15: 39–54. pmid:7752895 doi: 10.1111/j.1365-2958.1995.tb02219.x

24. Alsteens D, Beaussart A, Derclaye S, El-Kirat-Chatel S, Park HR, Lipke PN, et al. Single-Cell Force Spectroscopy of Als-Mediated Fungal Adhesion. Anal Methods. 2013;5: 3657–3662. doi: 10.1039/C3AY40473K. pmid:23956795

25. Alsteens D, Ramsook CB, Lipke PN, Dufrene YF. Unzipping a functional microbial amyloid. ACS Nano. 2012;6: 7703–7711. doi: 10.1021/nn3025699. pmid:22924880

26. Beaussart A, Alsteens D, El-Kirat-Chatel S, Lipke PN, Kucharikova S, Van Dijck P, et al. Single-molecule imaging and functional analysis of Als adhesins and mannans during Candida albicans morphogenesis. ACS Nano. 2012;6: 10950–10964. doi: 10.1021/nn304505s. ; 10.1021/nn304505s. pmid:23145462

27. Sawaya MR, Sambashivan S, Nelson R, Ivanova MI, Sievers SA, Apostol MI, et al. Atomic structures of amyloid cross-beta spines reveal varied steric zippers. Nature. 2007;447: 453–457. doi: 10.1038/nature05695. pmid:17468747

28. Morris KL, Zibaee S, Chen L, Goedert M, Sikorski P, Serpell LC. The structure of cross-beta tapes and tubes formed by an octapeptide, alphaSbeta1. Angew Chem Int Ed Engl. 2013;52: 2279–2283. doi: 10.1002/anie.201207699. pmid:23307646

29. Alsteens D, Dupres V, Klotz SA, Gaur NK, Lipke PN, Dufrene YF. Unfolding individual als5p adhesion proteins on live cells. ACS Nano. 2009;3: 1677–1682. doi: 10.1021/nn900078p. ; 10.1021/nn900078p. pmid:19534503

30. Gilchrist KB, Garcia MC, Sobonya R, Lipke PN, Klotz SA. New features of invasive candidiasis in humans: amyloid formation by fungi and deposition of serum amyloid P component by the host. J Infect Dis. 2012;206: 1473–1478. doi: 10.1093/infdis/jis464. ; 10.1093/infdis/jis464. pmid:22802434

31. Garcia-Sherman MC, Lysak N, Filonenko A, Richards H, Sobonya RE, Klotz SA, et al. Peptide detection of fungal functional amyloids in infected tissue. PLoS One. 2014;9: e86067. doi: 10.1371/journal.pone.0086067. ; 10.1371/journal.pone.0086067. pmid:24465872

32. Charm SE, Kurland GS. A comparison of couette, cone and plate and capillary tube viscometry for blood. Bibl Anat. 1969;10: 85–91. pmid:5407426

33. Bai G, Bee JS, Biddlecombe JG, Chen Q, Leach WT. Computational fluid dynamics (CFD) insights into agitation stress methods in biopharmaceutical development. Int J Pharm. 2012;423: 264–280. doi: 10.1016/j.ijpharm.2011.11.044. ; 10.1016/j.ijpharm.2011.11.044. pmid:22172288

34. Austin JW, Sanders G, Kay WW, Collinson SK. Thin aggregative fimbriae enhance Salmonella enteritidis biofilm formation. FEMS Microbiol Lett.

1998;162: 295–301. pmid:9627964 doi: 10.1111/j.1574-6968.1998. tb13012.x

35. Chapman MR, Robinson LS, Pinkner JS, Roth R, Heuser J, Hammar M, et al. Role of Escherichia coli curli operons in directing amyloid fiber formation. Science. 2002;295: 851–855. doi: 10.1126/science.1067484. pmid:11823641

36. Romero D, Aguilar C, Losick R, Kolter R. Amyloid fibers provide structural integrity to Bacillus subtilis biofilms. Proc Natl Acad Sci U S A. 2010;107: 2230–2234. doi: 10.1073/pnas.0910560107. ; 10.1073/ pnas.0910560107. pmid:20080671

37. Vidal O, Longin R, Prigent-Combaret C, Dorel C, Hooreman M, Lejeune P. Isolation of an Escherichia coli K-12 mutant strain able to form biofilms on inert surfaces: involvement of a new ompR allele that increases curli expression. J Bacteriol. 1998;180: 2442–2449. pmid:9573197

38. Ben Nasr A, Olsen A, Sjobring U, Muller-Esterl W, Bjorck L. Assembly of human contact phase proteins and release of bradykinin at the surface of curli-expressing Escherichia coli. Mol Microbiol. 1996;20: 927–935. pmid:8809746 doi: 10.1111/j.1365-2958.1996.tb02534.x

39. Olsen A, Jonsson A, Normark S. Fibronectin binding mediated by a novel class of surface organelles on Escherichia coli. Nature. 1989;338: 652– 655. doi: 10.1038/338652a0. pmid:2649795

40. Olsen A, Wick MJ, Morgelin M, Bjorck L. Curli, fibrous surface proteins of Escherichia coli, interact with major histocompatibility complex class I molecules. Infect Immun. 1998;66: 944–949. pmid:9488380

41. Sjobring U, Pohl G, Olsen A. Plasminogen, absorbed by Escherichia coli expressing curli or by Salmonella enteritidis expressing thin aggregative fimbriae, can be activated by simultaneously captured tissue-type plasminogen activator (t-PA). Mol Microbiol. 1994;14: 443–452. pmid:7885228 doi: 10.1111/j.1365-2958.1994.tb02179.x

42. Oli MW, Otoo HN, Crowley PJ, Heim KP, Nascimento MM, Ramsook CB, et al. Functional amyloid formation by Streptococcus mutans. Microbiology. 2012;158: 2903–2916. doi: 10.1099/mic.0.060855–0. ; 10.1099/mic.0.060855–0. pmid:23082034

43. Romero D, Vlamakis H, Losick R, Kolter R. An accessory protein required for anchoring and assembly of amyloid fibres in B. subtilis biofilms. Mol Microbiol. 2011;80: 1155–1168. doi: 10.1111/j.1365-2958.2011.07653.x. ; 10.1111/j.1365-2958.2011.07653.x. pmid:21477127

44. Coleman DA, Oh SH, Zhao X, Hoyer LL. Heterogeneous distribution of Candida albicans cell-surface antigens demonstrated with an Als1-specific monoclonal antibody. Microbiology. 2010;156: 3645–3659. doi: 10.1099/mic.0.043851–0. ; 10.1099/mic.0.043851–0. pmid:20705663

Chapter 7

TISSUE-SPECIFIC METABOLITE PROFILING AND QUANTITATIVE ANALYSIS OF GINSENOSIDES IN PANAX QUINQUEFOLI-UM USING LASER MICRODISSECTION AND LIQUID CHROMATOGRAPHY–QUADRUPOLE/TIME OF FLIGHT-MASS SPECTROMETRY

Yujie Chen[1,2], Liang Xu[3], Yuancen Zhao[1], Zhongzhen Zhao[1], Hubiao Chen[1], Tao Yi[1], Minjian Qin[2] and Zhitao Liang[1]

[1]School of Chinese Medicine, Hong Kong Baptist University, Kowloon, Hong Kong Special Administrative Region, People's Republic of China

[2] Department of Resources Science of Traditional Chinese Medicines, State Key Laboratory of Modern Chinese Medicines, College of Traditional Chinese Medicines, China Pharmaceutical University, Tongjiaxiang-24, Gulou District, Nanjing 210009, People's Republic of China

[3] School of Pharmacy, Liaoning University of Traditional Chinese Medicine, Dalian, China.

ABSTRACT

Background

The root of *Panax quinquefolium* L., famous as American ginseng all over the world, is one of the most widely-used medicinal or edible materials. Ginsenosides are recognized as the main bioactive chemical components responsible for various functions of American ginseng. In this study, tissue-specific chemicals of *P. quinquefolium* were analyzed by laser microdissection and ultra-high performance liquid chromatography- quadrupole/time-of-flight-mass spectrometry (UHPLC-Q/TOF–MS) to elucidate the distribution pattern of ginsenosides in tissues. The contents of ginsenosides in various tissues were also compared.

RESULTS

A total of 34 peaks were identified or temporarily identified in the chromatograms of tissue extractions. The cork, primary xylem or cortex contained higher contents of ginsenosides than phloem, secondary xylem and cambium. Thus, it would be reasonable to deduce that the ratio of total areas of cork, primary xylem and the cortex to the area of the whole transection could help to judge the quality of American ginseng by microscopic characteristics.

Conclusion

This study sheds new light on the role of microscopic research in quality evaluation, and provides useful information for probing the biochemical pathways of ginsenosides.

BACKGROUND

Microscopic authentication refers to examine the structure, cell and internal features of herbal medicines using a microscope and its derivatives. It has been recorded in many Pharmacopoeias as an authentication method, such as Chinese Pharmacopoeia, United States Pharmacopeia, European Pharmacopoeia, British Pharmacopoeia, Japanese Pharmacopoeia, and Korean Pharmacopoeia. Distinctly, microscopic authentication has been commonly used in the authentication of herbal medicines. As we know, the secondary metabolites of herbal medicine contribute to its effects. Nevertheless, the normal microscopic identification cannot provide the useful information of secondary metabolites in different herbal materials directly. Thus, microscopic method can identify the source species but not evaluate the quality of herbal medicines.

By using techniques of anatomy and histochemistry, some studies have demonstrated that there is a close relationship between microscopic characteristics and active components of herbal medicines. For example, the histochemical techniques and phytochemical methods have been applied in the distribution and accumulation of active components in *Sinomenium acutum, Aloe vera* var. *chinensis, Gynostemma pentaphyllum, Dioscorea zingiberensis* and *Macrocarpium officinacle* [1–5]. However, these studies used routine chemical reactions and thus the distribution of the detailed active components could not be identified. Moreover, those agents usually have poor specificity, which leads to the increase of false positive results. Also, it is noteworthy that these investigations lacked objective data and had not been validated by other methods yet. Recently, the combination of fluorescence microscopy, laser microdissection (LMD), and ultra-high performance liquid chromatography-quadrupole/time-of-flight-mass spectrometry (UHPLC-Q/

TOF–MS) has been successfully applied to explore the distribution pattern of secondary metabolites among different tissues from several Chinese medicinal materials (CMMs) [6–11]. This method can obtain the exact quantitative and qualitative data to profile the chemicals in tissues and cells of medicinal materials. American ginseng, the root of *Panax quinquefolium* L., is one of the most recognized herbal medicines all over the world. Also, American ginseng has become popular in oriental countries as dietary health supplements or additives to foods and beverages [12]. In the herbal markets, various specifications or grades of American ginseng can be found, including main root, rootlet and fibrous root. Production area also affects the grade or price of the commercial medicine. As we know, American ginseng contains the major bioactive triterpene saponins named ginsenosides, such as ginsenosides Rg_1, $20(S)$-Rg_2, Re, $20(S)$-Rh_1, Rb_1, Rb_2 and Rd, which possess a wide range of pharmacological effects, including cardiovascular, anti-diabetic, anti-inflammatory and anti-tumor properties [13–16].

To evaluate the quality of American ginseng, a number of analytical methods to determine the total ginsenoside content or the target compounds have been developed [17–19]. However, few of them focus on the distribution rules of ginsenosides among tissues or detect the relationship of the quality and the microscopic characteristics. Until now, ginsenosides in the rhizome and root of *P. ginseng* Meyer has already been located: the cork contained more kinds of ginsenosides than did the cortex, phloem, xylem and resin canals [8]. But whether this rule applies to *P. quinquefolium* or not still waits to be found out. Analyzing the distribution of ginsenosides in different anatomical structures will establish the relationship between microscopic features and active components. Then the microscopic features used for the quality evaluation and classification of different specifications or grades of American ginseng can be validated or clarified.

In this study, fluorescence microscopy, LMD and UHPLC-Q/TOF–MS were used to analyze and compare the spatial chemical profiles of various tissues from *P. quinquefolium* to correlate the relationship between microscopic features and active components for the quality evaluation of American ginseng, shedding new light on the role of microscopic research in quality evaluation.

RESULTS AND DISCUSSION

Microscopic Examination and Dissection by LMD

In this study, four fresh *P. quinquefolium* samples (Pq1–4) and nine dried commercial samples were collected for analysis (see Table 1; Fig. 1). As shown under the normal light and fluorescence mode (see Fig. 2), the transverse section of American ginseng was comprised of cork, cortex, phloem, cambium and xylem. The cork was consisted of several rows of densely-arranged flat cells. Red fluorescence was emitted from the cork while blue color was shown in other tissues. Cortex was narrow. Cracks could be seen in phloem. Resin ducts with orange red fluorescence were scattered in the cortex and phloem. Cambium was arranged in a ring, showing strong florescence. Xylem was broad, usually differentiated into primary xylem with strong florescence and secondary xylem with common florescence. Since our study on localization of ginsenosides in the rhizome and root of *P. ginseng* illustrated that the resin ducts contained few ginsenosides, the resin ducts of *P. quinquefolium* samples were not examined here. The cork, cortex, phloem, secondary xylem and primary xylem were dissected from the main roots of Pq1–4 and Pq5–13. For the branch roots of Pq1–4, the xylem was hardly seen differentiation, and was thus examined as a whole. Compared with other samples, the cambium in the cross sections of Pq6 and Pq8 was obvious with relative more layers of cells, hence, the cambium of Pq6 and Pq8 were also investigated. Therefore, various tissues possessed different features and could be recognized under fluorescence mode. According to previous reports [6–8], the size of about 2,500,000 and 1,000,000 μm^2 of each separated tissues of fresh and dried materials were dissected by LMD respectively which could detect the chemicals containing in tissues.

Table 1: Information of commercial samples of *Panax quinquefolium* materials

Sample no.	Commercial name	Specification	Harvest time	Harvest place
Pq1	American ginseng	–	September 12th, 2014	Cultivation in Mulin County, Mudanjiang City, Heilongjiang Province, China
Pq2	American ginseng	–	September 12th, 2014	Cultivation in Mulin County, Mudanjiang City, Heilongjiang Province, China
Pq3	American ginseng	–	September 12th, 2014	Cultivation in Mulin County, Mudanjiang City, Heilongjiang Province, China
Pq4	American ginseng	–	September 12th, 2014	Cultivation in Mulin County, Mudanjiang City, Heilongjiang Province, China
Pq5	Wild-mountain pao-shen no. 1	HK$ 66,137.57/1000 g	–	Wildlife in America

Pq6	Wild-mountain small pao-shen no. 3.5	HK$ 34,391.53/1000 g	–	Wildlife in America
Pq7	Wild-mountain small and rouond pao-shen	HK$ 25,873.02/1000 g	–	Wildlife in America
Pq8	Wild-mountain pao-mian no. 3.5	HK$ 76,190.48/1000 g	–	Wildlife in America
Pq9	Wild-mountain pao-mian no. 4	HK$ 52,645.5/1000 g	–	Wildlife in America
Pq10	Wild-mountain small and rouond pao-mian	HK$ 44,973.54/1000 g	–	Wildlife in America
Pq11	Cultivated big-branch Pao-shen	HK$ 1534.39/1000 g	–	Cultivation in Canada
Pq12	Cultivated middle-branch Pao-shen	HK$ 1428.57/1000 g	–	Cultivation in Canada
Pq13	Cultivated shen no. 4	HK$ 1111.11/1000 g	–	Cultivation in Canada

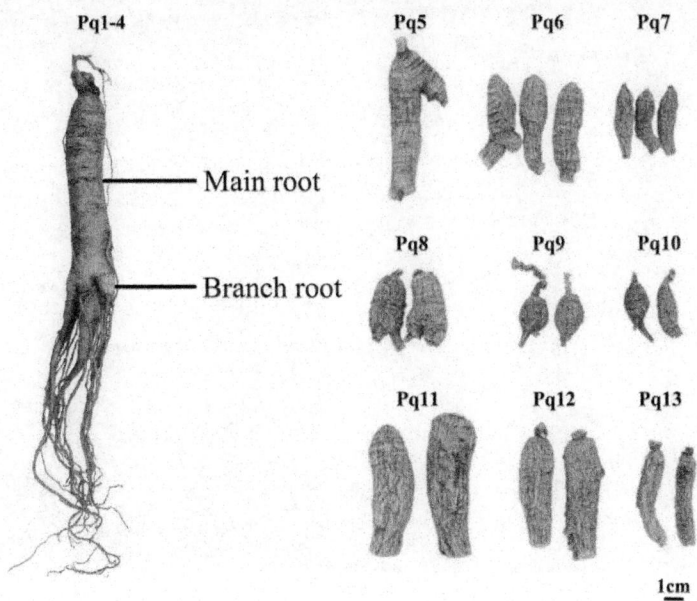

Figure. 1: Morphological features of *Panax quinquefolium* materials.

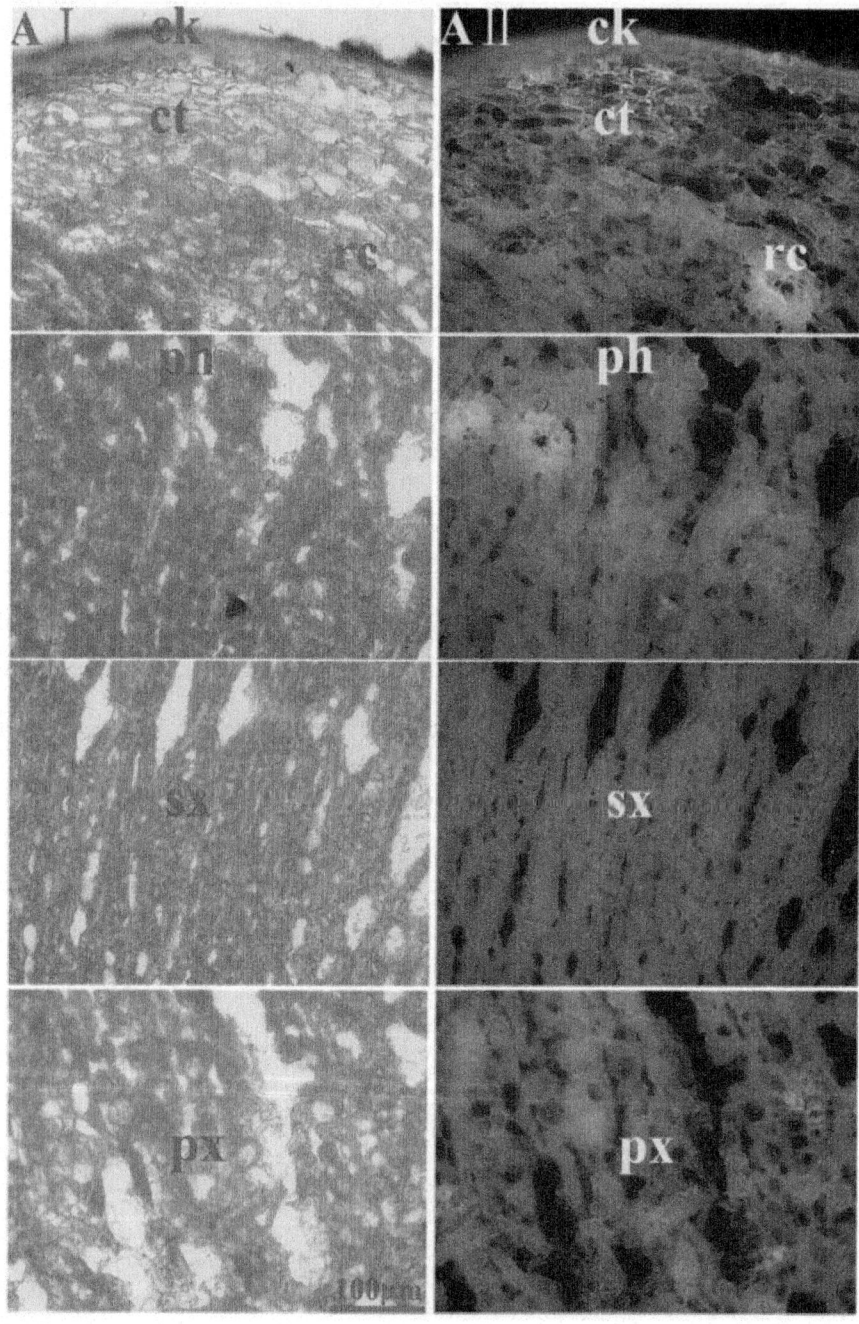

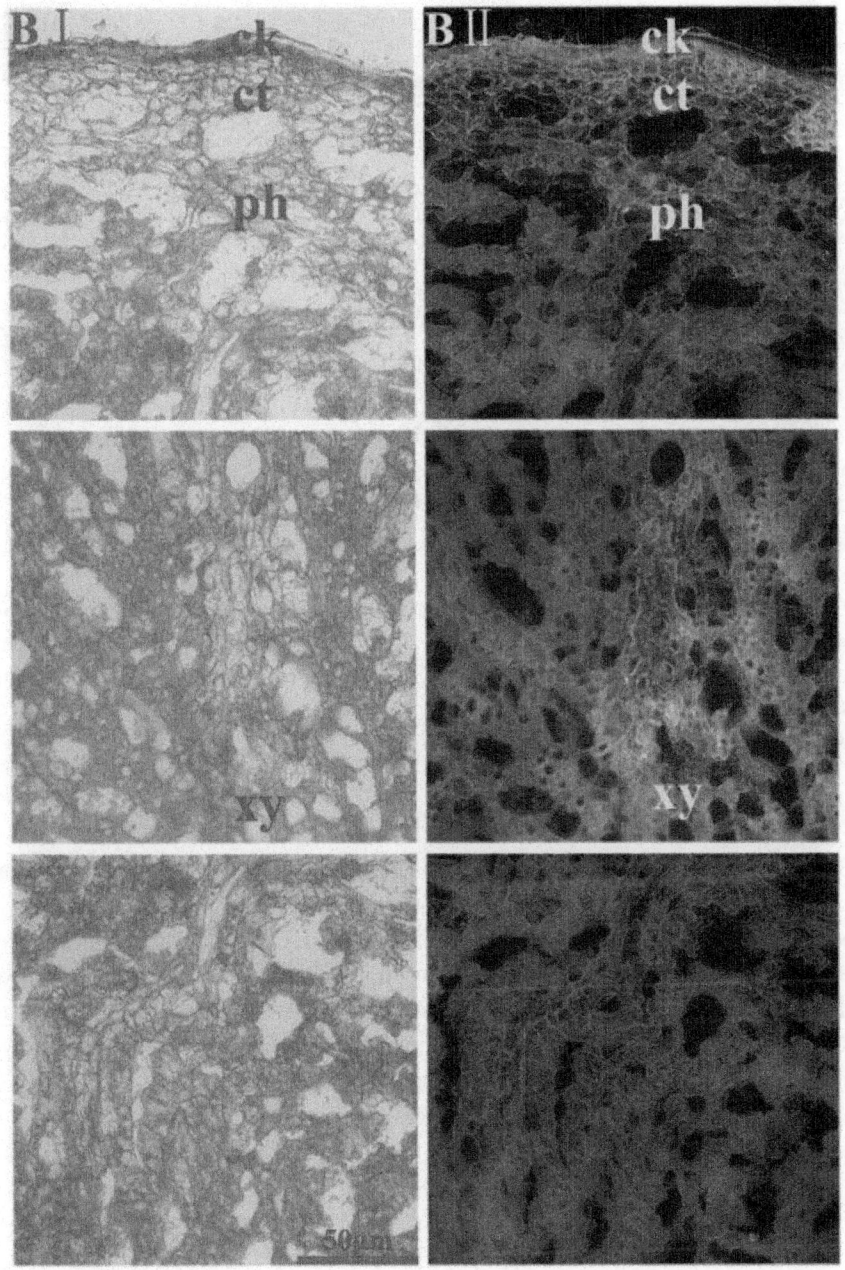

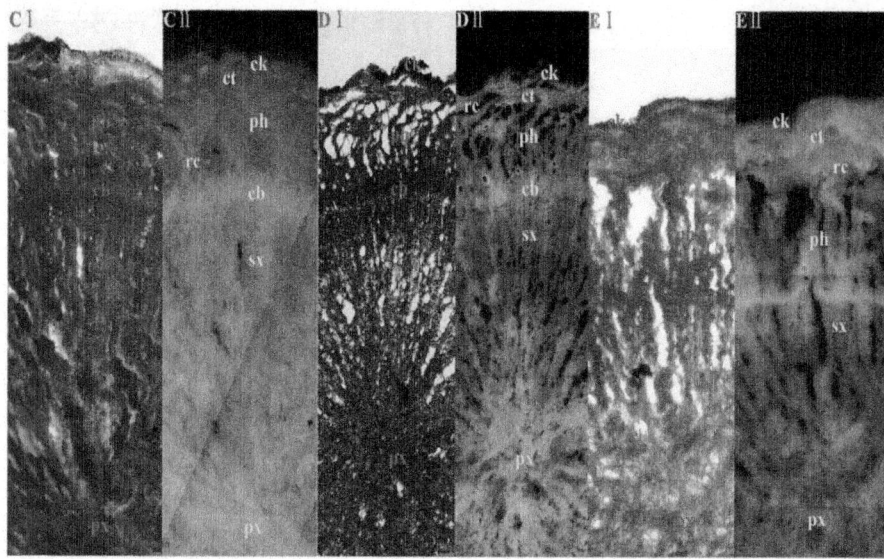

Figure. 2: Microscopic characteristics of *P. quinquefolium*. I Under normal light microscope, II under fluorescence mode with dichromatic mirror.**a, b** represented the main root and branch root of Pq1; **c–e** represented Pq6, Pq8 and Pq10 respectively. *ck* cork, *ct* cortex, *ph*phloem, *rc* resin canals, *cb* cambium, *xy* xylem, *sx* s econdary xylem, *px* primary xylem, *pt* pith.

Tissue-Specific Chemical Profiles

By UHPLC-Q/TOF–MS technique, tissue-specific chemical profiles of each sample were obtained as total ion chromatograms (see Figs. 3,4). A total of 34 peaks were detected in all the tissue extractions. By comparing retention times, accurate mass weights, and mass ions with the reference compounds, six peaks (Peaks 3, 4, 14, 15, 23, 29) were unambiguously identified as ginsenosides Rg_1, Re, 20(S)-Rg_2, 20(S)-Rb_1, Rb_2 and Rd. By matching those data with the components reported in the literature, 25 compounds were tentatively authenticated [12,20–24]. The identification result is shown in Table 2.

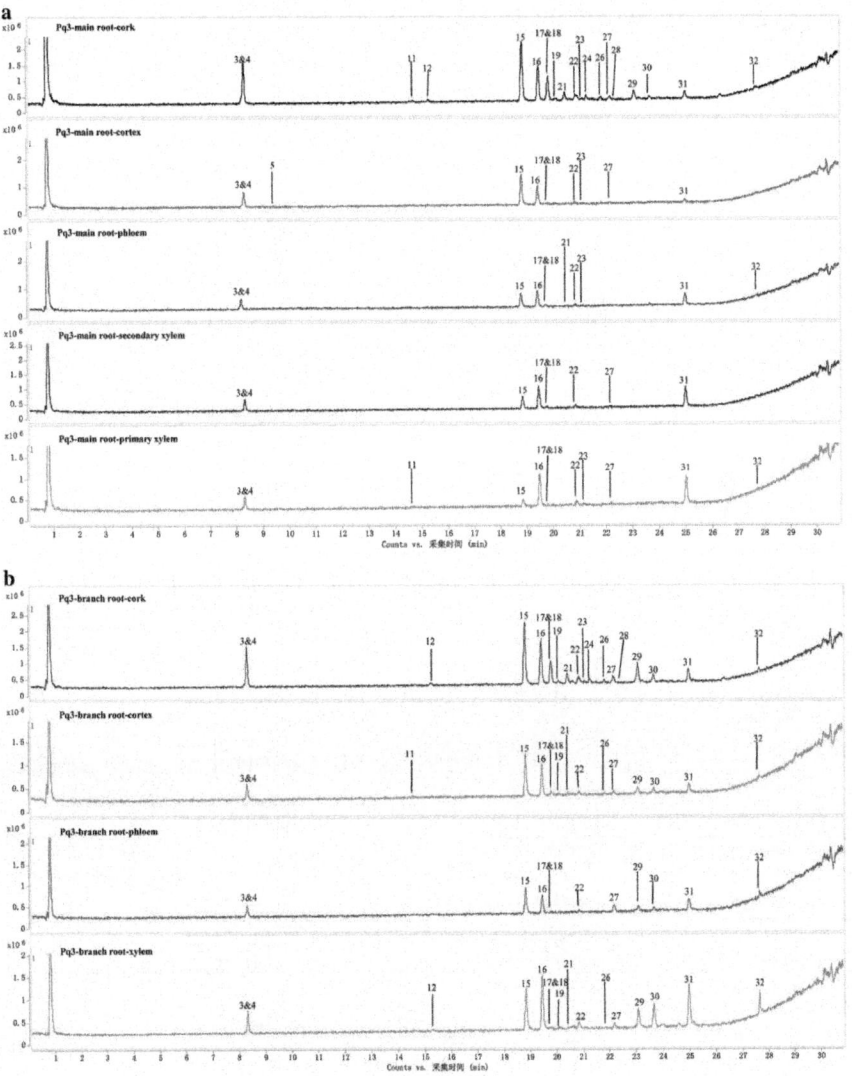

Figure. 3 The total ions current (TIC) chromatograms of microdissected tissues from main root (**a**) and branch root (**b**) of *P. quinquefolium* samples. The peak numbers referred to Table 2.

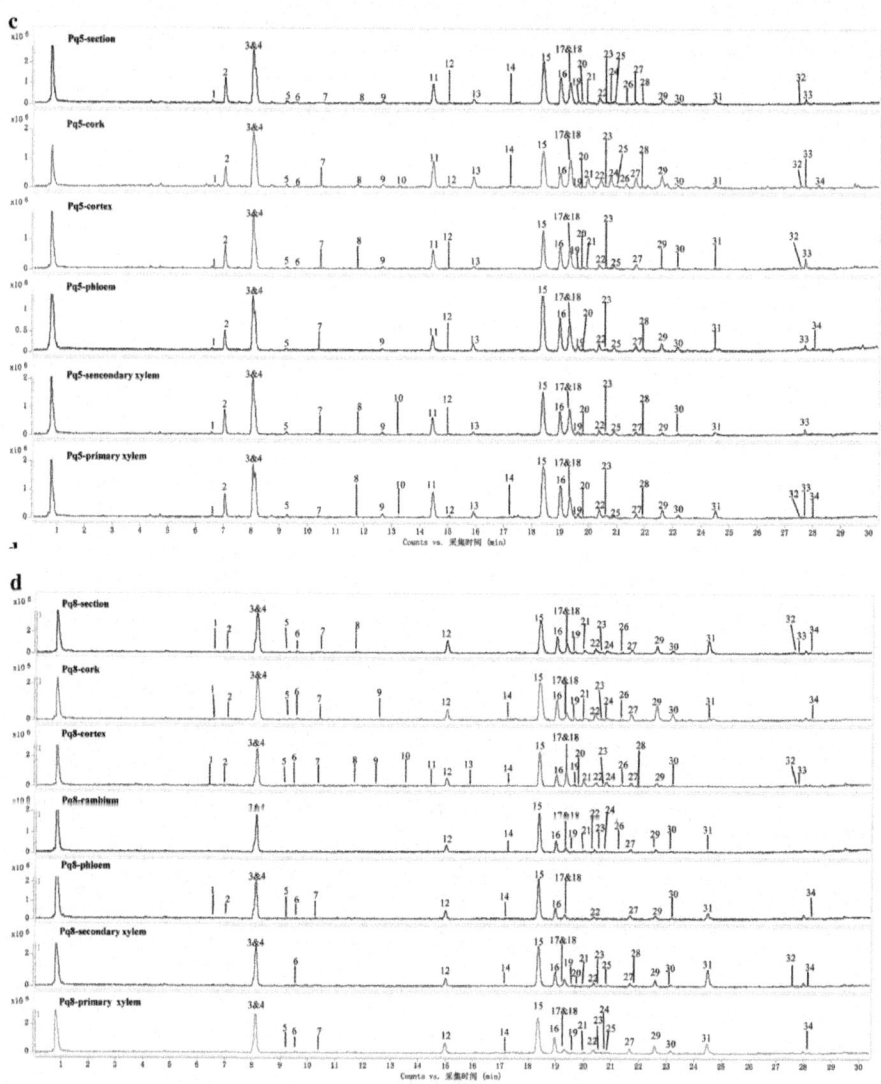

Figure. 4: The total ions current (TIC) chromatograms of microdissected tissues from *P. quinquefolium* samples of Pq5 (**c**) and Pq8 (**d**). The peak numbers referred to Table 2.

Table 2: Compounds identified from tissue extractions of *Panax quinquefolium* samples

Peak no.	Identity	t_R(min)	Molecular formular	[M−H]⁺ Mean measured mass (Da)	Theoretical exact mass (Da)	Mass accuracy (ppm)	[M−H+HCOOH]⁺ (mass accuracy, ppm)	Fragments of [M−H]⁺ (*m/z*)
1	20-Glc-G-Rf	6.58	$C_{48}H_{82}O_{19}$	961.5522	961.5378	14.98	1007.5578	799.5047 [M−H−Glc]⁻
2	Notoginsenoside R₁	7.04	$C_{47}H_{80}O_{18}$	931.5186	931.5278	−9.88	977.5465	799.5026 [M−H−Xyl]⁻; 637.4285[M−H−Glc−Xyl]⁻;
3	G-Rg₁ [a]	8.04	$C_{42}H_{72}O_{14}$	799.4975	799.4849	15.76	845.5026	637.4360 [M−H−Glc]⁻ 475.3785 [M−H−2Glc]⁻
4	G-Re[a]	8.12	$C_{48}H_{82}O_{18}$	945.5548	945.5428	12.69	991.5630	799.4935 [M−H−Rha]⁻ 783.5029 [M−H−Glc]⁻ 637.4407 [M−H−Rha−Glc]⁻
5	Malonyl-G-Rg₁	9.25	$C_{45}H_{74}O_{17}$	885.5082	885.4853	25.86	–	841.3240 [M−H−CO₂]⁻
6	Malonyl-G-Re isomer	9.56	$C_{51}H_{84}O_{21}$	1031.5547	1031.5432	11.15	–	987.5678[M−H−CO₂]⁻
7	Malonyl-G-Re	10.32	$C_{51}H_{84}O_{21}$	1031.5549	1031.5432	11.34	–	987.5644[M−H−CO₂]⁻
8	Floralquinquenoside B	11.73	$C_{42}H_{72}O_{15}$	815.4884	815.4793	11.16	–	637.4381[M−H−Rha−CH₃OH]⁻
9	Floralquinquenoside D	12.65	$C_{42}H_{72}O_{15}$	815.4882	815.4793	10.91	861.5002	653.4360 [M−H−Glc]⁻
10	Unknown	13.26	–	961.5559	–	–	1007.5580	–
11	Notoginsenoside Rw₂	14.43	$C_{41}H_{70}O_{14}$	785.4780	785.4687	11.84	831.4871	653.4361 [M−H−Xyl]⁻ 491.3674 [M−H−Xyl−Glc]⁻
12	Pseudoginsenoside F₁₁	14.99	$C_{42}H_{72}O_{14}$	799.4831	799.4844	−1.63	845.5015	653.4385 [M−H−Rha]⁻

13	Notoginsen-oside R_2	15.89	$C_{41}H_{70}O_{13}$	769.4573	769.4738	−21.44	815.4730	637.4392 [M−H−Xyl]⁻ 475.3839 [M−H− Xyl−Glc]⁻
14	20 (S)-G-Rg_2 [a]	17.23	$C_{42}H_{72}O_{13}$	783.5029	783.4900	16.46	829.5054	637.4394 [M−H−Rha]⁻ 475.3734 [M−H−Rha−Glc]⁻
15	G-Rb_1 [a]	18.38	$C_{54}H_{92}O_{23}$	1107.6097	1107.5957	12.64	−	945.5552[M−H−Glc]⁻ 783.5012 [M−H−2Glc]⁻
16	Malonyl-G-Rb_1	18.99	$C_{57}H_{94}O_{26}$	1193.6113	1193.5961	12.73	−	1149.6201[M−H−CO_2]⁻
17	G-Ro	19.33	$C_{48}H_{76}O_{19}$	955.5077	955.4908	17.69	−	793.2586[M−H−Glc]⁻
18	G-Rc	19.34	$C_{53}H_{90}O_{22}$	1077.5730	1077.5871	−13.08	−	945.5660 [M−H−Araf]⁻ 783.4980 [M−H−Araf −Glc]⁻
19	Malonyl-G-Rb_1 isomer I	19.63	$C_{57}H_{94}O_{26}$	1193.6142	1193.5961	15.16	−	1149.6185[M−H−CO_2]⁻
20	Unknown	19.80	−		1087.5461	−	−	−
21	Malonyl-G-Ra_2	19.97	$C_{56}H_{92}O_{25}$	1163.5993	1163.5855	11.86	−	1119.6041[M−H−CO_2]⁻
22	Malonyl-G-Rb_1 isomer II	20.38	$C_{57}H_{94}O_{26}$	1193.6101	1193.5961	11.73	−	1149.6192[M−H−CO_2]⁻
23	G-Rb_2 [a]	20.47	$C_{53}H_{90}O_{22}$	1077.5683	1077.5851	−15.59	1123.6337	945.5674 [M−H−Arap]⁻
24	G-Rb_3	20.79	$C_{53}H_{90}O_{22}$	1077.5977	1077.5851	11.69	1123.6637	945.5587 [M−H−Xyl]⁻ 915.5474 [M−H−Glc]⁻
25	Unknown	20.91	−		1119.6015	−	−	−	925.4844
26	Ma- Rb_2/ Rb_3 isomer	21.34	$C_{56}H_{92}O_{25}$	1163.5992	1163.5849	12.29	−	1119.6007[M−H−CO_2]⁻
27	O-acetyl-G-Rb_1	21.68	$C_{56}H_{94}O_{24}$	1149.6198	1149.6062	11.83	1195.6270	1107.6067 [M−H−Acetyl]⁻ 945.5466 [M−H−Ace-tyl−Glc]⁻
28	Zingibroside R_1	21.92	$C_{42}H_{65}O_{14}$	793.4479	793.4374	13.23	−	631.3332[M−H−Glc]⁻

29	G-Rd[a]	22.59	$C_{48}H_{82}O_{18}$	945.5548	945.5428	12.69	991.5613	783.4985 [M–H–Glc]⁻ 621.4432 [M–H–2Glc]⁻
30	Malonyl-G-Rd	23.18	$C_{51}H_{84}O_{21}$	1031.5614	1031.5432	17.64	–	987.5682[M–H–CO₂]⁻
31	G-Rd isomer	24.49	–	945.5543	945.5428	12.16	991.5069	783.4985 [M–H–Glc]⁻ 621.4432 [M–H–2Glc]⁻
32	20 (S)-G-Rg₃	27.55	$C_{42}H_{72}O_{13}$	783.4978	783.4900	9.96	829.5057	621.4375 [M–H–Glc]⁻ 459.4088 [M–H–2Glc]⁻
33	Chikusetsu-saponin IVa	27.69	$C_{42}H_{66}O_{14}$	793.4367	793.4380	–1.64	–	–
34	20 (R)-G-Rg₃	28.14	$C_{42}H_{72}O_{13}$	783.4982	783.4900	10.47	829.5065	621.4375 [M–H–Glc]⁻ 459.3964 [M–H–2Glc]⁻

G ginsenoside, *Glc* β-d-glucopyranosyl, *Rha* α-l-rhamnopyranosyl, *Xyl* β-d-xylopyranosyl, *Araf* α-l-arabinofuranosyl, *Arap* α-l-arabinopyranosyl

[a]Identified with chemical marker

As seen from Figs. 3, 4, the distribution differences of gensenosides in various tissues from American ginseng were not as distinct as Asian ginseng [8]. The cork extractions usually had the most peaks (20–34 peaks). The cortex and primary xylem took the second place, namely 11–31 peaks and 12–30 peaks respectively. The secondary xylem (9–28 peaks), phloem (11–27 peaks) and cambium (24 peaks for Pq6, 18 peaks for Pq8) possessed the least peaks. For example, the cork, cortex, phloem, secondary xylem and primary xylem of Pq1 showed 34, 29, 29, 28 and 30 peaks separately. The tissues above of Pq7 had 32, 19, 14, 19 and 21 peaks respectively. Thus, the cork, primary xylem and cortex possessed the most kinds of saponin compounds.

For most samples, the areas of Peaks 21–30 in the cork were larger than those in other tissues. Peaks 21–30 represented compounds with medium or low polarity, which might be concerned with the protection function of the cork. In the xylem, especially the primary xylem, the areas of Peaks 17–31 were larger than those in cortex, phloem and cambium, which might be relevant with the lignification, suberification and the channel function of xylem cells.

Quantification of Ginsenosides in Various Tissues

Ginsenosides Rg_1, Re, Rh_1, 20(S)-Rg_2, 20(S)-Rb_1, Rb_2 and Rd in various tissues of different samples were determined by UHPLC-Q/TOF–MS. The results are given in Table 3 and Fig. 5. For most samples (Pq1–5, Pq7–10), the cork contained the most ginsenosides compared with other tissues, with the content ranging from 1094.58 to 269944.16 $ng/10^5 \mu m^2$. Sometimes, the primary xylem possessed the highest level of ginsenosides (Pq6, Pq11–13), or possessed the second highest level (main root of Pq1, Pq5, Pq7–10), whereas sometimes low ginsenoside level was found in the primary xylem (main root of Pq2–4). The amounts of ginsenosides fluctuated in the cortex. It seemed that if the contents of ginsenosides were low in primary xylem, the contents would be high in cortex (main root of Pq2–4); and if the contents of ginsenosides were high in primary xylem, the cortex would have a medium (main root of Pq1, Pq5, Pq7, Pq8, Pq10) or low (Pq6, Pq9, Pq11–13) level of ginsenosides. The phloem, secondary xylem and cambium usually had fewer ginsenosides than other tissues. For the branch roots of Pq1-4, the cork, xylem and cortex occupied higher contents of ginsenosides than phloem did. Thus, the distribution pattern of ginsenosides in American ginseng was quite distinct from Asian ginseng. Distinctly, the cork, primary xylem or cortex had more ginsenosides than phloem, secondary xylem and cambium in American ginseng. Based on all the above, it was reasonable to deduce that the ratio of total areas of cork, primary xylem and the cortex to the area of whole transection could help to evaluate the quality of American ginsengs.

Table 3: Contents of ginsenosides in the tissues from *Panax quinquefolium* samples

Sample no.	Tissue	Amount in unit area ($ng/10^5 \mu m^2$)							
		Rg_1 [a]	Re	Rh_1	Rg_2	Rb_1	Rb_2	Rd	Sum
Pq1 main root	Cork	67.31	34.58	0.40	8.83	13,247.66	25.51	13.18	13,397.47
	Cortex	18.77	9.50	–[b]	1.83	5576.43	1.02	1.05	5608.60
	Phloem	11.53	7.46	0.25	1.87	4734.50	1.28	1.53	4758.42
	Secondary xylem	9.38	10.74	–	2.11	3176.85	2.20	1.69	3202.97
	Primary xylem	31.12	12.11	0.30	2.07	8104.59	2.15	1.49	8153.83
Pq1 branch root	Cork	74.68	50.67	–	1.16	16,897.70	31.86	7.00	17,063.07
	Cortex	10.84	9.88	–	2.81	7608.60	1.63	3.24	7637.00
	Phloem	5.50	5.46	0.35	2.17	5073.24	0.56	2.72	5090.00
	Xylem	8.10	9.36	–	3.97	7239.49	1.68	12.29	7274.89

Pq2 main root	Cork	130.53	69.38	0.27	3.45	16,012.69	42.06	27.09	16,285.47
	Cortex	36.33	19.63	–	0.75	5244.41	1.16	0.95	5303.23
	Phloem	16.39	8.66	–	2.57	3840.21	1.17	0.72	3869.72
	Secondary xylem	23.34	28.46	–	7.57	2344.29	2.51	1.50	2407.67
	Primary xylem	27.65	29.74	–	4.72	2688.51	3.21	0.93	2754.76
Pq2 branch root	Cork	62.44	46.66	0.30	13.93	17,558.77	52.02	29.52	17,763.64
	Cortex	11.64	8.78	0.36	2.84	3371.72	1.44	2.04	3398.82
	Phloem	11.57	8.17	0.37	4.20	3159.24	1.82	3.55	3188.92
	Xylem	23.09	18.41	0.34	9.31	5805.28	1.48	17.17	5875.08
Pq3 main root	Cork	41.66	18.92	0.39	4.15	269,855	16.80	7.24	269,944.16
	Cortex	18.67	7.19	0.59	1.61	145,606.6	3.61	1.15	145,639.42
	Phloem	11.31	6.40	0.51	0.98	67,598.38	–	1.21	67,618.79
	Secondary xylem	11.69	6.42	0.31	0.84	50,655.09	–	–	50,674.35
	Primary xylem	10.03	5.35	0.33	0.87	19,113.26	0.46	0.30	19,130.60
Pq3 branch root	Cork	23.16	20.68	0.32	4.98	252,865.9	12.32	17.92	252,945.28
	Cortex	6.69	6.20	0.39	2.06	114,430.5	3.50	4.44	114,453.78
	Phloem	5.17	4.43	0.32	1.56	85,663.43	0.95	3.07	85,678.93
	Xylem	6.18	5.95	0.27	0.30	134,882.3	0.70	9.10	134,904.80
Pq4 main root	Cork	48.13	23.62	0.32	0.56	11,972.59	20.01	8.35	12,073.58
	Cortex	11.50	5.10	0.31	0.84	4151.39	1.61	1.43	4172.18
	Phloem	11.85	5.45	0.33	0.52	1685.59	–	1.28	1705.02
	Secondary xylem	9.70	5.35	0.48	0.69	2659.33	1.51	1.45	2678.51
	Primary xylem	6.37	3.43	0.43	0.54	1766.77	0.74	0.47	1778.75
Pq4 branch root	Cork	19.34	20.25	0.30	3.61	20,298.81	19.70	23.67	20,385.68
	Cortex	7.09	7.63	0.40	1.48	12,388.83	5.48	7.41	12,418.32
	Phloem	2.94	4.66	0.32	1.07	5156.83	1.73	9.49	5177.04
	Xylem	7.50	8.57	0.35	2.25	15,479.39	2.95	9.38	15,510.39
Pq5	Cork	1723.58	838.53	10.24	11.41	869.15	167.96	229.08	3849.94
	Cortex	920.69	365.92	4.64	3.35	764.67	11.79	20.06	2091.13
	Phloem	527.99	390.62	2.07	1.50	885.82	6.30	70.71	1884.99
	Secondary xylem	798.60	434.04	0.95	2.28	821.14	6.09	26.04	2089.14
	Primary xylem	1028.47	924.56	1.19	5.86	1365.07	32.65	144.22	3502.02

Pq6	Cork	670.07	34.99	6.03	1.00	582.31	149.25	155.58	1599.24
	Cortex	320.81	13.99	3.79	0.84	364.01	41.22	47.94	792.61
	Phloem	417.83	18.60	2.80	0.94	432.25	5.95	25.73	904.10
	Cambium	605.12	26.43	7.10	1.03	600.16	6.08	40.85	1286.77
	Secondary xylem	906.45	35.99	4.26	0.85	814.07	7.77	59.66	1829.04
	Primary xylem	1501.30	74.73	5.11	1.32	1115.92	23.22	179.22	2900.82
Pq7	Cork	166.40	327.34	1.71	3.93	401.22	66.79	127.19	1094.58
	Cortex	174.18	207.77	1.30	2.98	163.75	16.12	24.81	590.91
	Phloem	119.12	131.49	0.65	1.40	191.36	4.16	21.26	469.43
	Secondary xylem	158.80	110.53	0.60	0.88	157.42	4.33	3.80	436.35
	Primary xylem	187.65	173.03	0.71	0.92	333.41	11.19	30.31	737.22
Pq8	Cork	149.28	1827.33	0.67	12.70	1347.97	41.12	429.50	3808.57
	Cortex	180.35	714.05	0.74	19.35	1173.10	80.68	82.68	2250.96
	Phloem	141.83	732.05	0.56	6.91	1002.23	9.38	49.03	1941.99
	Cambium	144.34	723.85	0.62	9.33	1154.96	5.40	61.17	2099.69
	Secondary xylem	144.52	987.34	0.80	9.24	1478.13	11.02	163.51	2794.55
	Primary xylem	145.17	1302.97	0.95	11.91	1365.79	12.07	218.33	3057.19
Pq9	Cork	907.61	14.06	2.08	0.88	799.16	195.43	170.42	2089.63
	Cortex	160.10	1.99	–	–	179.45	7.47	4.61	353.61
	Phloem	74.54	1.52	–	0.95	60.01	4.97	2.16	144.15
	Secondary xylem	392.43	2.41	–	–	430.97	3.61	22.49	851.91
	Primary xylem	676.25	2.78	0.84	–	1019.24	5.45	69.53	1774.09
Pq10	Cork	668.61	712.57	0.84	6.36	986.29	19.83	67.40	2461.89
	Cortex	139.10	669.75	0.54	14.92	635.65	78.61	39.19	1577.77
	Phloem	123.70	611.79	0.61	6.61	434.33	3.12	14.90	1195.07
	Secondary xylem	146.61	697.48	0.66	5.26	538.81	16.77	19.34	1424.93
	Primary xylem	147.68	743.10	0.62	7.22	714.65	2.12	20.17	1635.56

Pq11	Cork	62.97	537.33	0.88	5.07	511.33	65.37	188.27	1371.23
	Cortex	24.68	320.33	–	3.50	503.92	2.51	36.75	891.69
	Phloem	21.42	344.88	–	3.88	670.87	2.04	83.75	1126.83
	Secondary xylem	8.58	340.35	–	3.60	564.94	2.98	159.17	1079.62
	Primary xylem	22.13	619.29	0.56	7.31	916.71	3.81	364.94	1934.75
Pq12	Cork	115.07	518.18	1.85	6.69	634.75	104.45	319.73	1700.73
	Cortex	67.05	342.52	0.55	3.34	560.80	9.01	43.48	1026.75
	Phloem	69.79	375.80	–	3.56	871.87	3.74	48.98	1373.75
	Secondary xylem	45.97	610.82	0.61	4.39	1117.76	6.46	211.26	1997.28
	Primary xylem	147.28	871.49	0.47	5.95	1021.44	9.45	132.19	2188.28
Pq13	Cork	82.82	766.07	0.92	22.68	568.79	88.17	125.20	1654.66
	Cortex	33.11	428.44	–	6.46	453.02	7.45	37.26	965.74
	Phloem	34.04	547.62	–	5.32	526.29	3.31	41.08	1157.65
	Secondary xylem	41.02	453.95	0.52	8.36	772.36	5.45	104.84	1386.50
	Primary xylem	37.32	893.53	–	14.27	922.80	3.81	166.35	2038.07

[a]Ginsenoside

[b]Under detection limit

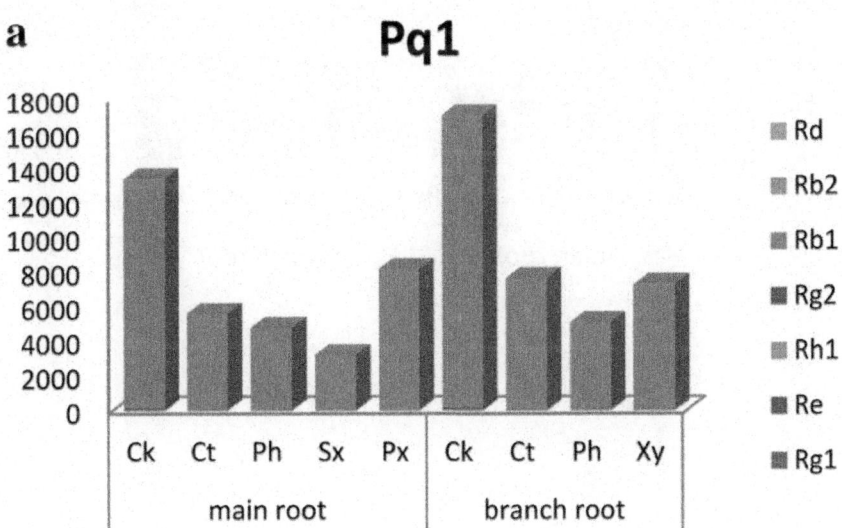

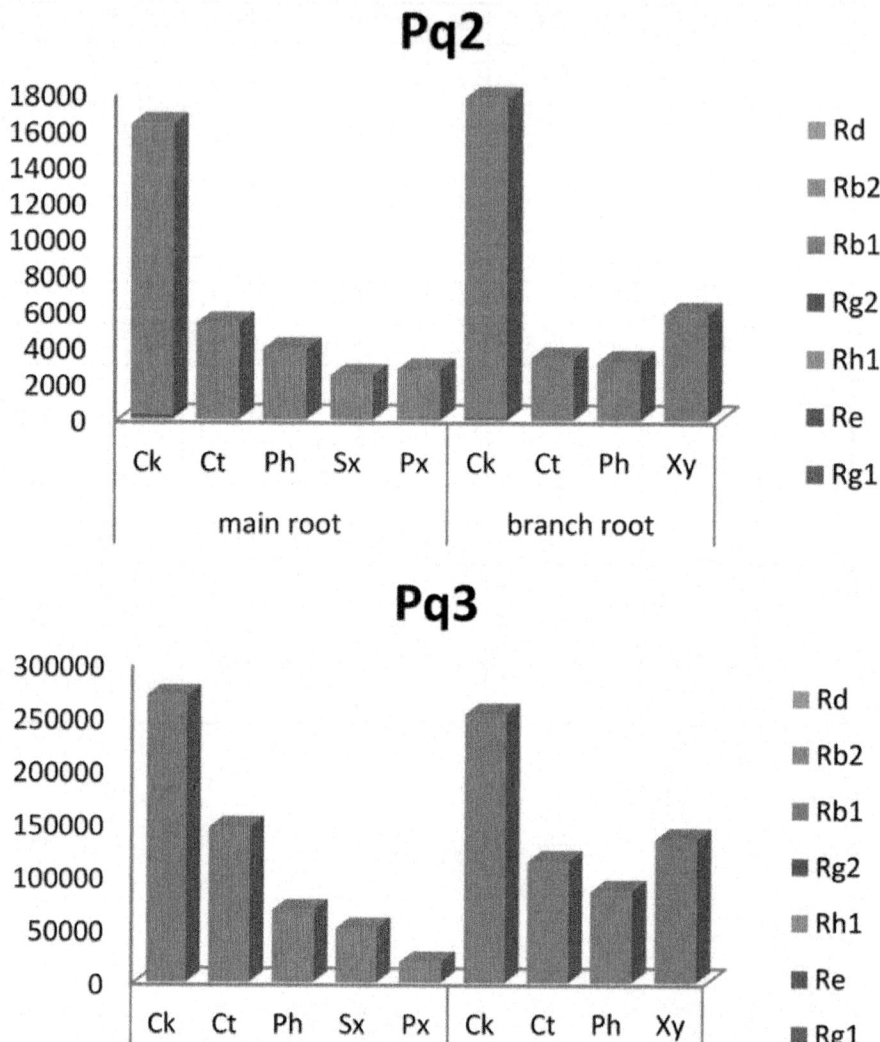

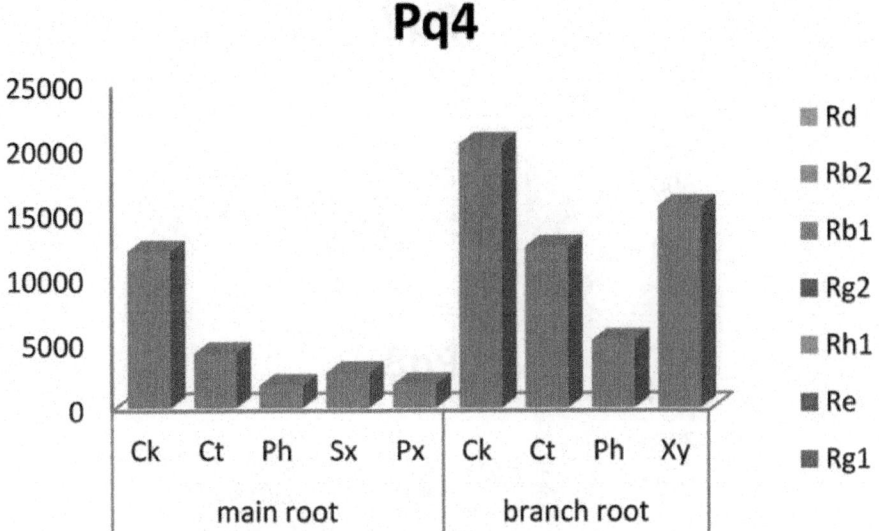

b

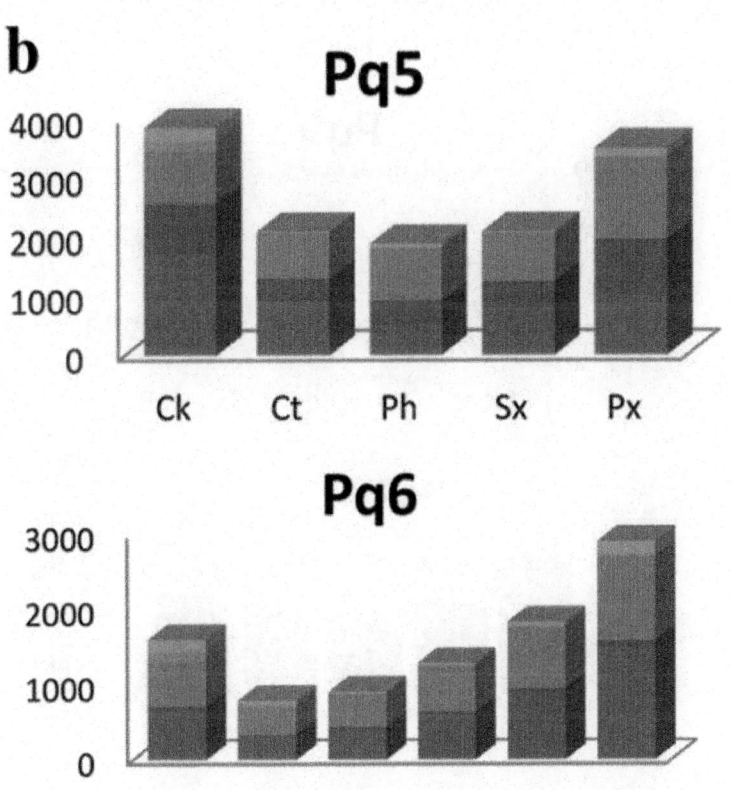

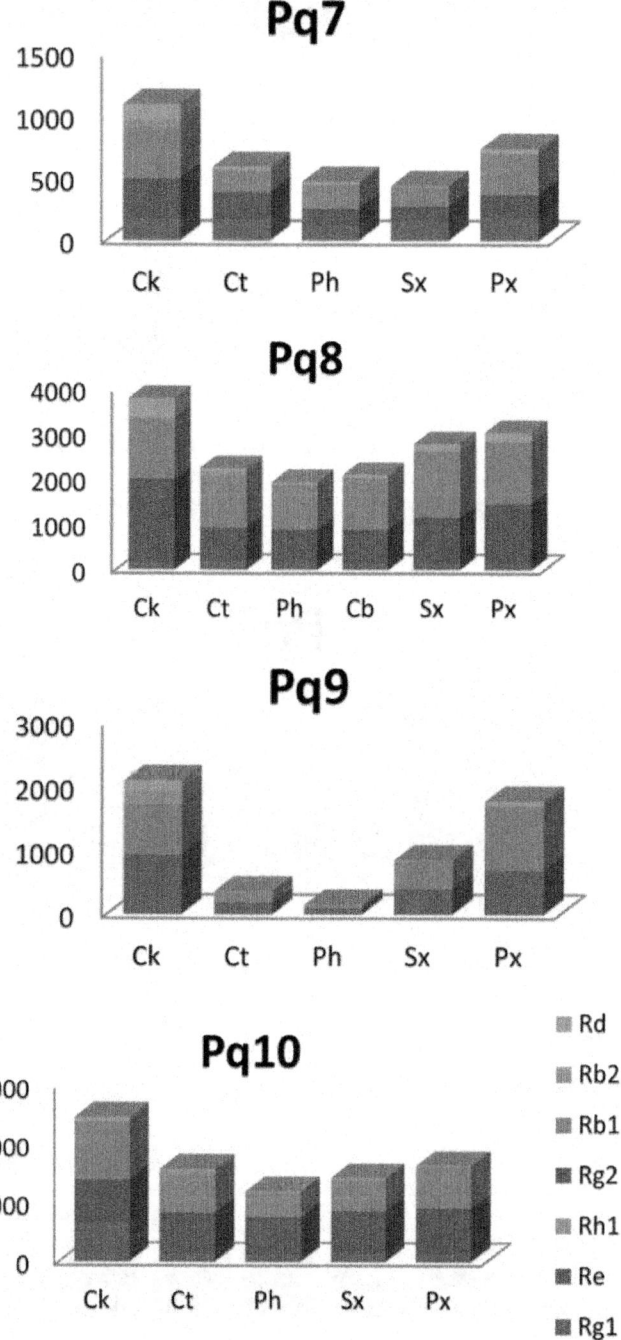

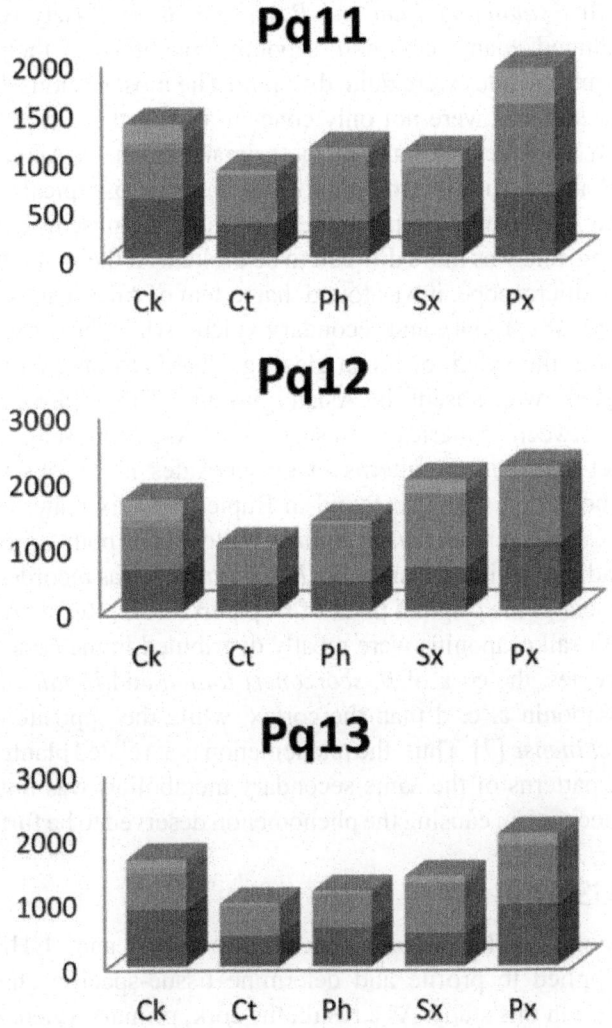

Figure. 5: Contents of ginsenosides in different tissues of Pq1-4 (**a**) and Pq5-13 (**b**). *Ck* cork, *Ct* cortex, *Ph* phloem, *Cb* cambium, *Sx* secondary xylem, *Px* primary xylem.

It was reported that the outer part of the *P. quinquefolium* root contained more ginsenosides than the center part [25]. However, another paper found that the peak areas of ginsenosides in the center part were larger than those of the outer part [26]. The outer part includes the cork and cortex, while the center part represented the primary xylem for most samples or xylem for branch roots. Our research illustrated that the both situations existed simultaneously in American ginseng.

Although *P. quinquefolium* and *P. ginseng* were closely related species which contained many common saponin constituents, their distribution patterns of ginsenosides were quite different. The most obvious difference was that the ginsenosides were not only concentrated in the cork and cortex, but also inclined to be accumulated in the primary xylem in American ginseng. This was identical with the morphological and microscopical characteristics of Asian and American ginseng. In detail, American ginseng was harder than Asian ginseng, and was more difficult to be broken. At the same time, under the fluorescence microscope, it was found that xylem of American ginseng usually differentiated into primary and secondary xylem, while the differentiation was scarely seen in the xylem of Asian ginseng. That is to say that the developed primary xylem was absent in Asian ginseng. The different microscopic structures between American ginseng and Asian ginseng may explain their distinct distribution patterns of ginsenosides in various tissues.. Such similar phenomenon was also found in Bupleuri Radix material. *Bupleurum chinense* DC. and *B. scorzoneri folium* Willd. were both original plants of Bupleuri Radix in China. Meanwhile, *B. falcatum* L. was recorded by Japanese Pharmacopoeia as the original plant of Bupleuri Radix. Recent research found that although saikosaponins were mostly distributed in the cork and cortex in the three species, the cork of *B. scorzoneri folium* and *B. falcatum* contained more saikosaponin a, c, d than the cortex, while the opposite situation was found in *B. chinense* [7]. Thus, the phenomenon that related plants had different distribution patterns of the same secondary metabolites was not an accident. The exact mechanism causing the phenomenon deserved to be further explored.

CONCLUSION

In conclusion, LMD, fluorescence microscopy, and UHPLC–Q/TOF–MS were applied to profile and determine tissue-specific chemicals of *P. quinquefolium* in this study. As a result, the cork, primary xylem or cortex had more ginsenosides than phloem, secondary xylem and cambium in American ginseng. Thus, the ratio of total areas of cork, primary xylem and the cortex to the area of the whole transection showed a potential to be used as a reference to judge the quality of American ginsengs.

EXPERIMENTAL

Plant Material

As seen from Table 1 and Fig. 1, four fresh *P. quinquefolium* samples (Pq1–4) were collected from Mulin County, Mudanjiang City, Heilongjiang Province, China. Nine dried samples (Pq5–13) of various commercial types were purchased from Hong Kong herbal markets. All of them were identified by Dr. Zhitao Liang from the School of Chinese Medicine, Hong Kong Baptist University. The voucher specimens were deposited in the Bank of China (Hong Kong) Chinese Medicines Centre of Hong Kong Baptist University. Collected samples were stored at −20 °C before use.

Chemicals and Reagents

Chemical standards of ginsenosides Rg_1, $20(S)$-Rg_2, Re, $20(S)$-Rh_1, Rb_1, Rb_2 and Rd were purchased from Shanghai Tauto Biotech Company (Shanghai, China). Acetonitrile and methanol of HPLC grade were from E. Merck (Darmstadt, Germany), and formic acid of HPLC grade was from Tedia (Fairfield, USA). Water was prepared by a Milli-Q system (Millipore, Bedford, MA, USA).

Laser Microdissection and Sample Solution Preparations

The dried materials were firstly softened by infiltrating with water-soaked-non-cellulose paper before frozen section. The softened and fresh roots were cut into small sections, embedded in cryomatrixTM (Thermo Shandon Limited, U.K.), and then placed on a cutting platform in the cryobar of a cryostat (Thermo Shandon As620 Cryotome, U.K.) at −20 °C. Serial slices of 40 μm in thickness were cut at −10 °C. Each sectioned tissue slice was mounted directly to a non-fluorescent PET microscope steel frame slide (76 mm × 26 mm, 1.4 μm thick, Leica Microsystems, Germany). The slide was observed with a Leica LMD 7000 microscope system (Leica, Benshein, Germany) in fluorescence mode with a dichromatic mirror. Microdissection was conducted by a DPSS laser beam at 349 nm wavelength, aperture of 12, speed of 10, power of 50–60 μJ and pulse frequency of 2895 Hz under a Leica LMD-BGR fluorescence filter system at 10x magnification. Tissue parts within an area of approximately 1×10^6 μm² were determined as the investigated size and dissected separately under fluorescence inspection mode. The microdissected tissues fell into caps of 500 μL microcentrifuge tubes (Leica, Germany) by gravity.

The separated tissue part in each cap was transferred to the bottom of the tube through centrifugation (Centrifuge 5415R, Eppendorf, Hamburg, Germany) at 12,000 rpm for 5 min. 100 μL methanol was added into each microcentrifuge tube. The tube was sonicated for 30 min (CREST 1875HTAG ultrasonic processor, USA). The microcentrifuge tube was centrifuged again for 10 min at 12,000 rpm, and 4 °C. 90 μL of the supernatant was transferred to a glass insert with plastic bottom spring (400 μL, Grace, USA) in a 1.5 mL brown HPLC vial (Grace, USA) and stored at 4 °C for analysis.

Qualitative and Quantitative Analysis

UHPLC-QTOF–MS analysis was performed on an Agilent 6540 ultra-high definition accurate mass quadrupole time-of-flight spectrometer with UHPLC (UHPLC-QTOF–MS, Agilent Technologies, USA). A UPLC C_{18} analytical column (2.1 mm × 100 mm, I.D. 1.7 μm, ACQUITY UPLC®BEH, Waters, USA) was used for separation, coupled with a C_{18} pre-column (2.1 mm × 5 mm, I.D. 1.7 μm, VanGuardTM BEH, Waters, USA) at room temperature of 20 °C. The mobile phase was a mixture of water (A) and acetonitrile (B), both containing 0.1 % formic acid, with an optimized linear gradient elution as follows: 0–3 min, 10–20 % B; 3–25 min, 20–38 % B; 25–30 min, 38–85 % B; 30–30.1 min, 85–100 % B; 30.1–32 min, 100 % B; 32–32.1 min 100–10 % B with 4 min of balance. The injection volume was 3 μL for tissue sample. The flow rate was set at 0.35 mL/min. The mass spectra were acquired in negative mode by scanning from 100 to 1700 in mass to charge ratio (m/z). The MS analysis was performed under the following operation parameters: dry gas temperature 300 °C, dry gas (N_2) flow rate 8 L/min, nebulizer pressure 45 psi, Vcap 3000, nozzle voltage 500 V, and fragmentor voltage 180 V. The energies for collision-induced dissociation (CID) were set at 30 and 45 eV respectively for the fragmentation information.

Data analysis was performed with Agilent MassHunter Workstation software-Qualitative Analysis and Q-TOF Quantitative Analysis (version B.04.00, Build 4.0.479.5, Service Pack 3, Agilent Technologies, Inc. 2011). By searching databases including PubMed of the US National Library Medicine and the National Institutes of Health, Scifinder Scholar of American Chemical Society and Chinese National Knowledge Infrastructure (CNKI) of Tsinghua University, all chemicals reported in the literatures as derived from *Panax* species were summarized in a Microsoft Office Excel table to establish a database, which includes the name, molecular formula, and molecular weight of each chemical. The "Search Database" in the "Identify Compounds" in Agilent MassHunter Workstation software-Qualitative Analysis was used to identify the chromatographic peaks.

To semi-quantitatively determine the spatial distributions of the individual metabolites in different tissue regions, the contents of chemical markers including ginsenosides Rg_1, 20(S)-Rg_2, Re, 20(S)-Rh_1, Rb_1, Rb_2 and Rd in various microdissected tissues were relatively determined using the above UHPLC-QTOF–MS method. Linearity was examined within selected concentration range with different levels and applied to calculate the amounts of these analytes in tissue extracts.

ABBREVIATIONS

LMD: laser microdissection

UHPLC-Q/TOF-MS: ultra-high performance liquid chromatography-quadrupole/time-of-flight- mass spectrometry

AUTHORS' CONTRIBUTIONS

ZL initiated and all authors designed the study. YC and YZ carried out the experimental study. YC drafted the manuscript. LX collected the herbal samples. All authors contributed to the data analysis and to finalizing the manuscript. All authors read and approved the final version.

ACKNOWLEDGEMENTS

This work is sponsored by the Faculty Research Grant of Hong Kong Baptist University (FRG2/12-13/030) and Innovation and Technology Fund (ITS/185/13FX). We are grateful to Mr. Alan Ho from the School of Chinese Medicine, Hong Kong Baptist University for his technical support.

REFERENCES

1. Cai X, Zhang AX, Wu H, Hu ZH (1999) Histochemistry of sinomenine in the stem of *Sinomenium acutum* and *Sinomenium acutum* var.*cinereum*. Acta BotBoreali-Occidentalia Sin 19:104–107

2. Shen ZG, Chauser-volfson E, Gutterman Y, Hu ZH (2001) Anatomy, histochemistry and phytochemistry of leaves in *Aloe vera* var.*chinensis*. Acta Bot Sin 43:780–787

3. Lin R, Cao YF, Hu ZH (2002) Anatomical structure of vegetative organs and histochemical localization of ginsenosides in *Gynostemma pentaphyllum*. Acta Bot Boreali-Occidentalia Sin 22:796–800

4. Cao YF, Lin R, Hu ZH (2003) Studies on the developmental anatomy of Rhizome of *Dioscorea zingiberensis* and its histochemistry. J Wuhan Bot Res 21:288–294

5. Qiao Q, Xiao YP, Wang ZZ (2004) Anatomical structure and histochemical localization of the drupe of *Macrocarpium officinacle*. Acta Bot Yunnanica 26:651–655

6. Liang ZT, Sham TT, Yang GY, Yi L, Chen HB, Zhao ZZ (2013) Profiling of secondary metabolites in tissues from *Rheum palmatum* L. using laser microdissection and liquid chromatography mass spectrometry. Anal Bioanal Chem 405:4199–4212

7. Liang ZT, Oh KY, Wang YQ, Yi T, Chen HB, Zhao ZZ (2014) Cell type-specific qualitative and quantitative analysis of saikosaponins in three *Bupleurum* species using laser microdissection and liquid chromatography–quadrupole/time of flight-mass spectrometry. J Pharm Biomed Anal 97:157–165

8. Liang ZT, Chen YJ, Xu L, Qin MJ, Yi T, Chen HB, Zhao ZZ (2015) Localization of ginsenosides in the rhizome and root of *Panax ginseng* by laser microdissection and liquid chromatography–quadrupole/time of flight-mass spectrometry. J Pharm Biomed Anal. 105c:121–133

9. Yi L, Liang ZT, Peng Y, Yao X, Chen HB, Zhao ZZ (2012) Tissue-specific metabolite profiling of alkaloids in Sinomenii Caulis using laser microdissection and liquid chromatography-quadrupole/time of flight-mass spectrometry. J Chromatogr A 1248:93–103

10. Chen YJ, Liang ZT, Zhu Y, Xie GY, Tian M, Zhao ZZ, Qin MJ (2014) Tissue-specific metabolites profiling and quantitative analyses of flavonoids in the rhizome of *Belamcanda chinensis* by combining laser-microdissection with UHPLC-Q/TOF-MS and UHPLC-QqQ-MS. Talanta 130:585–597

11. Jaiswal Y, Liang ZT, Ho A, Wong LL, Yong P, Chen HB, Zhao ZZ (2014) Distribution of toxic alkaloids in tissues from three herbal medicine Aconitum species using laser micro-dissection, UHPLC–QTOF MS and LC–MS/MS techniques. Phytochemistry 107:155–174

12. Sun BS, Xu MY, Li Z, Wang YB, Sang CK (2012) UPLC-Q-TOF-MS/MS analysis for steaming times-dependent profiling of steamed *Panax quinquefolius* and its ginsenosides transformations induced by repetitious steaming. J Ginseng Res 36:277–290

13. Attele AS, Wu JA, Yuan CS (1999) Ginseng pharmacology: multiple constituents and multiple actions. Biochem Pharmacol 58:1685–1693

14. Dou DQ, Hou WB, Chen YJ (1998) Studies on the characteristic constituents of Chinese ginseng and American ginseng. Planta Med 64:585–586

15. Qi LW, Wang CZ, Yuan CS (2011) Ginsenosides from American ginseng: chemical and pharmacological diversity. Phytochemistry 72:689–699

16. Yuan CS, Wang CZ, Wicks SM, Qi LW (2010) Chemical and pharmacological studies of saponins with a focus on American ginseng. J Ginseng Res 34:160–167

17. Corbit RM, Ferreira JFS, Ebbs SD, Murphy LL (2005) Simplified extraction of ginsenosides from American ginseng (*Panax quinquefolius* L.) for high-performance liquid chromatography-ultraviolet analysis. J Agric

Food Chem 53:9867–9873

18. Wang A, Wang CZ, Wu JA, Osinski J (2005) Determination of major ginsenosides in *Panax quinquefolius* (American ginseng) using high-performance liquid chromatography. Phytochem Anal 16:272–277

19. Qu CL, Bai YP, Jin XQ, Wang YT, Zhang K, You JY, Zhang HQ (2009) Study on ginsenosides in different parts and ages of *Panax quinquefolius* L. Food Chem 115:340–346

20. Zheng CN, Hao HP, Wang X, Wang XL, Wang GJ, Sang GW, Liang Y, Xie L, Xia CH, Yao XL (2009) Diagnostic fragment-ion-based extension strategy for rapid screening and identification of serial components of homologous families contained in traditional Chinese medicine prescription using high-resolution LC-ESI-IT-TOF/MS: Shengmai injection as an example. J Mass Spectrom 44:230–244

21. Qi LW, Wang CZ, Yuan CS (2011) Isolation and analysis of ginseng: advances and challenges. Nat Prod Rep 28:467–495

22. Li SL, Shen H, Zhu LY, Xu J, Jia XB, Zhang HM, Lin G, Cai H, Cai BC, Chen SL, Xu HX (2012) Ultra-high-performance liquid chromatography-quadrupole/time of flight mass spectrometry based chemical profiling approach to rapidly reveal chemical transformation of sulfur-fumigated medicinal herbs, a case study on white ginseng. J Chromatogr A 1231:31–45

23. Zhang HM, Li SL, Zhang H, Wang Y, Zhao ZL, Chen SL, Xu HX (2012) Holistic quality evaluation of commercial white and red ginseng using a UPLC-QTOF-MS/MS-based metabolomics approach. J Pharm Biomed Anal 62:258–273

24. Du XW, Wills RBH, Stuart DL (2004) Changes in neutral and malonyl ginsenosides in American ginseng (*Panax quinquefolium*) during drying, storage and ethanolic extraction. Food Chem 86:155–159

25. Liu WC, Zhang MP, Li CS, Sun CY, Jiang SC, Li C, Wang Y (2012) Determination of ginsenosides for different parts in *Panax quinquefolium* L. by HPLC. Chin J Ginseng Res. 22:20–23

26. Qu YX, Wang ZZ (2006) A study on saponins for different parts in *Panax quinquefolium* L. Chin J Prog in Modern Biomed 6:32–35

Chapter 8

QUANTITATIVE ANALYSIS OF THE MAJOR CONSTITUENTS IN CHINESE MEDICINAL PREPARATION SUOQUAN FORMULAE BY ULTRA FAST HIGH PERFORMANCE LIQUID CHROMATOGRAPHY/QUADRUPOLE TANDEM MASS SPECTROMETRY

Feng Chen , Hai-long Li , Yong-Hui Li, Yin-Feng Tan and Jun-Qing Zhang

School of Pharmacy, Hainan Medical University, Hainan Provincial Key Laboratory of R&D of Tropical Herbs, Haikou 571101, China

ABSTRACT

Background

The SuoQuan formulae containing Fructus *Alpiniae Oxyphyllae* has been used to combat the urinary incontinence symptoms including frequency, urgency and nocturia for hundreds of years in China. However, the chemical information was not well characterized. The quality control marker constituent only focused on one single compound in the current Chinese Pharmacopeia. Hence it is prudent to identify and quantify the main constituents in this herbal product. This study aimed to analyze the main constituents using ultra-fast performance liquid chromatography coupled to tandem mass spectrometry (UFLC-MS/MS).

Results

Fourteen phytochemicals originated from five chemical classes constituents were identified by comparing the molecular mass, fragmentation pattern and retention time with those of the reference standards. A newly developed UFLC-MS/MS was validated demonstrating that the new assay was valid, reproducible and reliable. This method was successfully applied to simultaneously quantify the fourteen phytochemicals. Notably, the content of these constituents showed significant differences in three pharmaceutical preparations. The major constituent originated from each of chemical class was isolinderalactone, norisoboldine, nootkatone, yakuchinone A and apigenin-4',7-dimethylther, respectively. The variation among these compounds was more than 1000 times. Furthermore, the significant content variation between the two different Suoquan pills was also observed.

Conclusion

The proposed method is sensitive and reliable; hence it can be used to analyze a variety of SuoQuan formulae products produced by different pharmaceutical manufacturers.

BACKGROUND

Urinary incontinence is common and costly. More than 200 million people worldwide live with this disorder, causing significant detrimental effects on their quality of life [1, 2]. Involuntary urine loss has been reported to occur in 30.9% of women and in 3%–10% of men in the mainland of China [3]. Whilst a conservative approach to the treatment of incontinence is justified in almost all cases drug therapy such as antimuscarinic agents remains integral in patients complaining of overactive bladder or bothersome stress urinary incontinence [4]. On the other hand, some traditional Chinese herbs, such as fruits of *Alpinia oxyphylla* Miq. (Zingiberaceae, known as Yizhi in Chinese), are recommended for improving frequent urination and/or enuresis symptoms in current Chinese pharmacopoeia [5]. *A. oxyphylla* Fructus is usually used in compound formulae. Among them, the most famous is SuoQuan pills, which was first described in Chinese canonical medicine about 800 years ago for the treatment of different urinary incontinence symptoms including frequency, urgency and nocturia. This formulae consists of three herbs: *A. oxyphyllae* Fructus, *Radix linderae* (Lauraceae) and *Dioscorea opposite* (Dioscoreaceae) and as one of essential herbal drugs has been documented in Chinese pharmacopeia [6]. In China, the State Food and Drug Administration (SFDA) has approved nine pharmaceutical manufacturers to produce this Chinese patent medicine in the dosage forms as pills and capsules (http://app1.sfda.gov.cn/datasearch/face3/base.jsp). However, the chemical information in these products is not well characterized. Identification of main constituents is helpful in optimizing manufacturing procedures, ensuring batch consistency and fostering understanding of the clinical effects of the herbal product.

According to the principle of *jun-chen-zuo-shi*, the most influential theory of traditional Chinese medicine, the *A. Oxyphyllae* Fructus is used as the *jun* (emperor) herb and the *R. linderae* as the *chen* (minister) herb in the SuoQuan pills formulae [7]. Chemical analysis reveal that *A. oxyphyllae* Fructus contains flavonoids, diarylheptanoids, sesquiterpenes, volatile oil, steroids and their glycosides, *etc.*[8, 9]. Isoquinoline alkaloids, sesquiterpene lactones, and flavonoids are the main bioactive components discovered in *R. linderae*[10, 11]. Modern pharmacological studies have demonstrated that the kernels of *A. oxyphylla* possess anti-inflammatory function [12, 13], and the diarylheptanoids (yakuchinone A and B) might be the active components [14–16]. Meanwhile, the alkaloids (*e.g.*, norisoboldine) [17–20] and lactones (*e.g.*, isolinderalactone) [21] from *R. linderae* can effectively alleviate inflammation. These constituents might have potential therapeutic roles for the prevention and treatment of incontinence. However, current chemistry quality control of SuoQuan pills focused only on quantitative analysis of linderane [6], and the results were inadequate for the whole quality assurance purposes. Therefore, a powerful analysis approach for identification and simultaneous determination of these multiple constituents occurred in SuoQuan formulae is urgently needed to ensure the quality control, as well as efficacy and safety, of the Chinese patent drug.

In our laboratory, we have systematically isolated the natural products of *A. oxyphylla* fruits [9] and characterized its nucleobases and nucleosides contents collected from different cultivation regions [8]. Pharmacological studies revealed that flavonol izalpinin could inhibit the muscarinic receptor-related detrusor contractile activity (published elsewhere). In the present work, a new UFLC-MS/MS method was developed and validated for simultaneous determination of 14 phytochemicals present in *A. Oxyphyllae* Fructus or *R. linderae*. Using the newly developed method, the content levels of 14 phytochemicals in SuoQuan pills were determined and compared. Notably, the content of these constituents demonstrated significant differences in three SuoQuan formulae products.

EXPERIMENTAL

Chemicals and Reagents

Reference standards of linderane (purity, 98%; similarly hereinafter) and norisoboldine (96.9%) were obtained from the National Institutes for Food and Drug Control (Beijing, China). Nootkatone (98%) and boldine (99%) were purchased from Sigma-Aldrich (St Louis, MO, USA). Atractylenoide III (98%) and isolinderalactone (98%) were obtained from Tianjin Bestbiotech Co. Ltd. (Tianjin, China). Yakuchinone A (98%), Yakuchinone B (98%) and oxyphyllacinol (98%) were purchased from Chenfun Medical Technology

(Shanghai) Co., Ltd. (Shanghai, China). Tectochrysin, izalpinin, chrysin, kaempferide and apigenin-4',7-dimethylther were separated and identified from *A. Oxyphyllae* by Prof. Zhang (Hainan Provincial Key Laboratory of R&D on Tropical Medicinal Plants, Haikou, China) and the purity of these compounds were > 98%. HPLC-grade methanol and acetonitrile were products of Sigma-Aldrich (St Louis, MO, USA). HPLC-grade formic acid was purchased from Aladdin Industrial Inc. (Shanghai, China). HPLC-grade water was prepared by double-distillation of deionized water. The other chemical reagents of analytical grade or better were obtained from Hainan YiGao Instrument Co., Ltd (Haikou, China). The utilized SuoQuan pills (SQP) and SuoQuan capsules (SQC) are commercially available *A. Oxyphyllae* Fructus products. SQP-1 were obtained from Guangdong YiHeTang Pharmaceutical Co., Ltd. (lot no.20111210, expiration: 2013/11; Chinese SFDA ratification no.Z44023146; Zhongshan, Guangdong Province, China) and SQP-2 were purchased from Guangdong HuaTianBao Pharmaceutical Group (lot no.1110103, expiration: 2014/09; Chinese SFDA ratification no. Z44023538; Foshan, Guangdong Province, China). Twenty pills weigh 1 g for both SQP-1 and SQP-2. The SQC were purchased from Hansen Pharm. (lot no.110602, expiration: 2013/05; Chinese SFDA ratification no.Z19991039; Yiyang, Hunan Province, China) and each capsule contains 0.3 g of solid.

Construction of Standard Curves

Blank solvent (acetonitrile or methanol) was spiked with the 14 test compounds to generate a nominal concentration of 1 mg/mL for each compound. Then, each stock solution was diluted step by step with methanol to prepare a sequence of standard solutions. All solutions were stored at -4°C before analysis. Standard curves were constructed using weighted ($1/X$ or $1/X^2$) linear regression of the peak areas of the phytochemical analyte (Y) against the corresponding nominal concentration of the analyte (X, ng/mL).

Sample Preparation

Samples of 0.2 g (accurately weighed) of dry and powdered SQP or SQC were macerated with 10 mL of methanol and then ultrasonicated twice or triple for 30 min every time. Between each ultrasonication, the residue was filtered and washed with 1 mL of methanol. The resulting extract solution were combined and then centrifuged. The extract solution was diluted with methanol from 10 to 100 times before analysis.

UFLC-MS/MS

An AB-SCIEX API 4000⁺ mass spectrometer (AB-Sciex, Toronto, Canada) interfaced via a Turbo V ion source with a Shimadzu Prominence UFLC chromatographic system (Shimadzu Corporation, Kyoto, Japan), which is equipped with two LC-20AD pumps, a model DGU-20A$_{3R}$ degasser unit, a SIL-20A HT autosampler and a CTO-20A column oven. The AB-SCIEX Analyst software packages were used to control the UFLC-MS/MS system, as well as for data acquisition and processing. The mass was commonly calibrated every one month using polytyrosine glycol as standard in our laboratory.

Chromatographic separations of prepared samples were achieved using a Shim-pack XR-ODS column (2.0 mm i.d × 100 mm) maintained at 40°C. The LC mobile phases included H$_2$O containing 0.1‰ formic acid for solvent A, and methanol containing 0.1‰ formic acid for solvent B. As shown in Figure 1, a specially designed LC binary gradient program was used to separate the phytochemicals and the effluent was delivered at 0.3 mL/min throughout the gradient program.

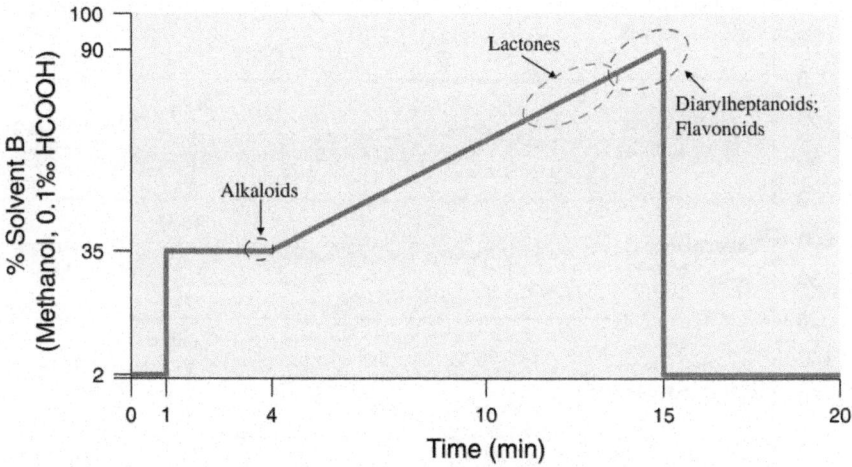

Figure 1: Solvent B (%) concentration-time profile of a 20-min LC binary gradient program. Solvent B containing 0.1‰ formic acid and the gradient program was as fallows: from 0% B to 2% B in 0.01 min, hold for 1 min; from 2% B to 35% B in 0.01 min, hold for 3min; from 35% B to 90% B in 11 min; back to 2% B in 0.01 min; maintain 4.99 min. The effluent was delivered at 0.3 mL/min throughout the gradient program and the column was maintained at 40°C. The dashed ovals showed the eluted analytes of different types of compounds (*i.e.*, alkaloids, lactones, flavonoids and diarylheptanoids) under this optimized UFLC-MS/MS conditions.

The mass spectrometer was operated in the positive ion ESI mode with multiple reaction monitoring (MRM) for all the analytes. The pneumatically nebulized ESI spraying was achieved by using inner coaxial nebulizer N_2 gas of 55 psi through a TurboIonSpray probe, a high voltage of + 5.5 kV applied to the sprayer tip, and heated dry N_2 gas of 55 psi at 550°C from two turbo heaters adjacent to the probe. To prevent solvent droplets from entering and contaminating the ion optics, a curtain N_2 gas of 25 psi was applied between the curtain plate and the orifice. The collision gas flow was set at level 4. The precursor-to-product ion pairs (Figure 2) used for MRM of nootkatone, yakuchinone A and B, oxyphyllacinol, tectochrysin, izalpinin, chrysin, kaempferide, apigenin-4',7-dimethylther, boldine, norisoboldine, linderane, isolinderalactone and atractylenoide III were m/z 219.2→163.0 (the optimal collision energy, 22 V), 313.2→136.9 (13 V), 311.2→117.0 (30 V), 315.3→137.0 (22 V), 269.1→226.0 (43.5 V), 285.0→242.0 (43 V), 255.1→152.9 (42 V), 301.1→286.0 (37 V), 299.2→256.0 (45 V), 328.2→237.2 (20 V), 314.1→265.0 (25 V), 261.1→173.0 (18 V), 245.1→156.0 (29 V) and 249.1→231.0 (28 V) respectively, with a scan time of 20 ms for each ion pair.

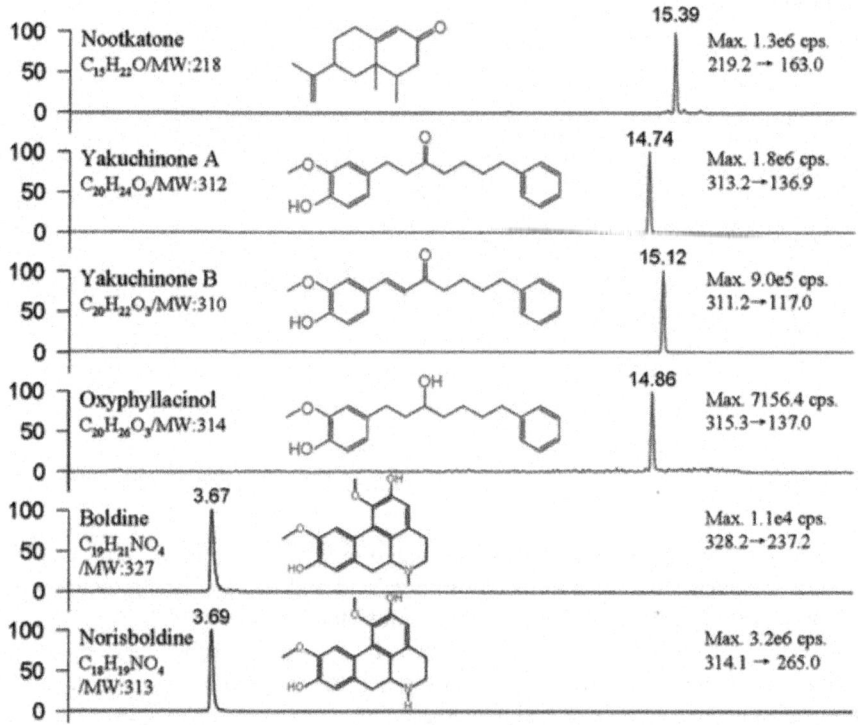

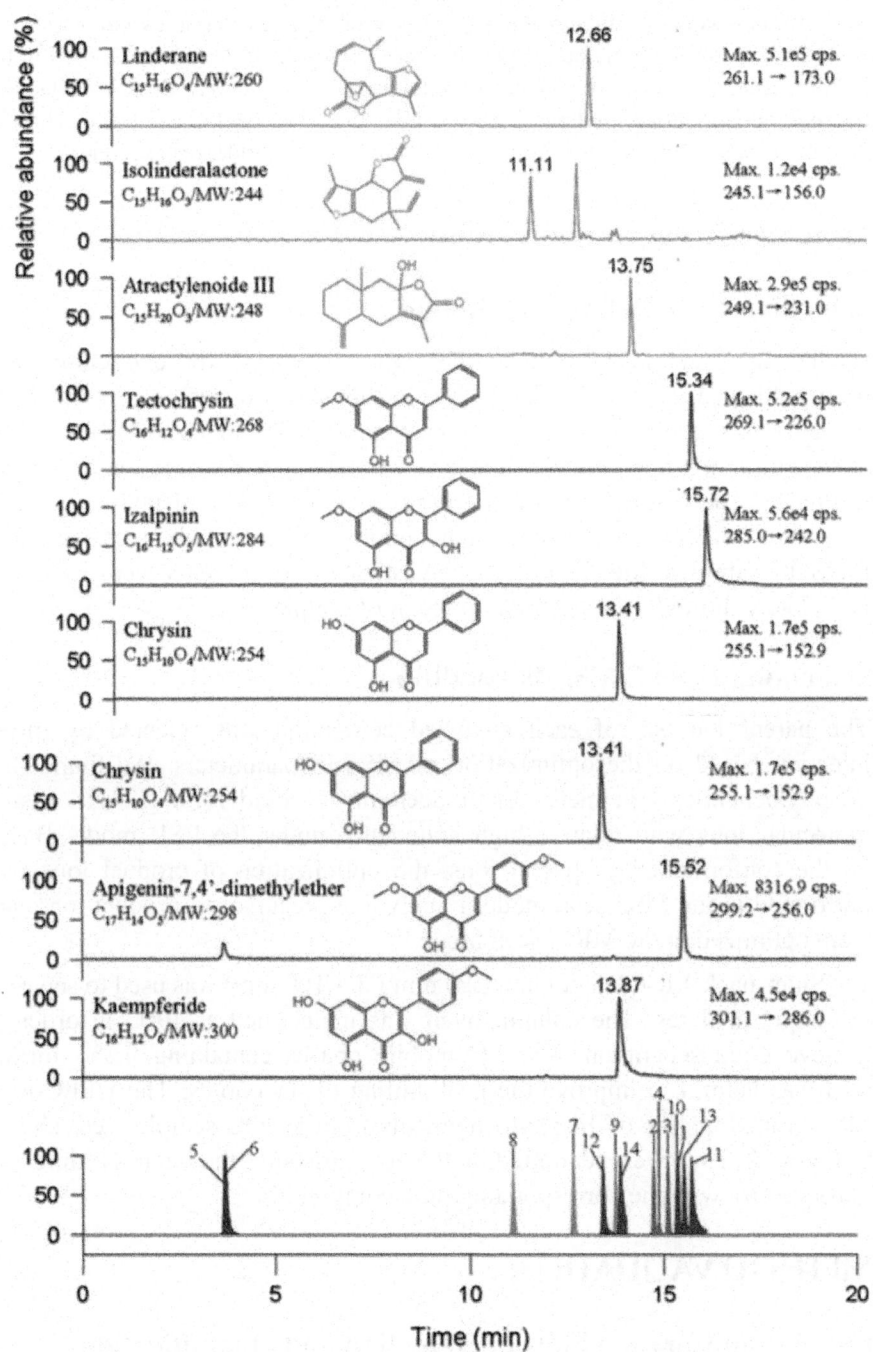

Figure 2: Typical UFLC-MS/MS chromatograms of 14 phytochemicals from a SQC

sample. They were identified by the comparison of the retention times and mass spectra (MRM mode) with the corresponding pure compounds. Peak from 1 to 14 in the bottom panel donates nootkatone, yakuchinone A, yakuchinone B, oxyphyllacinol, boldine, norisoboldine, linderane, isolinderalactone, atractylenoide III, tectochrysin, izalpinin, chrysin, apigenin-4',7-dimethylther and kaempferide, respectively.

RESULTS AND DISCUSSION

Selection of the Extraction Method

Studies designed to investigate the optimal method for extraction have been published previously [8, 11, 22]. In this study, therefore, we chose the methanol as the solvent and the ultrasonic extraction (40 KHz, 80 W) was used as the extraction method. The number of extraction and extraction time were evaluated. The results showed that the recoveries of the 14 phytochemicals were about 90% by single extraction but almost 100% through twice extraction for SQC samples. Meanwhile, the investigated constituents were extracted completely through triple extraction for SQP sample.

Selection of UFLC-MS/MS condition

The parent ion m/z of each analyzed compound was selected by direct injection based on the optimization of MS/MS parameters. We found that 14 phytochemicals in methanol or acetonitrile could form $[M+H]^+$ quasi-molecular ions with relative high abundance under the ESI^+ mode. Based on the conformation of parent ions, the optimization of product ions was performed in the MS2 scan mode. Finally, the precursor-to-product ion pairs were optimized in the MRM scan mode.

Shim-pack XR-ODS column (2.0 mm i.d × 100 mm) was used to separate the target analytes. The column oven was maintained at 40°C in order to achieve good separation. The LC mobile phases containing 0.1‰ formic acid was helpful to improve the peak tailing of flavonoids. The UFLC-MS/MS chromatograms of 14 phytochemicals from a SQC sample were shown in Figure 2. They were identified by the comparison of the retention time and mass spectra with the corresponding pure compounds.

METHOD VALIDATION

Calibration Curves, Limits of Detection and Quantification

Every calibration curve was performed with six different concentrations in quintuplicate and the calibration graphs were plotted after linear regression of the peak areas versus the corresponding concentration of each analyte. Calibration curves were linear with correlation coefficient (R^2) >0.99 for all analytes. Table 1 shows the characteristic parameters for linear dynamic ranges and coefficient. The limit of detection (LOD) and limit of quantification (LOQ) was the concentration of an analyte at which its signal-to-noise ratio were detected as 3:1 and 10:1, respectively. They were achieved by serial dilution of sample solution using the described UFLC-MS/MS conditions.

Table 1: Characteristic parameters of the investigated compounds analyzed with the UFLC-MS/MS system

Analytes	Linear ranges (ng/mL)	Linear equation (weight, 1/X or 1/X²)	R^2	LOD (ng/mL)[a]	LOQ (ng/mL)[b]
Nootkatone	2.0–1000	Y=3980X+13000	0.995	0.40	2.00
Yakuchinone A	2.0–2000	Y=7070X+10500	0.986	0.20	2.00
Yakuchinone B	2.0–1000	Y=7080X+5.450	0.993	0.27	2.00
Oxyphyllacinol	5.83–2330	Y=23.4X+800	0.998	5.83	11.7
Boldine	5.0–2000	Y=72.80X+59.40	0.996	2.00	5.00
Norisoboldine	1.94–1940	Y=6940X+19300	0.993	0.19	1.94
Linderane	2.16–2160	Y=3590X+630.0	0.995	0.43	2.16
Isolinderalactone	2.6–1040[c]	Y=0.604X+0.785	0.995	1.04[c]	2.60[c]
Atractylenoide III	2.0–2000	Y=1630X+1440	0.997	2.00	5.00
Tectochrysin	2.07–2070	Y=6470X+1190	0.998	0. 21	2.07
Izalpinin	2.0–2000	Y=1490X-2460	0.998	1.00	2.00
Chrysin	2.0–2000	Y=1560X+106.0	0.996	1.00	2.00
Apigenin-4',7-dimethylther	5.0–2000	Y=76.00X-134.0	0.994	2.00	5.00
Kaempferide	5.0–2000	Y=455.0X-710.0	0.999	2.00	5.00

Note: [a]Limits of detection; [b]Limits of quantification; [c]Concentration unit: μg/mL.

Precision, Repeatability and Stability

The precisions of the developed method were evaluated by analyzing the SQC solution in six replicates during a single day and by duplicating the experiments on three consecutive days. As shown in Table 2, the intra- and inter-day precision of the analytical method were quite good for all the investigated constituents, i.e., 0.26–5.54% and 2.06–5.32%, respectively. The repeatability was conducted by calculating the concentrations of six freshly prepared SQC samples. As a result, a good repeatability was achieved for the new method with RSD no more than 10% (Table 3). The stability of the analytes was evaluated under conditions mimicking situations likely to be encountered during sample storage and the analytical process. In this study, the freshly prepared SQC sample was stored at ambient temperature and injected into UFLC-MS/MS system at intervals 0, 6, 12, 24 and 48 h. The results (RSD <5%, Table 3) showed the sample solution was quite stable within two days.

Table 2: Intra-day and inter-day precision of the investigated compounds' in SQC sample

Analytes	Day 1 (n=6)		Day 2 (n=6)		Day 3 (n=6)		Inter-day (n=18)	
	Mean ± SD (ng/mL)	RSD (%)	Mean ± SD (ng/mL)	RSD (%)	Mean ± SD (ng/mL)	RSD (%)	Mean ± SD (ng/mL)	RSD (%)
Nootkatone	79.7 ± 2.0	2.51	75.9 ±1.0	1.32	75.0 ± 0.6	0.80	77.2±2.6	3.37
Yakuchinone A	820 ± 32	3.90	833 ± 23	2.76	786 ± 36	4.58	815 ± 34	4.17
Yakuchinone B	36.5 ± 0.9	2.47	36.8 ± 1.5	4.08	33.0 ± 1.3	3.94	35.7 ± 1.9	5.32
Oxyphyllacinol	613 ± 28	4.57	625± 12	1.92	620 ± 25	4.03	619 ± 22	3.55
Boldine	346 ± 12	3.47	338 ± 13	3.85	325 ± 18	5.54	338 ± 16	4.73
Norisoboldine	280 ± 10	3.57	279 ± 9	3.23	262 ± 8	3.05	270 ± 12	4.44
Linderane	20.8 ± 0.4	1.92	20.4 ± 0.7	3.43	19.1 ± 0.8	4.19	20.2 ± 0.9	4.46
Isolinderalactone[a]	121 ± 6	4.96	115 ± 4	3.48	114 ± 1	0.88	117 ± 5	4.27
Atractylenoide III	63.3 ± 2.2	3.48	61.2 ± 1.3	2.12	63.4 ± 2.8	4.42	62.6 ± 2.3	3.67
Tectochrysin	466 ± 17	3.65	465 ± 9	1.94	430 ± 20	4.65	456 ± 22	4.82
Izalpinin	78.3 ± 0.9	1.15	78.8 ± 1.2	1.52	75.9 ± 0.2	0.26	77.8 ± 1.6	2.06
Chrysin	29.9 ± 0.9	3.01	30.7 ± 1.5	4.89	29.3 ± 0.4	1.37	30.0 ± 1.1	3.67
Apigenin-4',7-dimethylther	672 ± 28	4.17	653 ± 17	2.60	610 ± 9	1.48	649 ± 33	5.08
Kaempferide	53.5 ± 1.8	3.36	53.5 ± 0.7	1.31	52.6 ± 1.6	3.04	53.3 ± 1.4	2.63

Note: [a]Concentration unit: μg/mL.

Table 3: Stability and repeatability of the investigated compounds in SQC sample

Analytes	Stability (n=6)		Repeatability (n=6)	
	Mean ± SD (ng/mL)	RSD (%)	Mean ± SD (ng/mL)	RSD (%)
Nootkatone	77.3 ± 2.8	3.62	71.2 ± 2.7	3.79
Yakuchinone A	818 ± 25	3.06	811 ± 42	5.18
Yakuchinone B	36.3 ± 1.7	4.68	36.4 ± 1.8	4.95
Oxyphyllacinol	1015 ± 44	4.33	637 ± 3	0.47
Boldine	342 ± 13	3.80	315 ± 10	3.17
Norisoboldine	271 ± 13	4.80	194 ± 17	8.76
Linderane	20.6 ± 0.58	2.82	190 ± 6	3.16
Isolinderalactone[a]	116 ± 4	3.45	105 ± 3	2.86
Atractylenoide III	63.7 ± 1.6	2.51	86.7 ± 3.1	3.58
Tectochrysin	457 ± 20	4.38	400 ± 15	3.75
Izalpinin	78.1 ± 1.6	2.05	85.1 ± 1.5	1.76
Chrysin	29.8 ± 1.3	4.36	256 ± 13	5.08
Apigenin-4',7-dimethylther	656 ± 32	4.88	576 ± 27	4.69
Kaempferide	53.3 ± 1.6	3.00	62.4 ± 1.1	1.76

Note: [a]Concentration unit: μg/mL.

Recovery

The recovery of known amounts of standards added to samples was used to assess the new method accuracy. Six portions of SQC or SQP sample were spiked with the mixed standards of 14 phytochemicals. Then the samples were pretreated as described in "Sample preparation"section, and the results are summarized in Table 4. All the recoveries were in the range of 98.2–105% for SQC sample, 95.6–102% for SQP sample, with the RSDs 0.80–4.93% and 1.00–4.76%, respectively.

Table 4: Recoveries of the investigated compounds in SQC or SQP samples

Analytes	SQC sample (n=6)				SQP sample (n=6)			
	Measured (µg) (Mean ± SD)	Added (µg)	Recovery (%, mean)	RSD (%)	Measured (µg) (Mean ± SD)	Added (µg) (Mean ± SD)	Recovery (%, mean)	RSD (%)
Nootkatone	5.00 ± 0.13	5.00	100	2.67	5.06 ± 0.19	5.00	101	3.84
Yakuchinone A	4.91 ± 0.14	5.00	98.2	2.80	4.77 ± 0.15	5.00	95.3	3.23
Yakuchinone B	4.97 ± 0.13	5.00	99.4	2.52	5.01 ± 0.24	5.00	100	4.76
Oxyphyllacinol	5.93 ± 0.22	5.83	102	3.64	5.71 ± 0.24	5.83	98.0	4.21
Boldine	5.00 ± 0.11	5.00	100.	2.13	5.04 ± 0.09	5.00	101	1.85
Norisoboldine	4.91 ± 0.24	4.85	101	4.93	4.81 ± 0.18	4.85	99.2	3.70
Linderane	5.37 ± 0.19	5.40	99.4	3.60	5.41 ± 0.21	5.40	100	3.84
Isolinderalactone	73.9 ± 2.2	75.0	98.5	3.02	74.8 ± 2.4	75.0	99.7	3.23
Atractylenoide III	5.23 ± 0.20	5.00	105	3.81	5.05 ± 0.15	5.00	101	3.04
Tectochrysin	3.96 ± 0.17	4.00	98.9	4.38	3.82 ± 0.08	4.00	95.6	2.18
Izalpinin	2.53 ± 0.07	2.50	101	2.77	2.51 ± 0.06	2.50	100	2.23
Chrysin	5.11 ± 0.13	5.00	102	2.45	5.13 ± 0.05	5.00	102	1.00
Apigenin-4',7-dimethylther	5.11 ± 0.15	5.00	102	2.89	5.04 ± 0.17	5.00	101	3.40
Kaempferide	7.37 ± 0.06	7.05	104	0.80	7.06 ± 0.21	7.05	100	2.92

Quantification of 14 Phytochemicals from three SuoQuan Formulae Preparations

The newly validated method was applied to simultaneously determine the 14 compounds in three commercial SuoQuan formulae products from different company. The results are summarized in Table 5 and Figure 3. Our results revealed that the contents of the 14 compounds are significantly different. The overall content level was as follows: lactones > alkaloids > naphthalenone (nootkatone) > diarylheptanoids > flavonoids, albeit the dosage forms (*i.e.*, capsules or pills). Accordingly, the highest content constituent of each type of compound was isolinderalactone, norisoboldine, nootkatone, yakuchinone A and apigenin-4',7-dimethylther, respectively. Moreover, the content variation was more than 1500 times among these phytochemicals .

Table 5: Contents of 14 phytochemicals from three SuoQuan formulae products

Analytes	Content (µg/g)		
	SQC	SQP-1	SQP-2
Nootkatone	712 ± 27	535 ± 12	461 ± 18
Yakuchinone A	811 ± 42	40.7 ± 2.0	181 ± 8.1
Yakuchinone B	36.4 ± 1.8	3.42 ± 0.09	16.5 ± 0.3
Oxyphyllacinol	63.7 ± 3.0	8.19 ± 0.38	23.8 ± 1.0
Boldine	315 ± 10	102 ± 5	18.4 ± 0.5
Norisoboldine	1940 ± 116	2360 ± 98	1036 ± 40
Linderane	19.0 ± 0.6	132 ± 1	266 ± 10
Isolinderalactone	10517 ± 264	34225 ± 1096	24875 ± 1228
Atractylenoide III	8.67 ± 0.31	12.2 ± 0.6	13.8 ± 0.6
Tectochrysin	40.0 ± 1.5	6.93 ± 0.29	21.9 ± 0.4
Izalpinin	8.51 ± 0.15	1.74 ± 0.07	5.13 ± 0.10
Chrysin	25.6 ± 1.3	3.15 ± 0.13	7.79 ± 0.40
Apigenin-4',7-dimethylther	57.6 ± 2.73	22.4 ± 1.0	56.2 ± 2.8
Kaempferide	6.24 ± 0.11	1.67 ± 0.09	4.97 ± 0.21

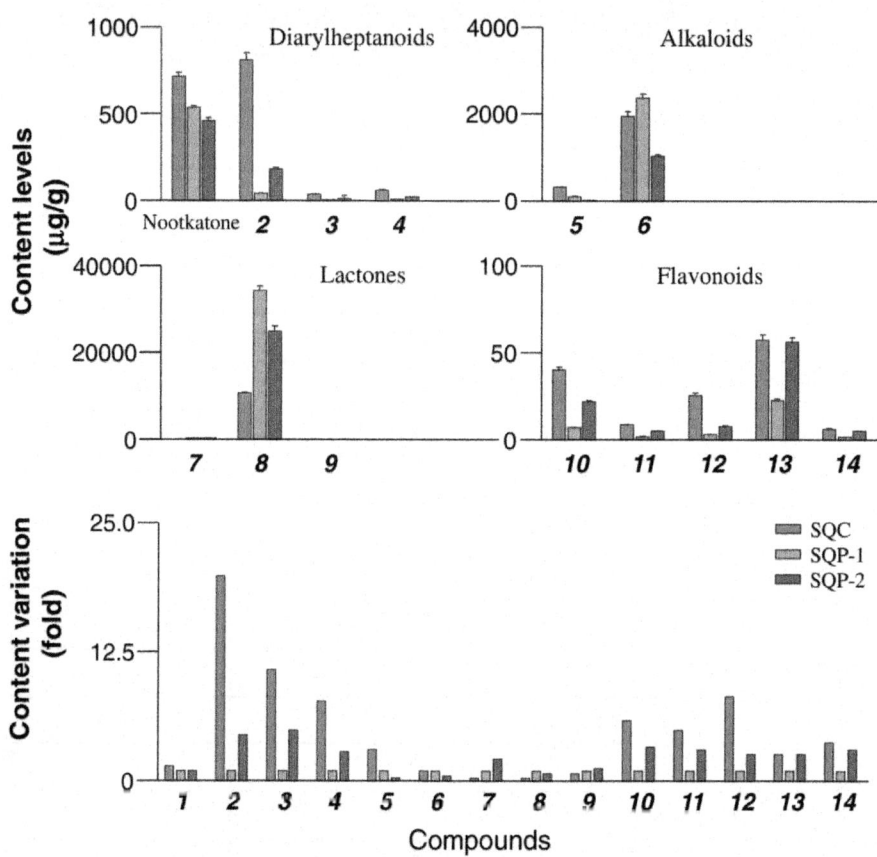

Figure 3: The content variation of 14 phytochemicals originated from SQC, SQP-1 and SQP-2 samples. In the bottom panel, the content level variation was normalized based on the contents in SQP-1 sample.

On the other hand, the manufacturing process influenced the content levels. Besides for lactones, the other types of compound had higher concentration in SQC samples than SQP samples. SQP is a Chinese traditional medical mixture containing the powder of *A. oxyphyllae* Fructus, *R. linderae* and *D. opposite* and the manufacturing process dose not involve the extraction technology. Therefore, all the constituents derived from three herbal powders are transformed into the final pills. On the contrary, the SQC is manufactured through refluxing extraction method with ethanol and the constituents that can dissolved in ethanol are differentially transformed into the final capsules according to the theory of "like dissolves like". As a result, most of the investigated constituents were concentrated and the content variation in SQC sample might smaller than the SQP sample (Figure 3).

As shown in Figure 3, it was worth noting that the significant content variation between the SQP-1 sample and SQP-2 sample was observed. As for linderane, a marker compound for quality control in current Chinese Pharmacopeia [6], the mean content levels (132 μg/g for SQP-1 and 266 μg/g for SQP-2) were superior to the mandatory criteria (*i.e.*, no less than 90 μg/g) albeit the 2-fold difference between the two products. However, for the most of the other constituents, the content levels in SQP-2 sample were larger than SQP-1 sample. The diarylheptanoids' content levels in SQP-2 were 3–5 times higher than SQP-1. Similarly, this data for flavonoids was 2–3 times. The possible explanation for these discrepancies could be that the herbal raw materials were different between the two companies. Thus, strict quality control for raw material and adequate in-process controls should be properly carried out. In the cases where the active principles are unknown, marker substance(s) should be established for analytical purposes.

In the present work, we also found other compounds occurred in the SQC and SQP sample including linderagalactone D, linderagalactone C, hydroxylindestenolide, neolinderalactone, reticuline and norboldine. These phytochemicals were well separated and their precursor-to-product ion pairs were set at $263.5 \rightarrow 227.2$, $263.5 \rightarrow 217.1$, $247.7 \rightarrow 229.2$, $245.5 \rightarrow 199.4$, $330.2 \rightarrow 192.2$ and $314.0 \rightarrow 265.3$ according to the data reported in literature [11, 22]. However, these compounds were not quantified because the corresponding reference standards were not available.

CONCLUSION

In summary, a new UFLC-MS/MS method was developed and validated. The proposed method was sensitive and reliable and then successfully applied to simultaneously quantify the 14 phytochemicals occurred in SuoQuan formulae. Significant content variations were observed. Manufacturing processes influenced the content levels between the different dosage forms. The quality of raw materials play a pivotal role in guaranteeing the consistently good quality of herbal preparations. Sufficient chemical information and assay for particular constituents or markers are helpful in understanding of the clinical effects of the herbal product.

ABBREVIATIONS

UFLC-MS/MS: Ultra fast liquid chromatography-tandem mass spectrometry

RSD: Relative standard variation

SQC: SuoQuan capsules

SQP: SuoQuan pills.

ACKNOWLEDGEMENTS

This work was supported by Grant 812189 from the Natural Science Fund of Hainan Province, Grant 2010ZY012,2011ZY004 and 2012ZY010 from the Hainan provincial special fund for the Modernization of Traditional Chinese Medicine and Grant 2011BA101B07 from the National Science & Technology Pillar Program during the 12th Five-Year Plan Period of China. The work was also financially supported by Grant HY2012-013 from the Hainan medical university for Young scholars.

AUTHORS' CONTRIBUTIONS

CF and LHL were the primary contributors to this manuscript. CF and LHL were responsible for preparing the first draft of the manuscript and performed most of the experimentation and analysis while also being involved heavily in data acquisition and interpretation. ZJQ was involved in design of the experiments and provided critical advice on operation of the analytical equipment due to previous expertise. LYH and TYF had a significant role in development of the experiments and interpretation of results. All authors read and approved the final manuscript.

REFERENCES

1. Rogers RG: Clinical practice. Urinary stress incontinence in women. *N Engl J Med* 2008,358(10):1029–1036.

2. Norton P, Brubaker L: Urinary incontinence in women. *Lancet* 2006,367(9504):57–67.

3. Wang YS, Chen YW: Prevalence of urinary incontinence among Chinese women. *Health News* 2012, 18:B002.

4. Thüroff JW, Abrams P, Andersson KE, Artibani W, Chapple CR, Drake MJ, Hampel C, Neisius A, Schröder A, Tubaro A: EAU guidelines on urinary incontinence. *Eur Urol* 2011,59(3):387–400.

5. Chinese Pharmacopoeia Commission: *Pharmacopoeia of the People's Republic of China-Part I*. Beijing: Chemical Industry Press; 2010:204–205.

6. Chinese Pharmacopoeia Commission: *Pharmacopoeia of the People's Republic of China-Part I*. Beijing: Chemical Industry Press; 2010:1218.

7. Qiu J: Traditional medicine: a culture in the balance. *Nature* 2007,448(7150):126–128.

8. Song WJ, Li YH, Wang J, Li ZY, Zhang JQ: Characterization of nucleobases and nucleosides in the fruit of *Alpinia oxyphylla*collected from different cultivation regions. *Drug Test Anal* 2013.

9. Qing ZJ, Yong W, Hui LY, Yong LW, Long LH, Ao DJ, Xia PL: Two new natural products from the fruits of *Alpinia oxyphylla* with inhibitory effects on nitric oxide production in lipopolysaccharide-activated RAW264.7 macrophage cells. *Arch Pharm Res*2012,35(12):2143–2146.

10. Wu YJ, Zheng YL, Luan LJ, Liu XS, Han Z, Ren YP, Gan LS, Zhou CX: Development of the fingerprint for the quality of *Radix Linderae* through ultra-pressure liquid chromatography-photodiode array detection/ electrospray ionization mass spectrometry.*J Sep Sci* 2010,33(17–18):2734–2742.

11. Wu YJ, Zheng YL, Liu XS, Han Z, Ren YP, Gan LS, Zhou CX, Luan LJ: Separation and quantitative determination of sesquiterpene lactones in *Lindera aggregata* (wu-yao) by ultra-performance LC-MS/MS. *J Sep Sci* 2010,33(8):1072–1078.

12. He ZH, Ge W, Yue GG, Lau CB, He MF, But PP: Anti-angiogenic effects of the fruit of*Alpinia oxyphylla*.*J Ethnopharmacol*2010,132(2):443–449.

13. Zhang ZJ, Cheang LC, Wang MW, Li GH, Chu IK, Lin ZX, Lee SM: Ethanolic extract of fructus Alpinia oxyphylla protects against 6-hydroxydopamine-induced damage of PC12 cells in vitro and dopaminergic neurons in zebrafish. *Cell Mol Neurobiol*2012,32(1):27–40.

14. Chun KS, Kang JY, Kim OH, Kang H, Surh YJ: Effects of yakuchinone A and yakuchinone B on the phorbol ester-induced expression of COX-2 and iNOS and activation of NF-kappaB in mouse skin. *J Environ Pathol Toxicol Oncol* 2002,21(2):131–139.

15. Chun KS, Park KK, Lee J, Kang M, Surh YJ: Inhibition of mouse skin tumor promotion by anti-inflammatory diarylheptanoids derived from Alpinia oxyphylla Miquel (Zingiberaceae). *Oncol Res* 2002,13(1):37–45.

16. Bayati S, Yazdanparast R: Antioxidant and free radical scavenging potential of yakuchinone B derivatives in reduction of lipofuscin formation using H_2O_2-treated neuroblastoma cells. *Iran Biomed J* 2011,15(4):134–142.

17. Wang C, Dai Y, Yang J, Chou G, Wang C, Wang Z: Treatment with total alkaloids from Radix Linderae reduces inflammation and joint destruction in type II collagen-induced model for rheumatoid arthritis. *J Ethnopharmacol* 2007,111(2):322–328.

18. Wei Z, Wang F, Song J, Lu Q, Zhao P, Xia Y, Chou G, Wang Z, Dai Y: Norisoboldine inhibits the production of interleukin-6 in fibroblast-like synoviocytes from adjuvant arthritis rats through PKC/MAPK/NF-κB-p65/CREB pathways. *J Cell Biochem*2012,113(8):2785–2795.

19. Luo Y, Liu M, Dai Y, Yao X, Xia Y, Chou G, Wang Z: Norisoboldine inhibits the production of pro-inflammatory cytokines in lipopolysaccharide-stimulated RAW 264.7 cells by down-regulating the activation of MAPKs but not NF-**κB**. *Inflammation*2010,33(6):389–397.

20. Wei ZF, Jiao XL, Wang T, Lu Q, Xia YF, Wang ZT, Guo QL, Chou GX, Dai Y: Norisoboldine alleviates joint destruction in rats with adjuvant-induced arthritis by reducing RANKL, IL-6, PGE2, and MMP-13 expression. *Acta Pharmacol Sin* 2013,34(3):403–413.

21. Wong SL, Chang HS, Wang GJ, Chiang MY, Huang HY, Chen CH, Tsai SC, Lin CH, Chen IS: Secondary metabolites from the roots of*Neolitsea daibuensis* and their anti-inflammatory activity. *J Nat Prod* 2011,74(12):2489–2496.

22. Han Z, Zheng Y, Chen N, Luan L, Zhou C, Gan L, Wu Y: Simultaneous determination of four alkaloids in *Lindera aggregata* by ultra-high-pressure liquid chromatography-tandem mass spectrometry. *J Chromatogr A* 2008,1212(1–2):76–81.

Chapter 9

QUANTITATIVE ANALYSIS OF TOTAL PETROLEUM HYDROCARBONS IN SOILS: COMPARISON BETWEEN REFLECTANCE SPECTROSCOPY AND SOLVENT EXTRACTION BY 3 CERTIFIED LABORATORIES

Guy Schwartz[1,2,3], Eyal Ben-Dor[2] and Gil Eshel[4]

[1]Porter School of Environmental Studies, Tel-Aviv University, Tel-Aviv 69978, Israel

[2]Remote Sensing Laboratory, Tel-Aviv University, Tel-Aviv 69978, Israel

[3]Geography and Human Environment Department, Tel-Aviv University, P.O. Box 39040, Tel-Aviv 69978, Israel

[4]The Soil Erosion Research Station, Ruppin Institute, Emeck Hefer 40250, Israel

ABSTRACT

The commonly used analytic method for assessing total petroleum hydrocarbons (TPH) in soil, EPA method 418.1, is usually based on extraction with 1,1,2-trichlorotrifluoroethane (Freon 113) and FTIR spectroscopy of the extracted solvent. This method is widely used for initial site investigation, due to the relative low price per sample. It is known that the extraction efficiency varies depending on the extracting solvent and other sample properties. This study's main goal was to evaluate reflectance spectroscopy as a tool for TPH assessment, as compared with three commercial certified laboratories using traditional methods. Large variations were found between the results of the three commercial laboratories, both internally (average deviation up to 20%), and between laboratories (average deviation up to 103%). Reflectance spectroscopy method was found be as good as the commercial laboratories in terms of accuracy and could be a viable field-screening tool that is rapid, environmental friendly, and cost effective.

INTRODUCTION

Among the chemicals that are relevant as environmental contaminants, petroleum hydrocarbons (PHC) are of particular significance. The widespread use of PHC for transportation, heating and industry has led to the release of these petroleum products into the environment through accidental spills, long-term leakage, or operational failures. Consequently, many soil and water areas are contaminated with PHC. PHC are well known to be neurotoxic to humans and animals. Several studies have been conducted in order to verify the effects of PHC on humans and animals [1–3]. For both the diagnosis of suspected areas and the possibility of controlling the rehabilitation process, there is a great need to measure correctly the amounts of PHC in soils.

Total petroleum hydrocarbons (TPH) is a commonly used gross parameter for quantifying environmental contamination originated by various PHC products such as fuels, oils, lubricants, waxes, and others [4]. Traditional wet chemistry methods for determining TPH level in soil samples is based on extracting the contaminant from the soil sample. The TPH level in the extracted solution is then determined by a gravimetric, FTIR, or GC measurement calibrated by an EPA calibration standard.

The TPH gross parameter is in use worldwide and facilitates an important stage of contaminated sites investigation; therefore, it is important to examine the effects of hydrocarbon type and soil properties on the extraction efficiency, as well as cross-lab repeatability.

The common method for assessing TPH in soil samples is based on a modified version of EPA method 418.1. This method is based on extraction with 1,1,2-Trichlorotrifluoroethane (Freon 113, GC 99.9%), although other extracting solvents are available (i.e., Carbon tetrachloride, N-Hexane, etc.). This method was originally introduced in 1978 [5] by the USEPA in order to assess TPH in waste water but was later adjusted in 1983 [6] for the assessment of TPH in soil samples. Newer methods are available for determining TPH in soil samples; these methods are based on extraction with other solvents and are usually followed by gas chromatograph analysis for THP determination. As these methods are more expensive, the EPA method 418.1 is in vast use as a screening tool [4, 7].

There are number of possible interactions between inorganic and organic soil components and organic pollutants, soil organic matter, and clays, having significant impact on solid-liquid extraction. Furthermore, the solvent extraction of compounds from soil or sludge samples is dependent on the moisture content in the soil [8]. There are some inherent problems with IR readings of the extracted solvent; all petroleum hydrocarbons do not respond equally to infrared analysis, and comparison of the unknown to a standard mixture may give results with high systematic errors [9]. The major problem with the adjusted EPA 418.1 method is that the extraction yields can be strongly matrix dependent, and the extraction method development and optimization may be quite complicated. These extraction-related problems mainly originate from the diversity of chemical and physical properties of petroleum hydrocarbons, which affect not only the solubility of hydrocarbons to the solvents, but also on the strength of analyte-soil matrix interactions, and therefore render the control of the extraction process of petroleum hydrocarbons from soil problematic.

In conclusion, it is clear that the adjusted EPA method 418.1 may overestimate TPH as a result of the following:

- differences in infrared molar absorptivity for calibration standards and petroleum products;
- detection of naturally occurring hydrocarbons;
- infrared dispersion by mineral particles.

Negative bias may also be introduced via

- poor extraction efficiency of Freon-113 for high-molecular-weight hydrocarbons;
- differences in molar absorptivity;
- removal of five to six-ring alkylated aromatics during the silica gel cleanup procedure [10].

Quality assurance in the area of TPH determination is under developed and actually, except in few cases [11–13], there have not been any attempts to estimate the uncertainty related to the analytical procedure of TPH determination.

Taking in consideration all the possible biases that can occur during the adjusted EPA 418.1 method, as well as the fact that each laboratory uses somewhat different protocols and equipment for the extraction process and TPH determination; a methodic cross-laboratory evaluation is needed.

In addition to the traditional analytical chemistry methods used for measuring TPH in the soil samples, a new novel method based on reflectance spectroscopy was applied. Reflectance spectroscopy is commonly applied for quantitative analysis in many disciplines. This method consists of measuring the reflected electromagnetic energy from the soil samples in the VIS-NIR-SWIR region (350–2500 nm), and modeling this spectral data against samples with known concentration levels. Extracting the information about the soil attributes that is hidden within the spectral information, is done by using multivariate statistical techniques, also called chemometrics. Essentially, this involves regression techniques coupled with spectral preprocessing. A more detailed description of the spectral preprocessing and the chemometrics process as well as an overview of reflectance spectroscopy as a tool for monitoring contaminated soils can be found in a recent publication by the authors [7].

The spectral properties of hydrocarbons were identified at the late 1980s, although it was argued that these properties are visible at concentrations of 4% wt and above [14]. Several studies were conducted during the past 20 years in the field of PHC and reflectance spectroscopy (ie., [15–24]) that showed the potential of reflectance spectroscopy as being used as a tool for predicting TPH content. For taking a step forward in acceptance of this tool by the environmental protection authorities, a validation study that includes a comparison of the results of commercial laboratories analysis and reflectance spectroscopy performance is needed. Therefore, The goals of this study are

- a comparison of the inner and interlaboratory TPH measuring capabilities,
- general accuracy of the measured TPH levels as compared to the known TPH levels of the contaminated soil samples, and
- Testing reflectance spectroscopy as a viable replacement for the traditional methods based on solvent extraction.

MATERIALS AND METHODS

Three certified laboratories in Israel were selected for this study. Analogue soils typical to Israel were artificially contaminated with PHC and sent at the same time and in the same conditions to all laboratories. In addition, the samples underwent a new NIRS procedure that we developed in TAU in which reflectance spectroscopy is used to determine TPH level [7].

Soils and Hydrocarbons

Three soils were selected for this study (defined according to Israeli naming system [25] as well as the USDA key to soil taxonomy [26]): Loess (Typic Xerofluvent), Hamra (Typic Xerocherept), and Gromosol (Typic Chromoxerert). These soils represent a wide range of soil properties as described in Table 1 and are significantly differ from each other. The soils were collected from areas that were assumed to have no PHC contamination and were air-dried and sieved through a 2 mm sieve twice. The soils properties were determined by the traditional methods in soil science as follows: hydroscopic moisture content was determined by weight loss after 24h at 105°C. pH level and electrical conductivity were determined with a laboratory bench top 86505 pH/Conductivity meter by M.R.C Ltd. in a 1:2 soil and DI water suspension (resp.) after reaching equilibrium (30 minutes). Specific surface area (SSA) was determined by the absorption of mono layer of ethylene glycol monoethyl ether (EGME) [27]. Particle size distributions were determined by Marvin Mastersizer 2000 following Eshel et al. methodology [28]. SOC, SIC, and Total N were determined by a flash CHN elemental analyzer (Thermo Scientific Flash 2000). The soils analogue contaminated samples were prepared by mixing a known weight of several PHC types including: octane fuel, diesel and kerosene with known quantities of soil. For making well-mixed low concentration samples, we initially mixed a batch of 98.5 gr of soil with 1.5 gr of the selected PHC; after mixing the initial batch, the batch was then mixed again with clean soil at three concentration levels. In order to minimize the loss of PHC components, we minimized exposure to open air as much as possible. Each sample was divided equally into 4 amber glass vials, capped with a PTFE lined cap, and kept at 4°C. Three of the vials were sent to the analytical laboratories for analysis, 1 vial was kept for reflectance spectroscopy analysis. Table 2 describes the samples contamination properties and presents the calculated concentration info.

Table 1: Major soil properties

Israeli local name	USDA classification	HM	Sand	Silt	Clay	SOC	SIC	Total N	pH[1]	EC[1]	SSA
		%	volume %			g kg^{-1}				mS m^{-1}	m^{2}g^{-1}
Loess	Typic xerofluvent	4.14	38.6	49.4	12	5.4	22.5	0.9	8.22	5.44	167
Hamra	Typic xerocherept	1.44	97.37	1.73	0.9	1.5	2.1	0.5	8.57	0.08	83
Gromosol	Typic chromoxerert	5.23	46.46	38.98	14.56	7.6	12.5	1.3	8.68	0.55	238

[1] 1 to 2 ratio.

Table 2: Soil samples calculated concentration, projected TPH, and laboratory TPH results

Sample	Soil name	Contaminant	Calculated concentration (ppm)	Projected TPH	Spectroscopy (TPH)	Lab A (TPH)			Lab B (TPH)			Lab C (TPH)		
						Min	Max	Avg	Min	Max	Avg	Min	Max	Avg
1	Hamra	None	0	0	411	6	8	7	10	10	10	10	10	10
2		Diesel	450	567	908	354	434	394	599	610	605	458	506	483
3			4500	5674	4617	4575	5288	4932	6179	6292	6236	3730	4480	4111
4			10500	13239	8693	8122	8175	8149	14534	15369	14952	9897	10217	10021
5		Kerosene	550	670	953	277	320	299	405	415	410	305	383	350
6			6000	7304	4871	5455	6039	5747	7441	7528	7485	3420	3814	3628
7			12000	14609	8567	8740	9608	9174	14078	14125	14102	9410	9880	9704
8		95% octane	600	365	511	39	43	41	52	66	59	47	56	51
9			5500	3348	1274	519	586	553	793	838	816	244	333	300
10			9500	5783	1800	1227	1816	1522	1142	2003	1573	260	312	279
11	Loess	None	0	0	10	9	9	9	10	10	10	10	15	12
12		Diesel	500	630	1378	252	275	264	615	625	620	483	510	498
13			2500	3152	2545	1139	2308	1724	3593	3601	3597	2816	3055	2936
14			9000	11348	6662	5984	7903	6644	12447	12958	12703	7970	8560	8313
15		Kerosene	400	487	909	128	145	137	345	354	350	210	255	236
16			4000	4670	3182	2606	3250	2928	4667	4698	4693	2145	2312	2219
17			11000	13391	6495	9435	9628	9532	13184	13411	13298	7264	7859	7533
18		95% octane	700	426	937	34	47	41	70	70	70	46	69	54
19			4500	2739	784	210	228	219	629	635	632	62	88	73
20			10000	6087	1100	1188	1193	1191	2674	3107	2891	578	629	601
21	Groomsool	None	0	0	635	6	6	6	10	10	10	78	110	91
22		Diesel	600	757	737	356	381	369	640	677	659	463	512	490
23			3500	4413	1419	2613	2917	2765	4441	4624	4533	2493	2706	2621
24			11000	13670	2376	11753	14593	13173	14705	14800	14753	10513	11219	10811
25		Kerosene	600	730	714	223	237	230	439	470	455	190	254	222
26			5000	6087	1728	3494	4169	3832	5588	5613	5601	1231	1376	1306
27			10000	12174	3320	5839	6209	6024	11245	11436	11341	7922	8510	8261
28		95% octane	500	304	739	20	20	20	51	62	57	10	10	10
29			5200	3165	1916	410	491	451	680	691	686	228	265	249
30			9000	5478	1883	938	1127	1043	1800	1852	1826	685	824	743

Extraction and TPH Measurement Method

The general methodology for the adjusted EPA 418.1 method is based on taking a representative soil sample (3–10 gr.), adding sodium sulfate (1–5 gr.) to absorb any water and adding an extracting solvent (usually Freon 113, 20–30 mL) to the mixture. This mixture is then kept in a sealed glass vial capped with a PTFE cap and placed in a sonic bath for assisting and hasting the extraction process (about 10–45 minutes). Silica gel is then added to the mixture to absorb any polar hydrocarbons (nonfuel-related soil organic matter and fatty acids), and the mixture is mixed well. The filtered extract is then measured in an FTIR spectrometer at 3.42 μm (some laboratories use other absorption peaks in the close region). A calibration curve is created by using the 418.1 EPA standard (consists of 31.5% isooctane, 35% hexadecane and 33.5% chlorobenzene) diluted in the same extracting solution at at least 3 concentrations. The absorption depth of the measured sample is then converted to TPH values by the calibration curve. As this method is an adjusted EPA method, it can vary slightly between analytical laboratories, depending on internal laboratory standards, procedures, and equipment. The three laboratories used for analyzing the samples prepared for this study are commercial laboratories, certified by the national laboratories certification authority, thus the exact procedure is confidential and not known to the authors, although the principal remains the same. All 30 contaminated samples prepared for this study as described above were sent to the three certified laboratories for chemical analysis determination of TPH levels, the results are summarized in Table 2.

IR Absorbance of Diesel, Kerosene, Octane 95, and 418.1 EPA Reference

PHC efficiency to absorb IR radiation depends on the PHC molecules structure. It was important to map these absorptions differences for the contaminants used in this study, relative to the 418.1 EPA reference that is usually used for TPH determination. Diesel, kerosene, octane 95, and the 418.1 EPA reference were mixed with Freon 113 at four different concentration levels each: ~50, ~100, ~150, and ~200 ppm. Each sample was then measured for its absorbance by a buck scientific 404 analyzer; the results are shown in Figure 1. Since the relation between the absorption and the concentration for each PHC is perfectly linear, (see Figure 1), the absorption was calculated for each PHC for the following concentrations: 50, 100, 150, 200, 250, 300, 350, 400, 450, 500 ppm. Each PHC was then plotted versus the 418.1 EPA reference as shown in Figure 2.

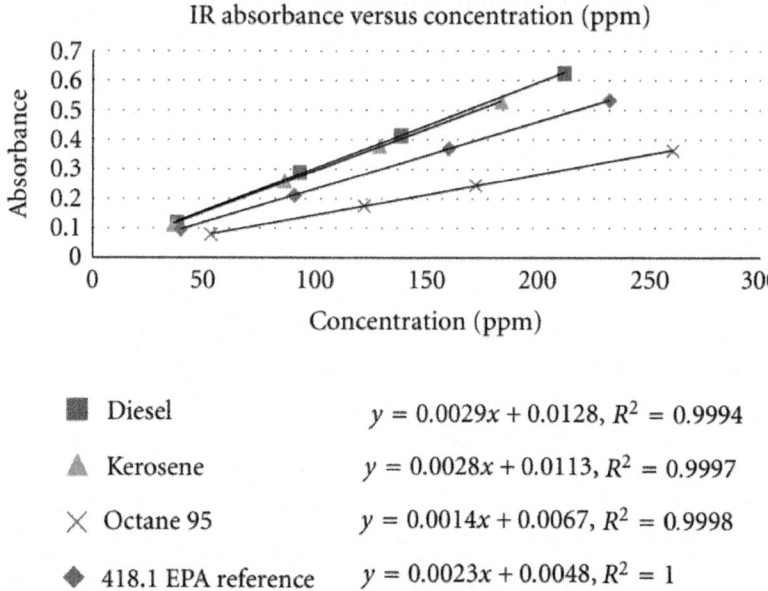

Figure 1: IR absorbance versus concentration (ppm).

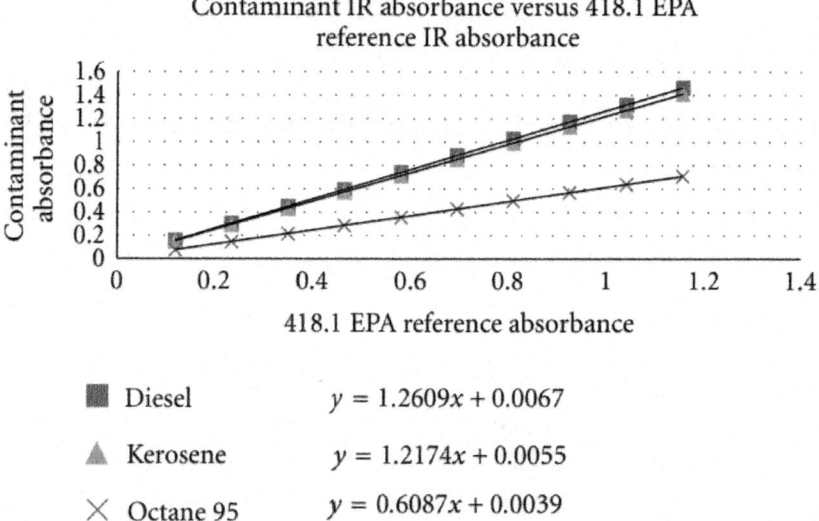

Figure 2: Contaminant IR absorbance versus 418.1 EPA reference IR absorbance.

Conversion of Specific PHC to TPH

Due to the fact that laboratories give results in TPH which is a gross parameter based on the EPA standard that represents a mixture of several PHC, and our soil samples were contaminated by a specific PHC, we need to apply a conversion factor from the specific PHC to the relative gross parameter TPH as seen is Figure 2. This resulted "Projected TPH" value should represent the contamination level of the contaminated samples if the laboratory process was flawless, thus eliminating one major bias factor, which is the difference between IR absorbance efficiency of the 418.1 EPA standard, relative to the specific PHC we used to contaminate the soil as described in the previous section. The conversion equations to project the specific PHC to TPH values in this study (Figure 2) are:

(1)TPH (ppm) = Diesel (ppm) * 1.2609 + 0.0067,(2)TPH (ppm) = Kerosene (ppm) * 1.2174 + 0.0055,(3)TPH (ppm) = Octane 95% (ppm) * 0.6087 + 0.0039,

The calculated projected TPH values are shown in Table 2, and are used for the rest of this study instead of the original specific PHC levels.

Intralaboratory Consistency Factors

The contaminated soil samples from each laboratory separately were divided into three groups: low, medium, and high, by the known concentration level, regardless of soil type or contaminant. The intralaboratory consistency was evaluated by four factors.

- Average delta: the difference between maximum TPH value and minimum TPH value of each sample in that group, followed by averaging the results of all the samples in that group.

- Average deviation: the difference between maximum TPH value and minimum TPH value of each sample in that group, then divided by the average TPH value for that sample, thus normalizing the results. Finally the normalized results of all samples were averaged for all samples in each group.

- Maximum delta: same as average delta, but instead of averaging the results for each group, only the maximum value was selected, portraying the "worst case scenario."

- Maximum deviation: same as average deviation, but instead of averaging the results for each group, only the maximum value was selected, portraying the "worst case scenario."

Results are shown in Table 3.

Table 3: Intralab repetition statistics for low, medium, and high TPH levels

Lab	A				B				C			
TPH level (calcu-lated)	AVG delta	AVG devia-tion	Max delta	Max devia-tion	AVG delta	AVG devia-tion	Max delta	Max devia-tion	AVG delta	AVG devia-tion	Max delta	Max devia-tion
Low (400–600)	24	12%	80	32%	15	7%	37	24%	38	17%	78	40%
Medium (2500–6000)	473	20%	1169	68%	54	2%	183	6%	229	16%	750	35%
High (9000–12000)	712	13%	2840	39%	361	10%	861	55%	390	9%	706	18%

Interlaboratory Consistency Factors

The interlaboratory consistency factors were calculated in the same way the intrafactors were calculated, but instead of taking the samples from each laboratory separately, all samples from all laboratories were joined together, as if they came from the same laboratory. The same four factors: average delta, average deviation, maximum delta, and maximum deviation were calculated as described in the intralaboratory consistency factors section. Results are summarized in Table 4.

Table 4: Interlab repetition statistics for low, medium, and high TPH levels

TPH level (calcu-lated)	AVG delta	AVG de-viation	Max delta	Max devia-tion
Low (400–600)	190	83%	373	199%
Medium (2500–6000)	2203	103%	4382	209%
High (9000–12000)	4564	90%	7247	178%

Spectroscopy TPH Measurements

The contaminated soil samples were measured according to TAU's protocol [29] by an ASD Fieldspec pro instrument with an ASD contact probe 3 times, each consisting of 30 measurements that have been averaged; the 3 resulting spectra for each sample were averaged. The average spectrum for each sample was used to predict the TPH level by a PLS model based on several soil types and PHC types, predeveloped in the last few years by the authors. The modeling procedure included five types of soils, three types of PHCs at 50 concentration levels, yielding 750 laboratory prepared samples. An "all possibilities" approach was used for generating robust NIRS models. This approach includes the evaluation of many preprocessing techniques (SNV, MSC, smoothing, absorbance, first and second derivatives, and continuum removal), as well as PLS and ANN modeling methods (i.e., [7, 22, 30–33]).

General Accuracy

In order to evaluate the reliability of the reflectance spectroscopy method as compared to the common EPA 418.1 method as an environmental monitoring tool, the general accuracy of both methods had to be examined. General accuracy is an important parameter as it determines not only the intra- and interperformances of the laboratories but also portrays the ability of the laboratory to determine the actual contaminant concentration in the sample. General accuracy of TPH measurements done by both reflectance spectroscopy and analytical laboratories, was measured by the same previously mentioned factors used for inter and intra groups as shown in Table 5 (average delta, average deviation, maximum delta, and maximum deviation). The average delta was calculated for each group; by first calculating the delta for each sample in that group (average TPH value-projected TPH value) followed by averaging the results of all the samples in that group. The average deviation was calculated for each group by first calculating the delta for each sample in that group (average TPH value-projected TPH value), then dividing the result with the projected TPH value for that sample, thus normalizing the results. Finally the normalized results of all samples were averaged for each group. The maximum delta and maximum deviation were calculated in the same manner, but instead of averaging the results for each group, only the maximum value was selected portraying the "worst case scenario."

Table 5: Accuracy of TPH determination by reflectance spectroscopy and three commercial laboratories

Lab	Spectroscopy				A				B				C			
TPH level (calculated)	AVG delta	AVG deviation	Max delta	Max deviation	AVG delta	AVG deviation	Max delta	Max deviation	AVG delta	AVG deviation	Max delta	Max deviation	AVG delta	AVG deviation	Max delta	Max deviation
Low (400–600)	325	68%	747	143%	349	68%	500	93%	192	42%	356	84%	283	57%	508	97%
Medium (2500–6000)	2055	47%	4360	74%	1956	51%	2795	92%	1010	30%	2532	78%	2590	60%	4781	97%
High (9000–12000)	6187	61%	11494	83%	4392	49%	6150	81%	1827	26%	4210	73%	4413	50%	5859	95%

RESULTS AND DISCUSSION

Inner laboratory consistency seems very acceptable with results of under 20% average deviation for all 3 labs with lab B having the best consistency of under 10% deviation (Table 3). Although the average deviation is low for all laboratories, in some cases high deviation can occur, even up to 68% as can be seen in Table 3 (medium concentration samples, Lab A). The interlaboratory consistency on the other hand is far from satisfactory. Average interlaboratory deviation is between 83% and 103% and can even reach values of ~200% in some cases, that is: a Hamra sample contaminated with diesel (Sample 4, Table 2) yielded an average value of 8149 TPH from Lab A and 14952 TPH from Lab B. Both intra and interlaboratory average deviation are presented in Figure 3, performance of Lab A and Lab C are about similar, with better performances by Lab B. General accuracy was also not satisfactory as seen in Table 5, average deviation ranged from 26% up to 68%. Many of the accuracy errors are in measuring 95% octane fuel; this could be a result of loosing most of the contaminant during the extraction process due to the high volatility nature of this PHC. Performance of all laboratories, including the reflectance spectroscopy method, are almost identical as shown in Figure 4, with Lab B being the most accurate laboratory. Although accuracy was not satisfactory, a good correlation appears when plotting the reflectance spectroscopy and laboratories TPH results against the projected TPH results as demonstrated in Figures 5, 6, and 7. This shows that both the spectroscopy and the laboratories TPH results are consistent and are good predictors of the contamination levels. Because it is clear that 95% octane fuel is a problematic contaminant due to its high volatility, when we examine the results while ignoring the 95% octane contaminated samples, almost perfect correlation coefficient appear (Figures 8, 9, and 10). These correlations between the reflectance spectroscopy and the laboratories TPH results shows consistency of Lab B being always over estimating the projected TPH values, and the reflectance spectroscopy, Lab A and Lab C always under estimating the projected TPH values at almost the same level. As this phenomena being so consistent, it can be corrected by the correlation factors specific for each Laboratory. The result of this study confirms the hypothesis of large variations between laboratories and methods, even though they are properly certified by the authorities. It is interesting to note that with a precise approach, it is possible to account for these variations, correct and calibrate the results to represent the contamination levels accurately, thus enabling reliable comparable results. Reflectance spectroscopy was found to be as good as the traditional method employed by the commercial certified laboratories. Reflectance spectroscopy is a nondestructive method that can be used for rapid, simple, and cost effective TPH determination both in the laboratory and in the field. Moreover, the resent advances in imaging spectroscopy field could enable the adding of a new spatial dimension for site investigation, opening new frontiers in monitoring PHC contamination in soil.

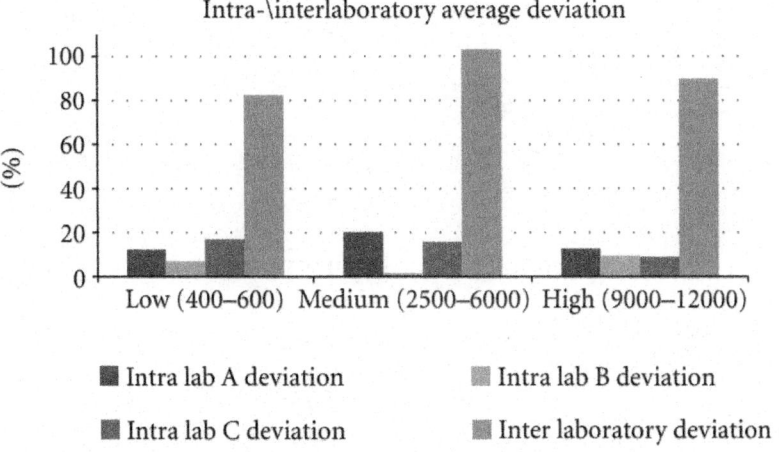

Figure 3: Intra-/interlaboratory deviation.

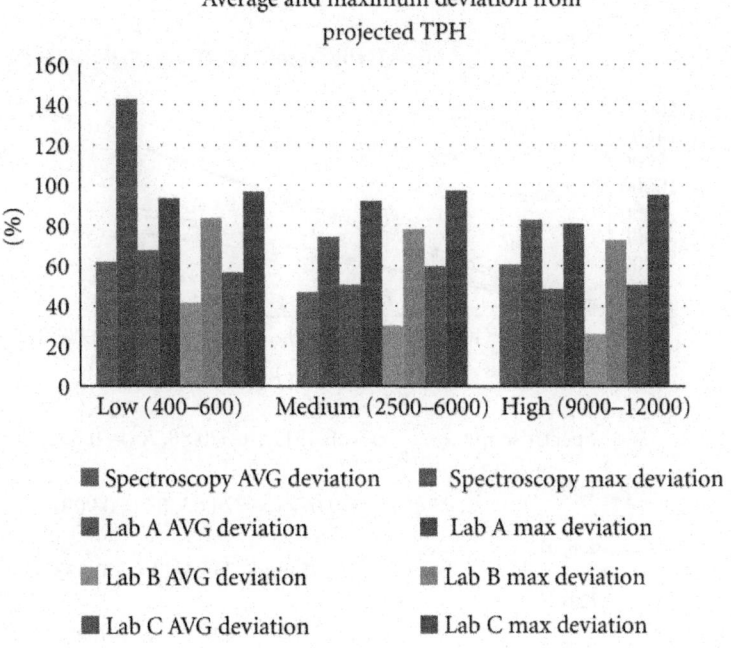

Figure 4: Average and maximum deviation from projected TPH.

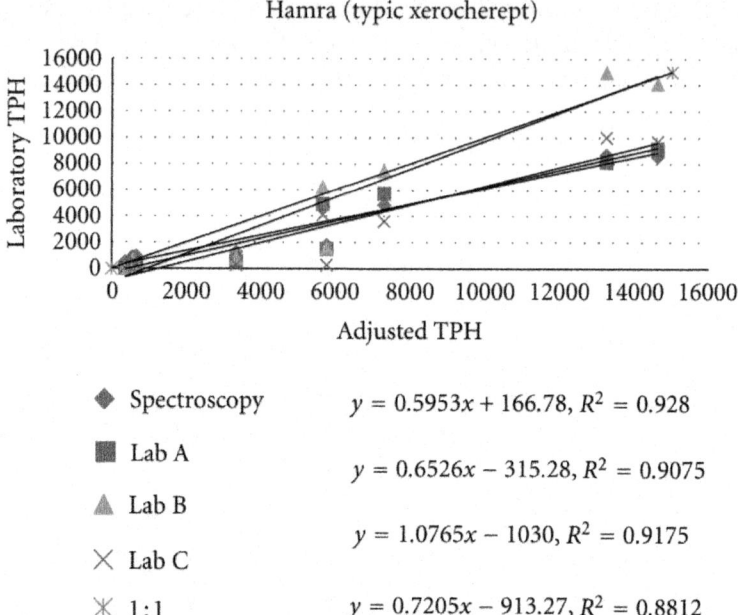

Figure 5: Hamra with all PHC types.

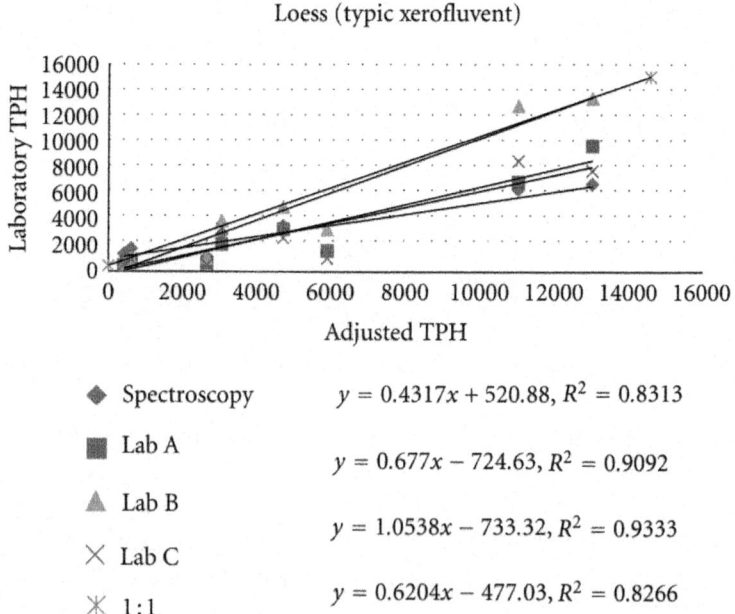

Figure 6: Loess with all PHC types.

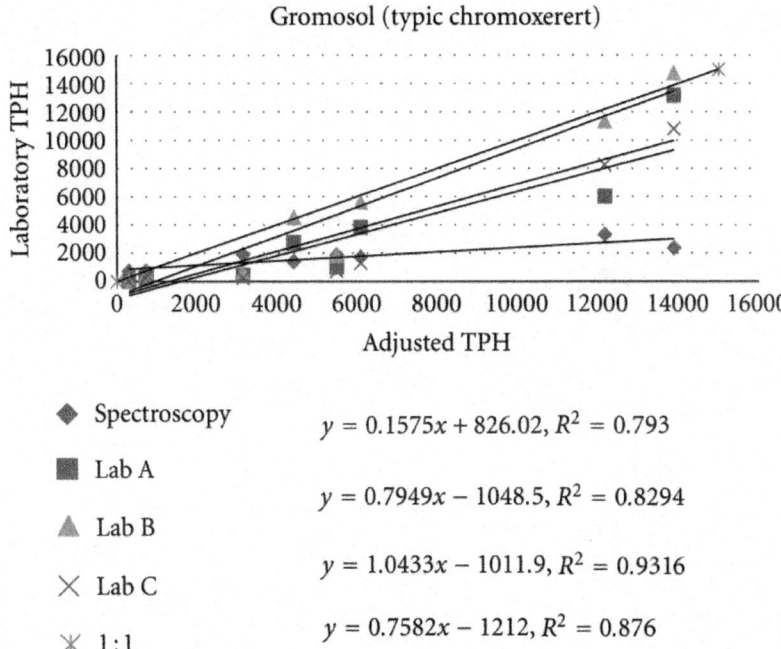

Figure 7: Gromosol with all PHC types.

For the Gromosol figure, the following relationships apply:

Spectroscopy: $y = 0.1575x + 826.02, R^2 = 0.793$

Lab A: $y = 0.7949x - 1048.5, R^2 = 0.8294$

Lab B: $y = 1.0433x - 1011.9, R^2 = 0.9316$

Lab C:

1:1: $y = 0.7582x - 1212, R^2 = 0.876$

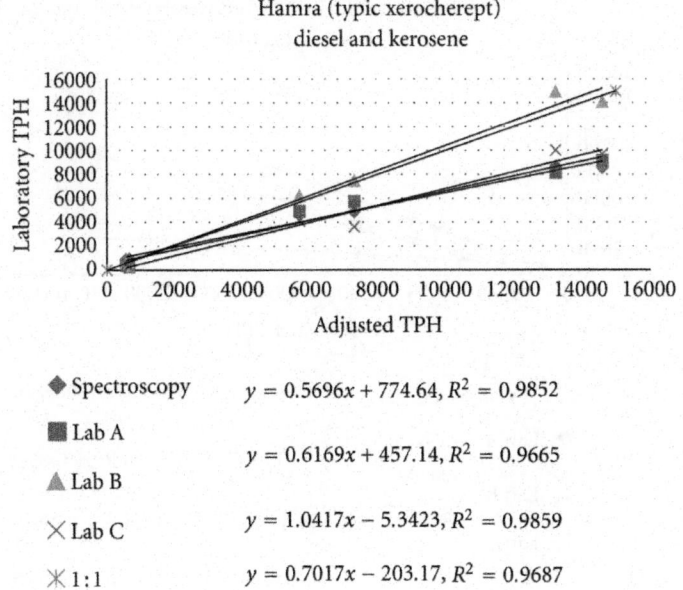

Figure 8: Hamra with diesel and kerosene.

For the Hamra figure, the following relationships apply:

Spectroscopy: $y = 0.5696x + 774.64, R^2 = 0.9852$

Lab A: $y = 0.6169x + 457.14, R^2 = 0.9665$

Lab B:

Lab C: $y = 1.0417x - 5.3423, R^2 = 0.9859$

1:1: $y = 0.7017x - 203.17, R^2 = 0.9687$

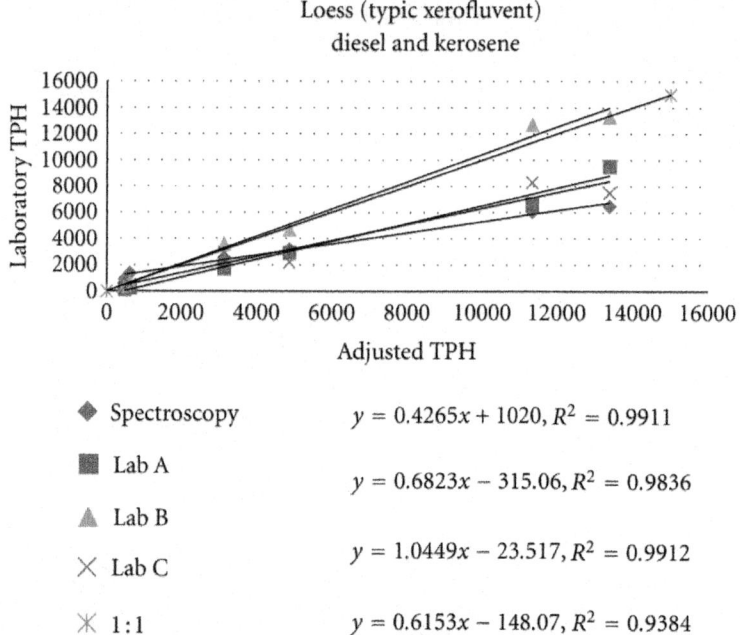

Figure 9: Loess with diesel and kerosene.

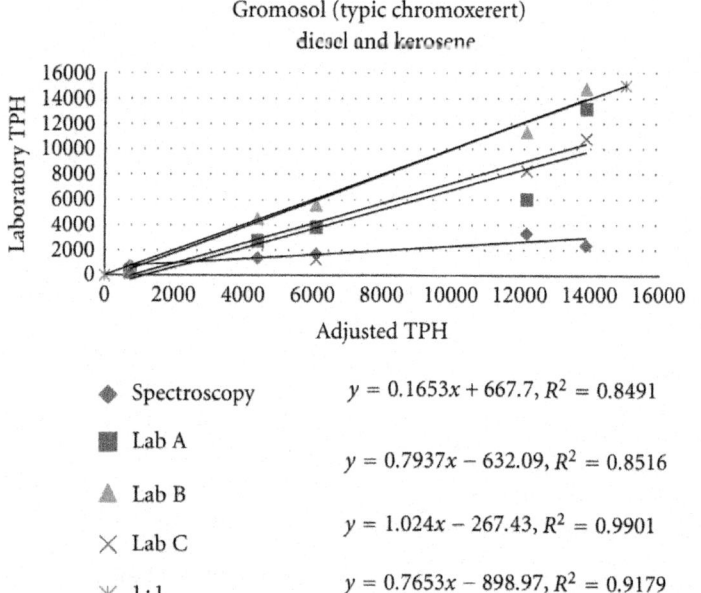

Figure 10: Gromosol with diesel and kerosene.

CONCLUSION

While accuracy level is affected by various elements such as laboratory protocols, equipment and personnel, results remain very consistent and can be corrected when certain factors specific for each laboratory are employed. When a new batch of samples needs to be evaluated, a sample of clean soil similar to the same batch, contaminated with the 418.1 EPA standard at two levels can be added to the batch, thus helping to model the bias for this batch and to calibrate the results. Due to the problematic nature of measuring the 95% octane TPH levels, a PID (Photo Ionization Detector) instrument should be used to accompany each sample to help measure the volatile PHC. Reflectance Spectroscopy performed very well in this study (almost the same as Lab A and Lab C), and should be considered as a tool for field screening due to its very low cost per sample, easy operation, ability to work in field conditions, and the possibility of fast measurements and instant results. Reflectance spectroscopy is a nondestructive environmental friendly method; that when coupled with a PID device (for volatile PHC detection) could be used as an excellent screening tool in the field. When using reflectance spectroscopy coupled with PID, contaminated samples should not elude detection. In general the 418.1 EPA method alone should not be used to grant a "clean bill of health" to any contaminated site, but only as a screening and decision-making tool before more expensive methods are employed. It is strongly recommended that any certified laboratory and method will be improved by using a standard protocol suggested in this study, for calibrating the laboratory results to the real contamination level of the soil. Applying these protocols will assure both intra- and interaccurate, consistent, and comparable results.

REFERENCES

1. M. S. Hutcheson, D. Pedersen, N. D. Anastas, J. Fitzgerald, and D. Silverman, "Beyond TPH: health-based evaluation of petroleum hydrocarbon exposures," Regulatory Toxicology and Pharmacology, vol. 24, no. 1, pp. 85–101, 1996

2. P. Boffetta, N. Jourenkova, and P. Gustavsson, "Cancer risk from occupational and environmental exposure to polycyclic aromatic hydrocarbons," Cancer Causes and Control, vol. 8, no. 3, pp. 444–472, 1997.

3. G. D. Ritchie, K. R. Still, W. K. Alexander et al., "A review of the neurotoxicity risk of selected hydrocarbon fuels," Journal of Toxicology and Environmental Health B, vol. 4, no. 3, pp. 223–312, 2001. ·

4. Environmental Sciences Division, Use of Gross Parameters for

Assessment of Hydrocarbon Contamination of Soils in Alberta, Oxford, UK, 1993.

5. United States Environmental Protection Agency (USEPA), Test Method for Evaluating Total Recoverable Petroleum Hydrocarbon, Method 418.1 (Spectrophotometric, Infrared), Government Printing Office, Washington, DC, USA, 1978.

6. United States Environmental Protection Agency (USEPA), Methods for Chemical Analysis of Water and Wastes, Government Printing Office, Washington, DC, USA, 1983.

7. G. Schwartz, G. Eshel, and E. Ben-Dor, "Reflectance spectroscopy as a tool for monitoring contaminated soils," in Soil Contamination, Intech, 2011.

8. R. S. G. Gómez, T. Pandiyan, V. E. A. Iris, V. Luna-Pabello, and C. D. de Bazúa, "Spectroscopic determination of poly-aromatic compounds in petroleum contaminated soils," Water, Air, and Soil Pollution, vol. 158, no. 1, pp. 137–151, 2004.

9. J. Krupcík, P. Oswald, D. Oktavec, and D. W. Armstrong, "Calibration of GC-FID and IR spectrometric methods for determination of high boiling petroleum hydrocarbons in environmental samples," Water, Air, and Soil Pollution, vol. 153, no. 1–4, pp. 329–341, 2004.

10. G. Xie, M. J. Barcelona, and J. Fang, "Quantification and interpretation of total petroleum hydrocarbons in sediment samples by a GC/MS method and comparison with EPA 418.1 and a rapid field method,"Analytical Chemistry, vol. 71, no. 9, pp. 1899–1904, 1999.

11. P. Lambert, M. Fingas, and M. Goldthorp, "An evaluation of field total petroleum hydrocarbon (TPH) systems," Journal of Hazardous Materials, vol. 83, no. 1-2, pp. 65–81, 2001.

12. E. Saari, P. Perämäki, and J. Jalonen, "A comparative study of solvent extraction of total petroleum hydrocarbons in soil," Microchimica Acta, vol. 158, no. 3-4, pp. 261–268, 2007.

13. M. Villalobos, A. P. Avila-Forcada, and M. E. Gutierrez-Ruiz, "An improved gravimetric method to determine total petroleum hydrocarbons in contaminated soils," Water, Air, and Soil Pollution, vol. 194, no. 1–4, pp. 151–161, 2008.

14. E. A. Cloutis, "Spectral reflectance properties of hydrocarbons: remote-sensing implications," Science, vol. 245, no. 4914, pp. 165–168, 1989

15. I. Schneider, G. Nau, T. V. V. King, and I. Aggarwal, "Fiber-optic near-infrared reflectance sensor for detection of organics in soils," IEEE Photonics Technology Letters, vol. 7, no. 1, pp. 87–89, 1995.

16. B. R. Stallard, M. J. Garcia, and S. Kaushik, "Near-IR reflectance spectroscopy for the determination of motor oil contamination in sandy loam," Applied Spectroscopy, vol. 50, no. 3, pp. 334–338, 1996

17. Z. Zwanziger and F. Heidrun, "Near infrared spectroscopy of fuel contaminated sand and soil. I. Preliminary results and calibration study," Journal of Near Infrared Spectroscopy, vol. 6, no. 1–4, pp. 189–197, 1998

18. D. F. Malley, K. N. Hunter, and G. R. B. Webster, "Analysis of diesel fuel contamination in soils by near-infrared reflectance spectrometry and solid phase microextraction-gas chromatography," Soil and Sediment Contamination, vol. 8, no. 4, pp. 481–489, 1999

19. B. Hörig, F. Kühn, F. Oschütz, and F. Lehmann, "HyMap hyperspectral remote sensing to detect hydrocarbons," International Journal of Remote Sensing, vol. 22, no. 8, pp. 1413–1422, 2001

20. F. Kühn, K. Oppermann, and B. Hörig, "Hydrocarbon index—an algorithm for hyperspectral detection of hydrocarbons," International Journal of Remote Sensing, vol. 25, no. 12, pp. 2467–2473, 2004.

21. K. H. Winkelmann, On the applicability of imaging spectrometry for the detection and investigation of contaminated sites with particular consideration given to the detection of fuel hydrocarbon contaminants in soil, Ph.D. thesis, Brandenburgische Technische Universität Cottbus, 2005.

22. G. Schwartz, G. Eshel, M. Ben-Haim, and E. Ben-Dor, "Rapid methods for classification and quantitative assessment of petroleum hydrocarbons pollution in soil samples using reflectance spectroscopy," EGU 2009-11441-2, Vienna, Austria, 2009.

23. S. Chakraborty, D. C. Weindorf, C. L. S. Morgan et al., "Rapid identification of oil-contaminated soils using visible near-infrared diffuse reflectance spectroscopy," Journal of Environmental Quality, vol. 39, no. 4, pp. 1378–1387, 2010.

24. T. Lammoglia and C. R. de S. Filho, "Spectroscopic characterization of oils yielded from Brazilian offshore basins: potential applications of remote sensing," Remote Sensing of Environment, vol. 115, no. 10, pp. 2525–2535, 2011.

25. J. Dan and H. Koyumdjisky, "The soils of israel and their distribution," European Journal of Soil Science, vol. 14, no. 1, pp. 12–20, 1963.

26. S. S. Staff, Keys to Soil Taxonomy, Government Printing Office, 2010.

27. D. L. Carter, M. M. Mortland, and W. D. Kemper, "Specific surface," in Methods of Soil Analysis Part I. Soil Science, A. Klute, Ed., pp. 413–422,

Society of America, Madison, Wis, USA, 1986.

28. G. Eshel, G. J. Levy, U. Mingelgrin, and M. J. Singer, "Critical evaluation of the use of laser diffraction for particle-size distribution analysis," Soil Science Society of America Journal, vol. 68, no. 3, pp. 736–743, 2004

29. A. Pimstein, E. Ben-Dor, and G. Notesco, "Performance of three identical spectrometers in retrieving soil reflectance under laboratory conditions," Soil Science Society of America Journal, vol. 75, no. 2, pp. 746–759, 2011.

30. G. Schwartz, G. Eshel, M. Ben-Haim, and E. Ben-Dor, Reflectance Spectroscopy as a Rapid Tool for Qualitative Mapping and Classification of Hydrocarbons Soil Contamination, Tel Aviv, Israel, 2009.

31. G. Schwartz, G. Eshel, M. Ben-Haim, and E. Ben-Dor, Quantitative Assessment of Petroleum Hydrocarbons in Situ by Diffused Reflectance Spectroscopy and a Penetrating Optical Sensor, GFZ, Potsdam, Germany, 2010.

32. G. Schwartz, G. Eshel, and E. Ben-Dor, An Operational Spectral Based Model to Predict Soil Petroleum Hydrocarbon Content in Field Samples, Edinburgh, Scotland, 2011.

33. G. Schwartz, Reflectance spectroscopy as a rapid tool for qualitative mapping and classification of hydrocarbons soil contamination, Ph.D. thesis, Tel Aviv University, 2012.

Chapter 10

QUANTITATIVE LC-MS/MS ANALYSIS OF SEVEN GINSENOSIDES AND THREE ACONITUM ALKALOIDS IN SHEN-FU DECOCTION

Na Guo[1], Mingtao Liu[2], Dawei Yang[3], Ying Huang[1], Xiaohong Niu[1], Ruifan Wu[4], Ying Liu[5], Guizhi Ma[4] and Deqiang Dou[6]

[1]Experimental Research Center, China Academy of Chinese Medical Sciences, Beijing 100700, China

[2] SRI International, Menlo Park, CA 94025, USA

[3] Key Laboratory of Biofuels, Qingdao Institute of Bioenergy and Bioprocess Technology, Chinese Academy of Sciences, Songling road 189, Qingdao 266101, China

[4]College of Pharmacy, Xinjiang Medical University, Urumqi 830011, China

[5] Key Laboratory of Bioactive Substances and Resource Utilization of Chinese Herbal Medicine, Ministry of Education, Institute of Materia Medica, Chinese Academy of Medical Sciences and Peking Union Medical College, Beijing 100050, China

[6] Department of Chinese Medicine Chemistry, Liaoning University of Traditional Chinese Medicine, Dalian 116600, China.

ABSTRACT

Background

Shen-Fu decoction is a traditional Chinese medicine prescription with a 3:2 ratio of *Radix Ginseng* and Fuzi (*Radix Aconiti lateralis praeparata*). Ginsenosides and alkaloids are considered to be the main active components of Shen-Fu decoction. However, no analytical methods have been used to quantitatively analyse both components in Shen-Fu decoction simultaneously.

RESULTS

We successfully developed a rapid resolution liquid chromatography coupled with tandem mass spectrometry (RRLC-MS/MS) method for the simultaneous

analysis of seven ginsenosides and three aconitum alkaloids in Shen-Fu decoction, the decoction of Radix ginseng and Fuzi (*Radix Aconiti lateralis praeparata*). Chromatogrpahic separation by RPLC was achieved using a reversed-phase column and a water/acetonitrile mobile phase, containing 0.05% formic acid and using a gradient system. The method was optimized to allow for simultaneous analysis of all analytes in 11minutes without the need for baseline resolution of the components. Furthermore, the separation demonstrated good linearity (r > 0.9882), repeatability (RSD < 7.01%), intra- and inter-day precisions (RSD < 5.06%) and high yields of recovery (91.13-111.97%) for ten major constituents, namely ginsenoside-Re, Rg_1, Rb_1, Rc, Rb_2, Rd, Rf, aconitine, hypacoitine and mesaconitine.

Conclusions

The developed method could be used as a rapid and reliable approach for assessment of the quantity of the major constituents in Shen-Fu decoction.

BACKGROUND

Decoction is the traditional prescription of traditional Chinese medicines (TCMs). Based on TCM theory, one single herb or several kinds of herbs combined are boiled in water to make the decoction. First documented in 1465, Shen-Fu decoction is a TCM prescription with a 3:2 ratio of *Radix Ginseng* and Fuzi (*Radix Aconiti lateralis praeparata*). Both components have been commonly used as herbal medicines in China for about 1800 years, predominantly used for folk treatment of diseases with the sign of Yangqi decline or Yang exhaustion. Shen-Fu decoction is also used to treat cardiovascular diseases such as circulatory collapse, shock, thoracic obstruction and acute thoracic pain. Shen-Fu Injection (SFI for intravenous medication), is a typical form of Shen-Fu decoction, that has been used for treatment of many kinds of diseases because of its cardiovascular protective effectiveness [1–3]. The main active components found in Shen-Fu decoction are ginsenosides and alkaloids. Ginsenosides are generally classified into four groups: protopanaxadiol, protopanaxatriol, ocotillol and oleanolic acid type [4–6], Currently, more than 150 ginsenosides have been isolated and identified in the literature. Among them, ginsenosides-Rb_1, Rb_2, Rc, Rd, Rg_1, Re and

Rf (Figure 1) are the most important compounds in chemical analysis of ginsengs. At present, about 224 alkaloids have been isolated and identified from Aconitum [7, 8]. These have been classified into four major groups, nonester alkaloids (NEAs), monoester diterpene alkaloids (MDAs), diester diterpene alkaloids (DDAs) and lipoalkaloids. *Aconitum* alkaloids are mainly constituted of three DDAs, diester-diterpence called aconitine (AC), measaconitine (MA) and hypaconitine (HA) (Figure 1). They are known for their high toxicity and pharmacological activity, as well as being the target markers of Fuzi. In general, the curative effect of traditional Chinese medicine is an integrative result of a number of ginsenosides and alkaloids. In order to minimize the variability of active ingredients in the decoction and ensure repeatable and reproducible therapeutic effects, it is very important to establish quality control methodology for the decoction. To this end, analysis of ginsenosides and *Aconitum* alkaloids is required to assess the quality of Shen-Fu decoction.

Compound	R_1	R_2	R_3
Protopanaxadiol-type			
Ginsenoside-Rb1	-O-Glc2-1Glc	-H	-O-Glc6-1Glc
Ginsenoside-Rb2	-O-Glc2-1Glc	-H	-O-Glc6-1Arap
Ginsenoside-Rc	-O-Glc2-1Glc	-H	-O-Glc6-1Araf
Ginsenoside-Rd	-O-Glc2-1Glc	-H	-O-Glc
Protopanaxadiol-type			
Ginsenoside-Re	-OH	-O-Glc2-1Rha	-O-Glc
Ginsenoside-Rg1	-OH	-O-Glc	-O-Glc
Ginsenoside-Rf	-OH	-O-Glc2-1Glc	-OH

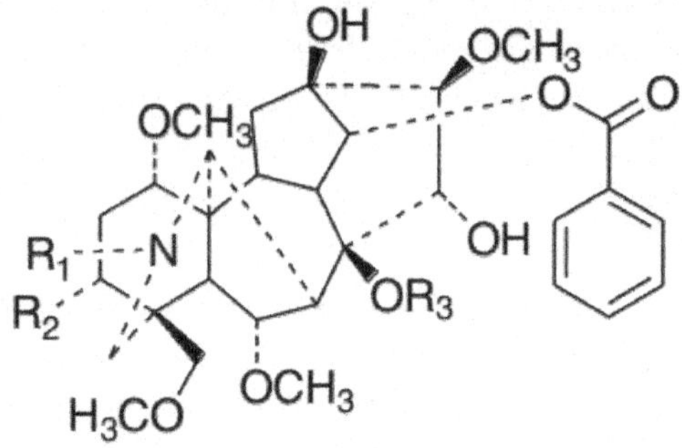

Compound	R_1	R_2	R_3
AC	C_2H_5	OH	acetyl
MA	CH_3	OH	acetyl
HA	CH_3	H	acetyl

Figure 1: Chemical structures of ginsenosides and Aconitum alkaloids analyzed in Shen-Fu decoction.

Previous methods that have been used to analyze ginsenosides and alkaloids include HPLC-DAD (ELSD), CE, GC-MS and LC-MS [9–19] and alkaloids [20–30]. In comparison with traditional HPLC, RRLC provides a higher peak capacity, greater resolution, increased sensitivity and higher speed of analysis. When coupled to a triple quadrupole tandem mass spectrometer (QQQ MS/MS), it can achieve high sensitivity and selectivity by using the multiple reaction monitoring (MRM) scan mode without the baseline chromatographic separation of target analytes. This method greatly facilitates the quantification of chemical markers in complex matrixes with only a small amount of sample. To date, there are no studies reporting the simultaneously quantitative determination of ginsenosides and Aconitum alkaloids in Shen-Fu decoction. The primary aim of the present study was to develop a direct and rapid RRLC-MS/MS method for simultaneously quantifying the ten constituents in Shen-Fu decoction, namely, ginsenosides-Rb1, Rb2, Rc, Rd, Rg1, Re and Rf and Aconitum alkaloids including AC, MA and HA.

RESULTS AND DISCUSSION

Chromatographic Conditions and MS/MS Method Development

Different mobile phases, including acetonitrile with 0.05%, 0.1% aqueous formic acid, acetic acid, 5 mM and 10 mM ammonium formate solutions were tested. The best peak shape and resolution was obtained with a mixture of acetonitrile and aqueous 0.05% formic acid solution. Using an optimized elution gradient, the main components were separately eluted within 11 min. The typical RRLC-QQQ MS/MS chromatograms of the marker chemicals in Shen-Fu decoction are shown in Figure 2. In order to increase sensitivity and specificity of quantification, multiple reaction monitoring was performed. All factors related with MS performance including ionization mode, capillary voltage, fragmentor voltage, collision energy, gas flow and desolvation temperature were analyzed. The optimum conditions were determined as follows: postive ion mode, capillary voltage 4000 V, drying gas, gas temperature 350°C and nebulizer pressure of 50 psi.

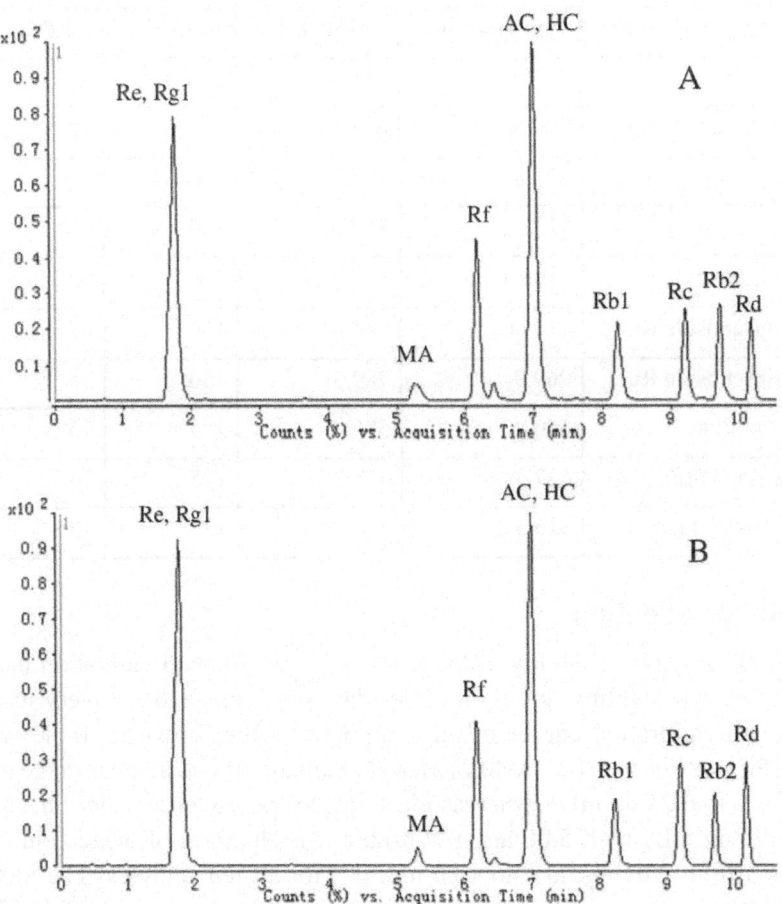

Figure 2: Typical RRLC-QQQ MS/MS chromatograms of marker chemicals in Shen-Fu decoction (A) standard mixture (B) Shen-Fu decoction.

Optimization of this MS/MS method produced highest achiveable response using the MRM pairs comprising of the precursor and product ions, which can achieve better quantitation than reported results using the selected ion monitoring (SIM) mode. After optimization, the precursor and product ions of the ten analytes were recorded (Table 1). The optimum collision energy was determined to be 50 eV for Ginsenoside Re, 40 eV for Rg_1, 55 eV for Rf and Rd, 65 eV for Rb_1, Rc and Rb_2. For alkaloids, they required a lower collision energy of 35 eV for MA, 40 eV for HA and 45 eV for AC (Table 1).

Table 1: Mass spectra properties of ten compounds in Shen-Fu decoction

Compound name	Precursor ion	Product ion	Frag(V)	CE(V)
Ginsenoside Re	969.6	789.5	150	50
Ginsenoside Rg_1	823.5	643.5	135	40
Ginsenoside Rf	823.3	365.3	140	55
Ginsenoside Rb_1	1131.6	365.0	150	65
Ginsenoside Rc	1101.7	335.0	150	65
Ginsenoside Rb_2	1101.6	334.8	150	65
Ginsenoside Rd	969.9	789.3	150	55
Aconitine	646.4	586.4	135	45
Mesaconitine	632.3	572.3	135	35
Hypacoitine	616.3	556.2	135	40

Method Validation

To determine the reliability of the test results, the method validation included linearity, repeatability, intra- and inter-day precisions and recovery test. The standard calibration curves of all compounds were shown in Table 2 with satisfactory linearity (r > 0.9882). *Aconitum*alkaloids had a linear range of 0.03 ng mL^{-1} to 6.24 ng mL^{-1}, whereas ginsenosides displayed a wider linear range of 3.90 ng mL^{-1} to 125.00 ng mL^{-1} (Table 2). The limit of dectection (LOD) ranged from 0.01 ng mL^{-1} to 1.25 ng mL^{-1} for all ten analytes. The intra-day and inter-day with RSD less than 5.06% are demonstrated in Table 2. The repeatability was satisfactory with RSD below 7.01%. Recovery of the ten compounds (Table 3) was within the range of 91.13-111.97% and showed no relevant difference in the percent yield recovered using with different concentrations of the compounds. Thus, the ten analytes can be quantitatively analyzed simultaneously in a relatively short-time using this optimized method.

Table 2: Calibration curves, LOD, LOQ, Precision and Repeatability for ten compounds in Shen-Fu decoction

Compound name	Calibration curve	r	Linear range (ng·mL⁻¹)	LOD (ng·mL⁻¹)	LOQ (ng·mL⁻¹)	Intra-day(n=6)	Inter-day (n=6)	Repeat-ability (n=5)
Ginsen-oside-Rb$_1$	Y=11.04X+181.44	0.9930	3.90~125.00	0.97	3.00	3.44	4.11	6.51
Ginsen-oside-Rb$_2$	Y=31.68X+246.74	0.9882	3.90~125.00	0.97	3.00	2.23	3.55	7.01
Ginsen-oside-Rc	Y=20.13X+25.79	0.9921	3.90~125.00	1.25	3.00	2.02	3.92	4.21
Ginsen-oside-Rd	Y=13.60X+69.00	0.9993	3.90~125.00	0.75	1.95	4.18	4.71	3.16
Ginsen-oside-Re	Y=18.52X+136.68	0.9973	3.90~125.00	0.48	1.95	3.34	5.06	4.67
Ginsen-oside-Rf	Y=37.66X+473.14	0.9952	3.90~125.00	0.97	3.00	2.05	3.12	5.03
Ginsen-oside-Rg$_1$	Y=52.38X+109.20	0.9994	3.90~125.00	0.48	1.50	2.29	2.49	4.12
Aconitine	Y=6193.52X-80.34	0.9996	0.03~1.25	0.01	0.04	2.76	3.51	4.45
Mesaconi-tine	Y=3617.22X-63.02	0.9939	0.03~1.25	0.01	0.04	3.97	3.28	4.96
Hypaconi-tine	Y=1207.82X+180.44	0.9960	0.19~6.24	0.01	0.04	2.55	2.42	4.81

Table 3: Analsysis of the recovery of ten compounds in Shen-Fu decoction

Compounds	Initial amount(ng)	Added amount(ng)	Detected amount(ng)	Recovery (%)	RSD/% (n=5)
Ginsenosid-Re	2074.12	1700	3673.54 ± 230.37	94.08	6.27
	2074.12	2100	3991.27 ± 117.73	91.29	2.95
	2074.12	2500	4475.56 ± 236.76	96.06	5.29
Ginsenoside-Rg$_1$	2260.51	1800	3959.88 ± 307.37	94.41	7.76
	2260.51	2250	4310.87 ± 229.80	91.13	5.33
	2260.51	2700	5176.34 ± 166.16	107.99	3.21
Ginsenoside-Rb$_1$	2423.46	2000	4299.39 ± 196.65	93.80	4.57
	2423.46	2500	4812.75 ±185.92	95.57	3.86
	2423.46	3000	5271.72 ± 130.37	94.94	2.47
Ginsenoside-Rc	2231.62	1800	3933.62 ± 155.38	94.56	3.95
	2231.62	2250	4644.8 ± 231.31	107.25	4.98
	2231.62	2700	5043.67 ± 244.21	104.15	4.84
Ginsenoside-Rb$_2$	1597.95	1200	2880.31 ± 83.82	106.86	2.91
	1597.95	1500	3277.51 ± 128.48	111.97	3.92
	1597.95	1800	3510.26 ± 169.97	106.24	4.84
Ginsenoside-Rd	816.73	640	1416.19 ± 86.39	93.67	6.10
	816.73	800	1642.99 ± 95.82	103.28	5.83
	816.73	960	1700.19 ± 80.93	92.03	4.76
Ginsenoside-Rf	2000.12	1600	3563.83 ± 252.68	97.73	7.09
	2000.12	2000	4173.63 ± 261.27	108.68	6.26
	2000.12	2400	4549.1 ± 256.57	106.21	5.64
Aconitine	2.91	2.4	5.52 ± 0.27	108.75	4.89
	2.91	3.0	5.67 ± 0.31	92.00	5.47
	2.91	3.6	6.7 ± 0.34	105.28	5.07
Mesaconitine	7.12	5.6	12.98 ± 0.66	104.64	5.08
	7.12	7.0	13.74 ± 0.68	94.57	4.95
	7.12	8.4	15.82 ± 0.56	103.57	3.54

Hypaconitine	100.06	80	174.69 ± 7.95	93.29	4.55
	100.06	100	204.9 ± 10.90	104.84	5.32
	100.06	120	225.07 ± 7.81	104.18	3.47

Sample Analysis

The described RRLC-QQQ-MS/MS method was subsequently applied to the analysis of Shen-Fu decoction, made by authenticated Radix ginseng and aconite root (see method part). The quantitative analytical results are shown in Table 4. The repeatability of the ten analytes in the Shen-Fu decoction was reliable (RSD<6.28%). From Table 4, Shen-Fu decoction showed higher amounts of ginsenosides than alkaloids. This result meant that the Shen-Fu decoction may have very low toxicity levels, as aconitine, hypacoitine and mesaconitine are the main toxicity source of some toxic herbal medicines [27]. Furthermore, ginsenoside-Rb1 was the most abundant of the ten compounds in Shen-Fu decoction. Conversely, aconitine was shown to be the least abundant of the ten compounds in Shen-Fu decoction. This method would allow for comparison of the quantity of ginsenosides and alkaloids between Shen-Fu decoction preparations and could therefore be used as a rapid and reliable approach for assessment of the quality of Shen-Fu decoction.

Table 4: Contents of ten compounds in Shen-Fu decoction

Samples	Content (µg/g)									
	Rb_1	Rd	Re	Rf	Rg_1	Rc	Rb_2	Aconitine	Mesaconitine	Hypaconitine
	247.17± 11.27	84.21± 4.31	210.64± 12.66	204.66± 10.24	231.22± 11.75	223.19± 14.01	121.16± 7.41	0.21± 0.01	0.76± 0.04	10.05± 0.48

MATERIALS AND METHODS

Chemicals, Standards and Samples

HPLC grade acetonitrile was purchased from Merck (Germany) and MS grade formic acid from Sigma-Aldrich. All other chemicals and solvents were of an analytical grade. Ultra-pure water (18.2MΩ) was prepared with a Milli-Q water purification system (Millipore, Bedford, MA, USA).

The standards reference samples of Ginsenosides Rb_1, Rb_2, Rc, Rd, Rg_1, Re, Rf, AC, HA and MA were purchased from the National Institute for Control of Pharmaceutical and Biological Products (Beijing, China). The purity of the standards was relatively high at no less than 98%. Radix ginseng was purchased from Liaoning luyuan Pharmaceutical Co., Ltd. in China. The processed aconite root was purchased from Tong-Ren-Tang Pharmaceutical store (Beijing, PR China). Panax ginseng and the prepared aconite root were authenticated by Professor Xirong, He, Insitute of traditional Chinese medicine, China Academy of Chinese Medical Sciences.

Sample Preparation

Reference Standards Solutions

Stock solutions were prepared by accurate measurement of ginsenoside Re, Rg_1, Rf, Rb_1, Rc, Rb_2, Rd, aconitine, hypacoitine and mesaconitine. They were dissolved with methanol respectively to get ten reference standards stock solutions (1.0 mg mL^{-1}), and were stored at 4°C.

Extracts of Shen-Fu Decoction

ShenFu Formula (SF) was prepared by combining of Radix ginseng and the processed aconite root (at a ratio of 3:2). Dried and pulverized white ginseng (18 g) and the processed aconite root (12 g) were ground and then refluxed three times with 300 mL of water for 60 min at 100°C. After cooling, the extracted solutions were filtered under vacuum. The solutions were condensed under decompression and finally were freeze-dried. The decoction extract was dissolved in a measured volume of water with a concentration equal to 10 mg of crude botanicals per milliliter. 1mL of the solution was precipitated with 8 mL ethanol allowed to sit for 24 h at 4°C. The solution was filtered under vacuum. The filtrate was transferred to a 50 mL volumetric flask. Prior to injection, all samples were filtered through a 0.22 μm membrane filter.

RRLC-MS Conditions

An Agilent-1200 RRLC/6410A QQQ system (Agilent, MA, USA) equipped with an electrospray ionization (ESI) source and operated in positive ion mode (data analysis software Masshunter version B.01.04) was used for the simultaneous determination of seven ginsenosides and three aconitum alkaloids in Shen-Fu decoction. The separation was performed on an Agilent ZORBAX C18 SB column (100 mm×2.1 mm, 1.8 μm). The gradient mobile phases consisted of (A) water containing 0.05% formic acid and (B) acetonitrile for gradient elution from the column at 40°C. The linear gradient conditions assessed for gradient optimization were as follows: 0–2 min, 28-34% B; 2–6 min, 34-35%; 6–10 min, 35-100%; 10–11 min, 100%. The flow rate was 0.35 ml/min. The column temperature was 40°C. The conditions for MS analysis were as follows: drying gas N_2 flow rate 12 L min^{-1}, gas temperature 350°C and nebulizer pressure was 50 psi. The capillary voltage was set to 4000 V. MRM was employed for quantification. The precursor-to-product ion pair, fragmentor voltage (Frag V) and collision energy (CE) for each analyte are described (Table 1). The dwell time of each ion pair was 200 ms.

Method Validation

An external calibration method was used for quantitative analysis with the linear calibration curves constructed using six different concentrations of the ten compounds. Each concentration was analyzed in triplicate and then the calibration curves were constructed by plotting the peak areas versus the concentrations of each analyte. The LOD and limit of quantification (LOQ) were measured with the signal-to-noise ratios of 3:1 and 10:1, respectively. The intra-day precision was determined by analysis of the standard solution at six times within 1 day. Inter-day precision on other hand, was determined by repeated analysis of the sample for three consecutive days. For the assessment of experimental repeatability test, five independent sample solutions were prepared by the procedures noted in *Extracts of Shen-Fu decoction*. The recovery of this method was determined using the standard addition method. Three different concentration levels (approximately equivalent to 0.8, 1.0 and 1.2 times of the concentration of the original amount in the matrix) of the references standards were added into the sample in triplicate. The average recoveries were determined by the following equation: Recovery(%) = (Observed amount - Original amount)/Spiked amount × 100%, RSD (%) = (SD/mean) × 100%.

CONCLUSIONS

This is the first report of the simultaneous determination of the major compounds in Shen-Fu decoction. By using RRLC coupled with an ESI triple quadrupole tandem spectrometer, we developed and validated a rapid, simple and reliable method to simultaneously determine ten marker chemicals (ginsenoside Re, Rg_1, Rb_1, Rc, Rb_2, Rd, Rf, aconitine, hypacoitine and mesaconitine) in the Shen-Fu decoction. This method provides an excellent quantitative tool for the quality assessments of TCM formulae because of its high capacity, high sensitivity, high selectivity and short analysis time.

ABBREVIATIONS

RRLC-MS/MS: Rapid resolution liquid chromatography coupled with tandem mass spectrometry

QQQ-MS/MS: Triple quadrupole tandem mass spectrometer

RSD: Relative standard deviations

TCMs: Traditional Chinese medicines

SFI: Shen-Fu Injection

SIM: Selected ion monitor

MRM: Multiple reaction monitor

ESI: Electrospray ionization

Frag V: Fragmentor voltage

CE: Collision energy

LOD: Limit of detection

LOQ: Limit of quantification

NEAs: Nonester alkaloids

MDAs: Monoester diterpene alkaloids

DDAs: Diester diterpene alkaloids

AC: Aconitine

MA: Measaconitine

HA: Hypaconitine.

ACKNOWLEDGMENTS

This work was financially supported by 2013 Program for Liaoning Innovative Research Team in University (LT2013020 the Autonomic Project of China Academy of Chinese Medicine Sciences (project number zz2012011) and the National Natural Science Foundation of China (Grant 81001597 and 81370095).

AUTHORS' CONTRIBUTIONS

GN, M-GZ and D-DQ conceived of the study, participated in its design and coordination, and drafted the manuscript. GN, L-MT and Y-DW performed experiments and analyzed results and helped to draft the manuscript. HY, N-XH, W-RF and LY helped to do experiments. All authors read and approved the manuscript.

REFERENCES

1. Wang YL, Wang CY, Zhang BJ, Zhang ZZ: Shenfu injection suppresses apoptosis by regulation of Bcl-2 and caspase-3 during hypoxia/reoxygenation in neonatal rat cardiomyocytes in vitro. *Mol Biol Rep* 2009, 36:365–370.

2. Zheng CD, Min S: Cardioprotection of Shenfu Injection against myocardial ischemia/reperfusion injury in open heart surgery.*Chin J Integr Med* 2008, 14:10–16.

3. Luo J, Min S, Wei K, Cao J: Ion channel mechanism and ingredient bases of Shenfu Decoction's cardiac electrophysiological effects. *J Ethnopharmacol* 2008, 117:439–445.

4. Chu S, Zhang J: New achievements in ginseng research and its future prospects. *Chin J Integr Med* 2009, 15:403–408.

5. Jia L, Zhao Y: Current evaluation of the millennium phytomedicine-ginseng (I): etymology, pharmacognosy, phytochemistry, market and regulations. *Curr Med Chem* 2009, 16:2475–2484.

6. Jia L, Zhao Y, Liang XJ: Current evaluation of the millennium phytomedicine—ginseng (II): collected chemical entities, modern pharmacology, and clinical applications emanated from traditional Chinese medicine. *Curr Med Chem* 2009, 16:2924–2942.

7. Judith S, Ming Z, Sonja P, Brigitte K: Aconitum in traditional Chinese medicine-a valuable drug or an unpredictable risk. j. *J Ethnopharmacol* 2009, 126:18–30.

8. Gao F, Li YY, Wang D, Huang X, Liu Q: Diterpenoid Alkaloids from the Chinese Traditional Herbal "Fuzi" and Their Cytotoxic Activity. *Molecules* 2012, 17:5187–5194.

9. Liu Y, Yang J, Cai Z: Chemical investigation on Sijunzi decoction and its two major herbs Panax ginseng and Glycyrrhiza uralensis by LC/MS/MS. *J Pharm Biomed Anal* 2006, 41:1642–1647.

10. Qi LW, Wang CZ, Yuan CS: Isolation and analysis of ginseng: advances and challenges. *Nat Prod Rep* 2011, 28:467–495.

11. Xie GX, Plumb R, Su MM, Xu ZH, Zhao AH, Qiu MF, Long XB, Liu Z, Jia W: Ultra☐performance LC/TOF MS analysis of medicinal Panax herbs for metabolomic research. *J Sep Sci* 2008, 31:1015–1026.

12. Toh DF, New LS, Koh HL, Chan ECY: Ultra-high performance liquid chromatography/time-of-flight mass spectrometry (UHPLC/TOFMS) for time-dependent profiling of raw and steamed Panax notoginseng. *J Pharm Biomed Anal* 2010, 52:43–50.

13. Hu P, Luo GA, Wang Q, Zhao ZZ, Wang W, Jiang ZH: The Retention Behavior of Ginsenosides in HPLC and Its Application to Quality Assessment of Radix Ginseng. *Acta Pharmacol Sin* 2009, 32:667–676.

14. Popovich DG, Hu C, Durance TD, Kitts DD: Retention of ginsenosides in dried ginseng root: Comparison of drying methods. *J Food Sci* 2005, 70:s355-s358.

15. Fuzzati N: Analysis methods of ginsenosides. *J Chromatogr B* 2004, 812:119–133.

16. Li L, Luo GA, Liang QL, Hu P, Wang YM: Rapid qualitative and quantitative analyses of Asian ginseng in adulterated American ginseng preparations by UPLC/Q-TOF-MS. *J Pharm Biomed Anal* 2010, 52:66–72.

17. Lai CM, Li SP, Yu H, Wan JB, Kan KW, Wang YT: A rapid HPLC-ESI-MS/MS for qualitative and quantitative analysis of saponins in "XUESETONG" injection. *J Pharm Biomed Anal* 2006, 40:669–678.

18. Qi LW, Wang HY, Zhang H, Wang CZ, Li P, Yuan CS: Diagnostic ion filtering to characterize ginseng saponins by rapid liquid chromatography with time-of-flight mass spectrometry. *J Chromatogr A* 2012, 1230:93–99.

19. Yu Q, Yu B, Yang H, Li X, Liu S: Silver (I)-assisted enantiomeric analysis of ginsenosides using electrospray ionization tandem mass spectrometry. *J Mass Spectrom* 2012, 47:1313–1321.

20. Wang JS, van der Heijden R, Spijksma G, Reijmers T, Wang M, Xu GW, *et al*.: Alkaloid profiling of the Chinese herbal medicine Fuzi by combination of matrix-assisted laser desorption ionization mass spectrometry with liquid chromatography-mass spectrometry. *J Chromatogr A* 2009, 1216:2169–2178.

21. Tang L, Gong Y, Lv C, Ye L, Liu L, Liu Z: Pharmacokinetics of aconitine as the targeted marker of Fuzi (Aconitum carmichaeli) following single and multiple oral administrations of Fuzi extracts in rat by UPLC/MS/MS. *J Ethnopharmacol* 2012, 141:736–741.

22. Chen JH, Lee CY, Liau BC, Lee MR, Jong TT, Chiang ST: Determination of aconitine-type alkaloids as markers in fuzi (Aconitum carmichaeli) by LC/(+) ESI/MS3. *J Pharm Biomed Anal* 2009, 48:1105–1111.

23. Liu H, Su J, Yang X, He YJ, Li HY, Ye J, Zhang WD: A novel approach to characterize chemical consistency of traditional Chinese medicine Fuzi Lizhong pills by GC-MS and RRLC-Q-TOFMS. *Chin Nat Med* 2011, 9:267–273.

24. Wang XJ, Wang HY, Zhang AH, Lu X, Sun H, Dong H, Wang P: Metabolomics study on the toxicity of aconite root and its processed products using ultraperformance liquid-chromatography/electrospray-ionization synapt high-definition mass spectrometry coupled with pattern recognition approach and ingenuity pathways analysis. *J Proteome Res* 2011, 11:1284–1301.

25. Wang ZH, Wen J, Xing JB, He Y: Quantitative determination of diterpenoid alkaloids in four species of Aconitum by HPLC. *J Pharm Biomed Anal* 2006, 40:1031–1034.

26. Yan GL, Sun H, Sun WJ, Zhao L, Meng XC, Wang XJ: Rapid and global detection and characterization of aconitum alkaloids in Yin Chen Si Ni Tang, a traditional Chinese medical formula, by ultra performance liquid chromatography–high resolution mass spectrometry and automated data analysis. *J Pharm Biomed Anal* 2010, 53:421–431.

27. Lu GH, Dong ZQ, Wang Q, Qian GS, Huang WH, Jiang ZH, Leung KS, Zhao ZZ: Toxicity assessment of nine types of decoction pieces from the daughter root of Aconitum carmichaeli (Fuzi) based on the chemical analysis of their diester diterpenoid alkaloids. *Planta Med* 2010, 76:825–830.

28. Wu W, Liang ZT, Zhao ZZ, Cai ZW: Direct analysis of alkaloid profiling in plant tissue by using matrix□assisted laser desorption/ionization mass spectrometry. *J Mass Spectrom* 2007, 42:58–69.

29. Ito K, Ohyama Y, Hishinuma T, Mizugaki M: Determination of Aconitum alkaloids in the tubers of Aconitum japonicum using gas chromatography/selected ion monitoring. *Planta Med* 1996, 62:57–59.

30. Sun H, Ni B, Zhang AH, Wang M, Dong H, Wang XJ: Metabolomics study on Fuzi and its processed products using ultra-performance liquid-chromatography/electrospray-ionization synapt high-definition mass spectrometry coupled with pattern recognition analysis. *Analyst* 2012, 137:170–185.

Chapter 11

QUANTITATIVE STRUCTURE-ANTIOXIDANT ACTIVITY MODELS OF ISOFLAVONOIDS: A THEORETICAL STUDY

Gloria Castellano [1] and Francisco Torrens [2]

[1]Departamento de Ciencias Experimentales y Matemáticas, Facultad de Veterinaria y Ciencias Experimentales, Universidad Católica de Valencia *San Vicente Mártir*, Guillem de Castro-94, E-46001 València, Spain

[2]Institut Universitari de Ciència Molecular, Universitat de València, Edifici d'Instituts de Paterna, E-46071 València, Spain

ABSTRACT

Seventeen isoflavonoids from isoflavone, isoflavanone and isoflavan classes are selected from *Dalbergia parviflora*. The ChEMBL database is representative from these molecules, most of which result highly drug-like. Binary rules appear risky for the selection of compounds with high antioxidant capacity in complementary xanthine/xanthine oxidase, ORAC, and DPPH model assays. Isoflavonoid structure-activity analysis shows the most important properties (log P, log D, pK_a, QED, PSA, NH + OH $\approx$ HBD, N + O $\approx$ HBA). Some descriptors (PSA, HBD) are detected as more important than others (size measure Mw, HBA). Linear and nonlinear models of antioxidant potency are obtained. Weak nonlinear relationships appear between log P, *etc.* and antioxidant activity. The different capacity trends for the three complementary assays are explained. Isoflavonoids potency depends on the chemical form that determines their solubility. Results from isoflavonoids analysis will be useful for activity prediction of new sets of flavones and to design drugs with antioxidant capacity, which will prove beneficial for health with implications for antiageing therapy.

INTRODUCTION

Flavonoids and isoflavonoids influence intercellular redox status to interact with specific proteins in intracellular signaling pathways and present antioxidant properties [1]. Antioxidants are chemical entities that function breaking free-radical chain reaction and metal ion chelation, which would catalyze free-radical-induced systemic damage. The molecules are polyphenolic and electron-rich, potentially acting as substrate inhibitors for the cytochrome P450 (CYP) enzymes and inducing detoxification enzymes, e.g., CYP-dependent monooxygenases (MOs) [2]. Some polyphenols penetrate the blood-brain barrier (BBB) into regions mediating cognitive behavior [3]. Because of flavonoids structural diversity, quantitative structure-activity relationships (SARs) (QSARs) were studied via antioxidant capacity assays [4]. Flavonoids potency depends on their chemical structure, which is influenced by the number and position of hydroxyl groups (OH) attached to both aromatic rings [5]. Isoflavonoids QSARs are scarce [6,7,8,9]. Isoflavonoids antioxidant activity depends on the redox properties of their hydroxyphenolic groups and structural relationship among the different moieties of the chemical structure, which allows many substitution patterns and variations on ring C (Table 1). Promden et al. evaluated antioxidant activities of 24 isoflavonoids from Dalbergia parviflora via three complementary in vitro antioxidant-based assay systems [10]: xanthine/xanthine oxidase (X/XO) [11], oxygen radical absorbance capacity (ORAC) [12] and 2,2-diphenyl-1-picrylhydrazyl (DPPH) [13]. The isoflavonoids consist of three subgroups. The isoflavones exhibited the highest antioxidant potency based on all three assays. The additional presence of an OH in ring B at either R3′ or R5′ from the basic structure of R7-OH in ring A, and R4′-OH or -OMe of ring B increased the antioxidant activities of all isoflavonoid subgroups.

Modeling via QSAR became important in the drug candidate (new chemical entity, NCE) design, environmental fate modeling, toxicity and property prediction of chemicals, since they offer an economical and time-effective alternative to the medium-throughput in vitro and low-throughput in vivo assays [14,15]. A QSAR model is a simple mathematical equation, which is evaluated from a set of molecules with known activities, properties and toxicities via computational approaches. Hypothesis of QSAR supports the replacement, refinement and reduction (3Rs) in animals in the research paradigm as an alternative for untested NCEs [16]. Tropsha and co-workers reviewed QSAR [17]. A QSAR model is limited to query chemicals structurally similar to the training compounds in the applicability domain (AD). Robust validation of QSAR relationships is key for a predictive model, which may be considered for forecasting molecules via interpolation (true prediction) inside

AD or extrapolation (less reliable guess) outside AD. A test molecule that is similar to those in the training set is predicted by QSAR model developed on the corresponding training set. On the contrary, a molecule quite dissimilar to the training ones will never be predicted with the same efficacy, since it is impossible for a single QSAR model to capture the property of an entire universe of chemicals. Relationships of QSAR present applications in drug discovery, environmental fate modeling, risk assessment and chemicals property prediction. The addition of descriptors to a model leads to a rise in the correlation coefficient but this does not always indicate an improvement in predictability. Models of QSAR were used for developing drugs. An objective of QSAR modeling is to predict absorption, distribution, metabolism, excretion (ADME), activity, property and toxicity (ADMET) of NCEs falling within developed-models AD. Chemical qualification (QSAR) programs depend on quantification of physicochemical and physiochemical properties, which facilitate selectivity towards antioxidant capacity.

In earlier publications, quantitative structure-property relationships (QSPRs) allowed prediction of chromatographic retention times of phenylurea herbicides [18] and pesticides [19]. This study aimed to investigate isoflavonoids QSARs via X/XO (pH 9.4), ORAC (blood-serum physiological pH 7.4) and DPPH (methanol, MeOH) assays via different solvents: inhibitions of water-soluble superoxide radical $O_2^{\cdot-}$ formation and peroxyl radical $HO_2^{\cdot}$-induced oxidation, and water-insoluble DPPH, respectively. Antioxidant capacities were derived from Promden et al. [10]. The improvements with regard to this qualitative work have been illustrated and discussed. In our QSARs, the different activity trends for the three complementary assays are explained.

RESULTS AND DISCUSSION

The molecular structures of 17 isoflavonoids, viz. eight isoflavones, six isoflavanones and three isoflavans, from the heartwood (duramen) of D. parviflora are displayed in Table 1. However, the obtained results are limited to the 17 substances contained in the ChEMBL database.

Isoflavonoids antioxidant activities in ORAC, X/XO and DPPH model assays were derived from Promden et al. [10]. However, no QSAR analysis was provided. For inactive Entries 12–14, 14 and 3–7–8–12–13–14 in Table 2, ORAC Trolox™ (a water-soluble vitamin-E analogue) equivalent antioxidant capacity (TEAC) was taken as minimum (minimum log ORAC), X/XO and DPPH concentration for 50% radical-trapping (scavenging, SC_{50}) were taken as maximum. Notice the opposite trends of ORAC and X/XO-DPPH results.

Table 1. Molecular structure of isoflavonoids from *Dalbergia parviflora*

Molecular Structure	Entry	Isoflavones	R_5	R_7	R_2'	R_3'	R_4'	R_5'
	1	Khrinone C	OH	OH	OMe	OH	OMe	H
	2	Calycosin	H	OH	H	OH	OMe	H
	3	Genistein	OH	OH	H	H	OH	H
	4	3'-*O*-Methylo-robol	OH	OH	H	OMe	OH	H
	5	Cajanin	OH	OMe	OH	H	OH	H
	6	Khrinone B	OH	OH	OH	H	OMe	OH
	7	Biochanin A	OH	OH	H	H	OMe	H
	8	Formononetin	H	OH	H	H	OMe	H
Molecular Structure	**Entry**	**Isoflavanones**	R_5	R_7	R_2'	R_3'	R_4'	R_5'
	9	3(*R,S*)-Violanone	H	OH	OMe	OH	OMe	H
	10	3(*S*)-Secundiflo-rol H	OH	OH	OMe	OH	OMe	H
	11	3(*R,S*)-Dalparvin	H	OH	OMe	H	OMe	OH
	12	3(*R,S*)-Onogenin	H	OH	OMe	H	OCH$_2$O	
	13	3(*S*)-Sativanone	H	OH	OMe	H	OMe	H
	14	3(*R,S*)-3'-*O*-Methylviolanone	H	OH	OMe	OMe	OMe	H
Molecular Structure	**Entry**	**Isoflavans**	R_7	R_8	R_2'	R_3'	R_4'	R_5'
	15	3(*R*)-Vestitol	OH	H	OH	H	OMe	H
	16	3(*R*)(+)-Mucro-nulatol	OH	H	OMe	OH	OMe	H
	17	3(*S*)-8-Demeth-ylduartin	OH	OH	OMe	OH	OMe	H

Table 2. Antioxidant activity (X/XO, ORAC, DPPH assays) of isoflavonoids from *D. parviflora* and ChEMBL physico/physiochemical descriptors

Entry	X/XO Assay SC$_{50}$ [μM] [a]	ORAC Assay TE [μM] [b]	DPPH Assay SC$_{50}$ [μM][a]	Log X/XO	Log ORAC	Mw [Da][c]	ALog P[d]	ACD Log P[e]	ACD LogD[f]	ACD pK$_a$[g]	RBN[h]	QEDw[i]	PSA [Å^2][j]	NH + OH	HBD[k]	N + O	HBA[l]
1	0.64	43.5	61.7	−0.194	1.638	330	2.11	2.25	0.82	6.32	3	0.79	105	3	3	7	7
2	0.25	37.8	96.2	−0.602	1.577	284	2.37	1.33	0.75	6.95	2	0.89	76	2	2	5	5
3	9.0	37.8	300	0.954	1.577	270	2.14	3.11	1.93	6.51	1	0.74	87	3	3	5	5
4	36.7	35.7	81.2	1.565	1.553	300	2.12	2.63	1.25	6.35	2	0.79	96.2	3	3	6	6
5	54.3	34.7	70.8	1.735	1.540	369	3.52	3.88	3.86	8.93	8	0.54	96.2	3	3	6	6
6	0.60	34.2	133.6	−0.222	1.534	316	1.88	1.71	0.37	6.38	2	0.63	116	4	4	7	7
7	203.3	26.6	300	2.308	1.425	284	2.37	3.34	2.11	6.5	2	0.89	76	2	2	5	5
8	116.92	2.8	300	2.068	0.447	268	2.61	6.99	2.86	2.31	2	0.91	55.8	1	1	4	4

9	43.7	31.1	89.7	1.640	1.493	286	2.48	7.69	2.63	2.44	2	0.89	76	2	2	6	5
10	247.2	27.4	74.3	2.393	1.438	302	2.24	2.76	2.34	7.5	2	0.79	96.2	3	3	7	6
11	48.2	21.8	80.4	1.683	1.338	332	2.22	7.48	3.01	2.58	3	0.79	105	2	3	6	7
12	56.9	0.0	300	1.755	0.0	330	2.25	4.52	4.1	7.48	2	0.87	94.4	1	2	6	7
13	59.3	0.0	300	1.773	0.0	270	2.72	3.48	3.31	7.7	2	0.91	55.8	1	1	5	4
14	300	0.0	300	2.477	0.0	330	2.69	2.93	2.74	7.67	7	0.93	74.2	1	1	6	6
15	6.4	40.1	204.1	0.806	1.603	272	3.2	3.26	3.25	9.53	2	0.88	58.9	2	2	4	4
16	10.0	39.8	75.41	1.000	1.600	302	3.18	2.84	2.84	9.87	3	0.91	68.2	2	2	5	5
17	13.4	27.0	115.4	1.127	1.431	318	2.94	1.65	1.65	9.75	3	0.75	88.4	3	3	6	6

[a] SC_{50}: concentration providing 50% inhibition; [b] Expressed as Trolox equivalents (TE, μM Trolox)/10 μM isoflavonoid; [c] Mw: molecular weight; [d] ALog P: decimal logarithm of the 1-octanol-water partition coefficient (log P) calculated by the method ALog P; [e] ACD Log P: log P calculated by ACD/ Log P; [f] ACD Log D: decimal logarithm of the 1-octanol-water distribution coefficient (log D) calculated by ACD/Log D at pH 7.4; [g] ACD Acidic pK_a: pK_a calculated by ACD/pK_a; [h] RBN: rotatable bonds; [i] QEDw: weighted quantitative estimate of drug-likeness; [j] PSA: topological polar surface area; [k] HBD: hydrogen-bond donor; [l] HBA: hydrogen-bond acceptor.

Isoflavonoids (IfOH) scavenge free radicals R$^•$ according to three possible reducing pathways.

(i) H-atom transfer (HAT) from the molecule to the radical (direct O–H bond breaking):

$$\text{IfOH (antioxidant)} + R^• \text{(free radical)} \rightarrow \text{IfO}^• + RH \tag{1}$$

High HAT rate is expected for a low O–H bond dissociation enthalpy (BDE).

(ii) Electron transfer (ET) from molecule to radical, leading to indirect H-abstraction or proton transfer (PT) (ET-PT):

$$\text{IfOH (antioxidant)} + R^• \text{(free radical)} \rightarrow \text{IfOH}^{•+} + R^- \rightarrow \text{IfO}^• + RH \tag{2}$$

(iii) Sequential proton-loss-electron-transfer (SPLET). Since antioxidants primarily function by HAT, which involves formation of an H-bond with the harmful free radicals [20], a rise in the count of OH substituents facilitates interaction with the toxic radicals (Fujita-Ban analysis) [21].

Correlations between the Different Methods, and Physicochemical and Physiochemical Properties

Physicochemical and physiochemical properties of isoflavonoids were calculated (NH + OH, N + O) or taken from ChEMBL database: *steric* (molecular weight, Mw), lipophilic (log P/D, topological polar surface area,

PSA), acid (pK_a), flexibility (rotatable bond, RBN), drug-likeness (weighted quantitative estimate of drug-likeness, QEDw, QED) and H-bond donor/ acceptor (HBD/A) [22]. All Mw < 400 Da were in agreement with the rule of five (RO5). Cajanin (Entry 5 in Table 2)Mw = 369 Da and its log P/PSA could be decreased. All ACD log P < 5 according to RO5 with the exception of Entries 8, 9 and 11. However, these results should be taken with care because atom type summation log P (Alog P) < 3 and log D < 4. All log D = 0–3 predicting high oral bioavailability (OB) except Entries 5, 11–13 and 15. All pK_a = 2–10 and isoflavonoids are weak acids in water, most resulting anionic while they are neutral without separation of charges in organic solvents (MeOH). Entries 8, 9 and 11 present maximum ACD log P ~7 and minimum pK_a ~2. All RBN ≤ 8 and N + O ≤ 7 forecasting OB. All QED > 0.7 (highly drug-like, HD) except Entries 5 and 6: QED = 0.5–0.7 (drug-like, D). It decays with Mw, *etc.* Entries 2, 3, 7, 8, 13, 15 and 16 with N + O = 4–5 present a chance of entering BBB. Entries 8, 13, 15 and 16 show PSA < 70 Å^2, and are foreseen with OB and to penetrate BBB in agreement with N + O = 4–5. Entries 15 and 16 with Alog P > 3 when PSA < 70 Å^{2}carry toxicity risk. Entries 1–7, 9–12, 14 and 17 show 70 < PSA < 120 Å^2 and are envisaged with high/middle OB. All NH + OH ≈ HBD ≤ 4 and N + O ≈ HBA ≤ 7 following RO5. The PSA trends are similar to HBA.

Xanthine/Xanthine Oxidase Assay

Most isoflavonoids exhibited high antioxidant activity in X/XO assay. The role of ring C is confirmed in the presence of the 2,3-double bond. Fragment =O environment primarily dictates its contribution to the antioxidant capacity profile of isoflavonoids. The class of planar isoflavones showed the highest potency. The activity of the different divisions were confirmed comparing the capacity of compounds with the same substitution pattern: planar, ring-C-unsaturated isoflavone khrinone C was detected much more potent than nonplanar, ring-C-saturated isoflavan 3(S)-8-demethylduartin and isoflavanone 3(S)-secundiflorol H (Entries 1, 17 and 10, respectively). The X/XO correlated with PSA and HBD properties. Conversion of X/XO to its logarithm got a better relationship with log D and pK_a descriptors. The best linear fit turns out to be:

−Log X/XO = − (0.494±0.647) − (0.570±0.167) ACD Log D + (0.0769 ± 0.0765) ACD pKan = 17, r = 0.683, s = 0.725, F = 6.1, MAPE = 38.48%, AEV = 0.5340, q = 0.553   (3)

where *n* is the number of points, *s* standard deviation, *F* Fischer ratio, MAPE mean absolute percentage error, AEV approximation error variance and *q*, leave-1-out cross-validated (CV) correlation coefficient. The pK_a correlates positively, while log D associates negatively, with $-\log$ X/XO. The positive coefficient for pK_a implies that activity rises for weaker-acids isoflavonoids in agreement with the fact that the assay prefers isoflavans ($pK_a \approx 10$) to isoflavanones ($pK_a \sim 6$). The negative coefficient for log D signifies that capacity rises for isoflavonoids more stable in the aqueous than in the organic phase. If a quadratic term is included in the fit, the model is improved:

$-$Log X/XO=(2.97±1.31)$-$(2.44±0.63)ACD Log D+(0.410±0.136)ACD (Log D)2$-$(0.229±0.156)HBAn = 17, r = 0.833, s = 0.570, F = 9.8, MAPE = 30.42%, AEV = 0.3104, q = 0.726   (4)

and AEV decays by 42%. Log D correlates negatively with $-\log$ X/XO in agreement with Equation (3). However, (log D)^{2}correlates positively with $-\log$ X/XO in a model passing via a minimum, in agreement with log P parabolic models of *in vitro*penetration of xenobiotics across artificial lipoidal/biomembranes [23]. Its small absolute coefficient indicates a weak nonlinear relationship. Linear Equation (3) has only two variables and is better appropriated for extrapolation than nonlinear Equation (4).

Oxygen Radical Absorbance Capacity Assay

Most isoflavones showed high antioxidant activity in ORAC assay, which correlated with PSA and HBD properties. The conversion of ORAC to its logarithm got better relationship with the same descriptors. The best linear fit results:

Log ORAC=(1.37±0.49)$-$(0.0335±0.0101)PSA+(1.12±0.21)HBDn = 17, r = 0.842, s = 0.361, F = 17.1, MAPE = 22.58%, AEV = 0.2903, q = 0.758

$$(5)$$

The HBD $\approx$ NH + OH correlates positively with log ORAC in agreement with Fujita-Ban analysis. However, PSA associates negatively with log ORAC. Adding two quadratic terms, fit is improved:

Log   ORAC=$-$(0.966±0.455)+(0.00435±0.00268)ACD   Log   P$^2-$(0.0975±0.0536)ACD Log D$-$(0.122±0.060)N + O +(2.40±0.25)NH + OH $-$(0.411±0.057) (NH + OH)^{2}n = 17, r = 0.972, s = 0.177, F = 38.1, MAPE = 9.37%, AEV = 0.0600, q = 0.870   (6)

and AEV decays by 79%. The NH + OH $\approx$ HBD correlates positively with log ORAC in agreement with Fujita-Ban analysis and Equation (5). However, log D and N + O associate negatively with log ORAC. Quadratic ACD log P^2 correlates positively with log ORAC in a parabola with a minimum, while (NH + OH)^2 associates negatively in a parabola with a maximum. Linear Equation (5), with only two variables, results better suited for extrapolation than nonlinear Equation (6).

2,2-Diphenyl-1-picrylhydrazyl Assay

Most isoflavones displayed high antioxidant activity in DPPH assay, which correlated with properties Alog P, QED, N + O and HBD. Best linear fit is:

$$-DPPH=-(1390\pm513)+(134\pm54)ALog\ \ P+(458\pm347)QED+(40.2\pm27.2)N+O$$
$$+(120\pm47)HBD n = 17, r = 0.790, s = 73.881, F = 5.0, MAPE = 32.53\%, AEV$$
$$= 0.3938, q = 0.400 \hspace{2cm} (7)$$

All descriptors correlate positively with $-$DPPH, and HBD $\approx$ NH + OH is in agreement with Fujita-Ban analysis and Equations (5) and (6). A positive coefficient for log P implies that antioxidant activity in the assay rises for isoflavonoids more soluble in the organic than in the aqueous phase. As DPPH assay is in MeOH (not water), the corresponding interpretation is that water, compared to MeOH, presents the capacity of forming a number of H-bonds (nets), while MeOH affinity for creating H-bonds is smaller because of the *steric* interference of the CH_3 group and inability to receive-give more H atoms. This is in concordance with the positive sign of log P and N + O terms. If quadratic pK_a^2 is included in the fit, the correlation is improved:

$$-DPPH=-(2090\pm501)+(235\pm68)ALog\ P-(1.58\pm0.83)ACD\ pKa2+(800\pm317)$$
$$QED\ \ +(70.1\pm23.6)N+O+(151\pm44)HBD n = 17, r = 0.865, s = 63.201, F = 6.5,$$
$$MAPE = 22.79\%, AEV = 0.2693, q = 0.655 \hspace{1cm} (8)$$

and AEV decays by 32%. All linear descriptors correlate positively with DPPH in agreement with Equation (7), and HBD $\approx$ NH + OH is in concordance with Fujita-Ban analysis. Quadratic pK_a^2 associates negatively with DPPH in a parabolic model with a maximum. Linear Equation (7) with only four variables is better appropriated for extrapolation than nonlinear Equation (8). The use of log DPPH as dependent variable does not improve the models.

Comparison between the Three Methods

The log X/XO can be estimated from log ORAC:

$$Log\ X/XO=2.18-0.729Log\ ORAC \hspace{2cm} (9)$$

The log X/XO can be approximated from DPPH:

$$Log\ X/XO=0.721+0.00347DPPH \hspace{2cm} (10)$$

The DPPH can be calculated from log ORAC:

$$DPPH=(308\pm41)-(117\pm31)Log\ ORAC n = 17, r = 0.702, s = 76.797, F = 14.6,$$
$$q = 0.638 \hspace{2cm} (11)$$

in agreement with the opposite trends of X/XO-DPPH and ORAC. The correlation is poor (Equation (11)). However, when a correction is made for the fact that ORAC assay is in water while DPPH assay is in MeOH, by adding a term in log N + O, a better fit is obtained:

DPPH = (725±164) − (107±26) Log ORAC − (574±221) Log N + On = 17, r = 0.811, s = 65.310, F = 13.4, q = 0.748 $\qquad$ (12)

where the term in log N + O ≈ log HBA corrects for the fact that in the ORAC assay, water presents greater ability to H-bond transfer than MeOH in the DPPH test.

The physicochemical and physiochemical properties used in Table 2 are simple to calculate, and their use gained widespread acceptance but the bulk physical properties of molecules are correlated [24]. One issue in using these properties is the potential redundancy, which is illustrated simply among isoflavonoids, where all four RO5 parameters are clearly linked:

ALog $\qquad$ P=−(0.418±0.540)+(0.0219±0.0027)MW−(0.697±0.086)HBA +(0.0745±0.0704)HBDn = 17, r = 0.934, s = 0.181, F = 29.5, q = 0.824 (13)

in agreement with the data for oral drugs taken from literature (N + O ≈ HBA, NH + OH ≈ HBD) [25]:

CLog P=0.19+0.018MW−0.64N+O−0.40NH+OHn = 1193, r = 0.79 (14)

The standard errors of the coefficients show that all ones in Equations (3)–(13) are acceptable.

Leave-m-out (1 ≤ m ≤ 14) CV correlation coefficient r_{cv} calculated for isoflavonoids ($q = r_{cv}$ ($m = 1$), cf. Table 3) show that r_{cv} decays with m except −DPPH (Equation (7)) and Alog P (Equation (13)), which indicate possible outliers. In particular, both antioxidant activity models log ORAC vs. {PSA, HBD} (Equation (5)) and vs. {(ACD log P)², log D, N + O, NH + OH, (NH + OH)²} (Equation (6)) give the greatest r_{cv}. The interpretation is that these are the most predictive descriptors sets for modeling isoflavoniods antioxidant activity. However, models −log X/XO vs. {log D, (log D)², HBA} (Equation (4)) and −DDPH vs. {Alog P, (pK_a)², QED, N + O, HBD} (Equation (8)) give smaller r_{cv}. Equation (6) is more predictive than Equations (4) and (8).

Drug design, discovery and development are complex and difficult because drug action is much more than binding affinity. A successful, efficacious and safe drug must present a balance of properties, e.g., activity against its intended target, appropriate ADME and acceptable safety profile. Based on the obtained results, new definitions of (stringent) drug-likeness, tractability and central nervous system (CNS)-active are proposed. Drug-likeness evaluates the suitability of the molecule under RO5, etc. The CNS-active is stricter. However, tractability is under more relaxed conditions. A summary of physicochemical and physiochemical descriptors was selected for every property (*cf.* Table 4). Properties Mw, Clog P(estimated as Alog P), RBN, QEDw, N + O, NH + OH, HBD, HBA, and no metal, sugar and carbohydrates fulfill drug-likeness for all 17 isoflavonoids. The only exception is khrinone B, which presents a Clog $P - (N + O) \approx Alog\, P - (N + O) = -5.12 \le -5$ and PSA = 116 > 105 Å² but it fulfills tractability: Alog $P - (N + O) > -8$ and PSA ≤ 140Å².

Table 3. Cross-validation correlation coefficient in a leave-*m*-out procedure for iso-flavonoids

m	−Log X/ XO Equation (3)	−Log X/XO Equation (4)	Log ORAC Equation (5)	Log ORAC Equation (6)	−DPPH Equation (7)	−DPPH Equation (8)	DPPH Equation (11)	DPPH Equation (12)	ALog PEquation (13)
1	0.553	0.726	0.758	0.870	0.400	0.655	0.638	0.748	0.824
2	0.552	0.725	0.757	0.870	0.405	0.653	0.638	0.748	0.828
3	0.550	0.724	0.756	0.869	0.409	0.651	0.638	0.747	0.832
4	0.549	0.722	0.755	0.867	0.415	0.648	0.638	0.746	0.836
5	0.546	0.720	0.753	-	0.422	0.645	0.637	0.745	0.839
6	0.544	0.717	0.751	-	0.431	0.641	0.637	0.744	-
7	0.540	0.713	0.749	-	0.442	0.636	0.635	0.743	-
8	0.537	0.707	0.746	-	-	-	0.633	0.742	-
9	0.532	0.698	-	-	-	-	0.629	0.741	-
10	0.527	0.682	-	-	-	-	0.624	0.740	-
11	0.521	0.650	-	-	-	-	0.615	0.741	-
12	0.516	-	-	-	-	-	0.600	-	-
13	0.509	-	-	-	-	-	0.565	-	-
14	0.436	-	-	-	-	-	-	-	-

Table 4. Summary of physicochemical and physiochemical descriptors selected for every property

Property	Mw [Da][a]	CLog P[b]	RBN[c]	QEDw[d]	N + O	CLog P − (N + O)	NH + OH	PSA [Å²][e]	HBD[f]	HBA[g]	Others
Tractability	200–800 [i]	≤8 [i]	≤16	>0.2	≤16	>−8	≤8	≤140	≤8	≤15	No metal, sugar, carbohydrates
Drug-likeness	100–500 [i]	≤5 [i]	≤10	>0.5	≤10	>−5	≤5	≤105	≤5	≤10	No metal, sugar, carbohydrates
Stringent drug-likeness	100–450 [i]	≤4 [i]	≤10	>0.5	≤8–9	>−4.5	≤3	≤105	≤3	≤8–9	No metal, sugar, carbohydrates

CNS-active [h]	100–400	≤3.5 [i,j]	≤7	>0.7	≤5	>0	≤4	≤70 [j]	≤4	≤5	No metal, sugar, carbo-hydrates

[a] Mw: molecular weight; [b] CLog P: decimal logarithm of 1-octanol-water partition coefficient (log P) calculated by CLog P; [c] RBN: rotatable bonds; [d] QEDw: weighted quantitative estimate of drug-likeness; [e] PSA: topological polar surface area; [f] HBD: hydrogen-bond donor; [g] HBA: hydrogen-bond acceptor; [h]Central nervous system (CNS)-active: penetrating the blood-brain barrier (BBB); [i] Mw > 400 Da when CLog P > 4 carries toxicity risk; [j] CLog P > 3 when PSA < 75 Å^2 carries toxicity and promiscuity risks.

Discussion

This study is in agreement with Promden *et al.* [10], providing an extension and further discussion. It would be expected that the results of the present work had not change if the larger set of 24 compounds were considered. However, the obtained results are limited to the 17 substances contained in the ChEMBL database. The novelty finding in comparison to Promden *et al.* [10] is described in the following paragraphs, essentially: in the present study, a comparative analysis of the three assays in different solvents and pHs is illustrated and analyzed. The main difference is that the work of Promden *et al.* [10] is qualitative SAR while this study is QSAR. A possibility exists of integrating parameters sets but the structural data of Promden *et al.* would be only indicators of functional-groups absence/presence. The predictability of the approach would be qualitative but not quantitatively improved.

There are two main types of empirical QSAR models: linear models and nonlinear ones. The linear models provide an appropriate representation of the activity in a small neighborhood of a set of molecular properties. However, when the molecules are tried outside this constrained region, the model predictions will not be accurate. On the other hand, the quadratic models tend to capture more precisely the capacity behavior, making the adequate for predicting a real potency in a wide region of properties. Weak nonlinear relationships were detected between some physicochemical and physiochemical properties, especially log P, and isoflavonoids antioxidant activity in X/XO, ORAC and DPPH assays. Key strengths of the obtained descriptors follow:

- easy to understand and apply;
- compounds with *non-drug-like* properties lie in the regions of property space with poor precedence; and

- good guide to avoid potential pitfalls.

Considering the structure of isoflavonoids, some parameters {log P, log D, PSA, HBD} are used. A simple linear correlation is proved to be a good model for the antioxidant activity of the molecules; other properties are redundant information. Procedure CV leave-m-out shows that {PSA, HBD} and {(ACD log P)^2, ACD log D, N + O, NH + OH, (NH + OH)^2} are the most predictive sets of descriptors for linear and nonlinear modeling isoflavonoids antioxidant capacity, respectively, according to the criterion of maximization of CV correlation coefficient. Both sets contain the essential characters of the antioxidant potency for isoflavonoid structures. The proposed method allows rapid estimation of the antioxidant activity for these molecules. The linear methods require that fewer parameters be estimated and, therefore, may be more parsimonious (Occam's razor). Linear and nonlinear correlation models were obtained for isoflavonoids antioxidant capacity, pointing, not only to a homogeneous molecular structure of these molecules, but also to the ability to predict and tailor drug properties. The latter is nontrivial in pharmacology.

EXPERIMENTAL SECTION

The 1-octanol-water *partition coefficient P* is the ratio of concentrations of compound S:

$$P = \frac{[S]_{\text{1-octanol}}}{[S]_{\text{water}}} \qquad (15)$$

Its decimal logarithm log P measures lipophilicity. The ALog P is calculated from a regression based on the hydrophobicity contribution of 115 atom {H, B-F, Si-Cl, Se-Br, I} kinds [26]. Every atom in every structure is classified into one of 115 sorts. Log P results:

$$A \log P = \sum_i n_i a_i \qquad (16)$$

where n_i is the number of the atoms of type i and a_i is hydrophobicity constant. Codes ACD/Log P and calculated log P(CLog P) [27] predict it from structure.

Distribution coefficient D is the ratio of sum of the concentrations of all forms of compound (unionized/ionized) in each phase; e.g., for a weak acid HA:

$$D = \frac{[HA]_{\text{1-octanol}}}{[A^-]_{\text{water}} + [HA]_{\text{water}}} \qquad (17)$$

As log$D$ is pH dependent, aqueous phase pH is buffered, e.g., blood-serum physiological pH 7.4 in ORAC assay. For unionizable compounds, log P = log D. 0 < log D < 3 enhances OB [28]. Code ACD/Log D predicts it understanding ionizable-molecules lipophilicity from structure. Programs ACD/Log $P - D$ are modules of ACD/Percepta (ACD/Labs).

An *acid dissociation constant* K_a measures the strength of an acid in solution. It is the equilibrium constant for acid-base dissociation reaction. The larger K_a, the more there is dissociation of the molecules in solution. Acids and neutrals present decreased toxicity risks related to bases [29]. Code ACD/pK_a predicts dissociation constants from structure.

An RBN is any single non-ring bond, bounded to nonterminal *heavy* (non-H) atom. Amide C–N bonds are not considered because of their rotational energy barrier. The count of RBNs measures the molecular flexibility.

An H atom attached to a relatively electronegative (EN) atom is an HBD [30]. The EN atom usually ranges from N to F atoms. The count NH + OH ≈ HBD. An EN atom, e.g., N to F atoms, is an HBA, whether it is bonded to an H atom or not (e.g., HBD ethanol presents an H atom bonded to an O atom, HBA O atom in diethyl ether does not show an H atom bonded to it). The count N + O ≈ HBA. The solvatochromic parameters are: dipolarity-polarizability π^*, HBD acidity α and HBA basicity β [31].

The PSA of an organic is calculated by Ertl *et al.* method as a sum of fragment contributions [32]. The N/O-centered polar fragments are considered [33]. The PSAs are similar to HBA trends. The PSA describes drug absorption (e.g., OB, human carcinoma of colon cell line type-2 (Caco-2) permeability, BBB penetration). In order to enter BBB, most CNS drugs show PSA $\leq$ 70 Å^2 but PSA $\leq$ 75 Å^2 when Clog P > 3 carries toxicity and promiscuity risks [34]. When Mw > 400 Da, Clog P > 4 presents some toxicity risk [35].

The QED combines eight characteristics: Mw, ALog P, HBD/A, PSA, RBN, number of aromatic rings (AROM), and count of alerts for undesirable substructures (ALERT) [36]. It avoids the pitfalls of hard cut-offs, providing a single metric for *similarity* of a compound to known oral drugs [37]. Based on QED, molecules can be classified: nondrug-like (ND), poorly drug-like (PD), D and HD for QED in 0.0–0.2, 0.2–0.5, 0.5–0.7 and 0.7–1.0, respectively. The RO5 predicts OB when HBD $\leq$ 5, HBA $\leq$ 10, Mw $\leq$ 500 Da and log $P \leq$ 5 [38]. Most OB compounds present RBN $\leq$ 10 and PSA $\leq$ 140 Å^2 [39]. Drugs with OB show N + O $\leq$ 10. Rules predict CNS activity: (1) if N + O $\leq$ 5, the molecule presents a high chance of entering BBB; (2) if log $P-$ (N + O) > 0, the compound is CNS-active [40]. The Mw, log P and PSA decline with Mw > 340 Da [41].

The correlation coefficient between CV representatives and the property values r_{cv} has been calculated with the leave-m-out procedure [42]. The process furnishes a new method for selecting the best set of descriptors: leave-m-out selects the best set of descriptors according to the criterion of maximization of the value of r_{cv}.

The statistics r, s and F were calculated with Microsoft Excel (Microsoft Office 2015); MAPE and AEV were computed with Knowledge Miner Insights for Excel; CV correlation coefficients (q, etc.) were evaluated with leave-m-out [42].

CONCLUSIONS

From the present results and discussion, the following conclusions can be drawn.

- Seventeen isoflavonoids from *Dalbergia* were selected from ChEMBL database representing *medicinal chemistry*compounds. Most are detected highly drug-like. Binary rules for compounds selection result risky: filters neglect valuable opportunities. Structure-antioxidant activity analyses indicate most important properties: log D-pK_a, PSA-HBD and log P-QED-N + O-HBD for X/XO, ORAC and DPPH assays, respectively. Capacity in X/XO prefers weaker-acids isoflavonoids more soluble in water than in 1-octanol, in agreement with X/XO (pII 9.4) favoring neutral isoflavans (p$K_a \approx 10$) rather than anionic isoflavanones (pK_a ~6). However, DPPH chooses isoflavonoids more soluble in 1-octanol with greater N + O count because this test is in methanol with H-bond transfer ability smaller than water. Models of QSAR provide quantitative information that filters drugs based on log D, etc. suggesting strategies for priority. Some descriptors (PSA, HBD) are more important than others (size, HBA). An advantage of our QSARs is that they detect weak nonlinear relationships between log$P$, etc. and potency. Simple, consistent analyses are described, improving our general understanding of activity. The rules are consistent with the literature.

- Isoflavonoid ring-C role was confirmed in the presence of isoflavones 2,3-double bond, explaining their greatest activity. Capacity gave preferences: Planar unsaturated isoflavones greater than non-planar saturated isoflavans and isoflavanones because unsaturation and planarity stabilize the phenoxyl radical. On comparing isoflavanones with isoflavans, this study demonstrates different favorites of X/XO, ORAC and DPPH: X/XO (pH 9.4) prefers neutral isoflavans (p$K_a \approx 10$) liking better phenoxyl-radical stabilization, which is not the case of anionic isoflavanones (pK_a ~6); in DPPH (methanol), an intramolecular H-bond $R_4 = O...HO-R_5$ can be formed in isoflavanones, but not in isoflavans lacking this moiety; and ORAC (pH 7.4) liking is intermediate. Isoflavonoids potency depends on the chemical form determining its solubility, which is modified by changing pH or solvent. Models of QSAR may predict activity of new series of isoflavonoids and design strong drugs.

ACKNOWLEDGMENTS

The authors acknowledge Dr. E. Besalú for providing us his full-linear leave-many-out program before publication. One of us, Francisco Torrens, thanks financial support from the Spanish Ministerio de Economía y Competitividad (Project No. BFU2013-41648-P) and EU ERDF.

AUTHOR CONTRIBUTIONS

Gloria Castellano and Francisco Torrens conceived and designed the experiments, performed the experiments, analyzed the data, contributed materials/analysis tools and wrote the paper.

REFERENCES

1. Williams, R.J.; Spencer, J.P.; Rice-Evans, C. Flavonoids: Antioxidants or signalling molecules? *Free Radic. Biol. Med.* **2014**, *36*, 838–849.

2. Stahl, W.; Ale-Agha, N.; Polidori, M.C. Non-antioxidant properties of carotenoids. *Biol. Chem.* **2002**, *383*, 553–558.

3. Jäger, A.K.; Saaby, L. Flavonoids and the CNS. *Molecules* **2011**, *16*, 1471–1485.

4. Yang, J.G.; Liu, B.G.; Liang, G.Z.; Ning, Z.X. Structure-activity relationship of flavonoids active against lard oil oxidation based on quantum chemical analysis. *Molecules* **2009**, *14*, 46–52.

5. Harsa, A.M.; Harsa, T.E.; Bolboacâ, S.D.; Diudea, M.V. QSAR in flavonoids by similarity cluster prediction. *Curr. Comput. Aided Drug Des.* **2014**, *10*, 115–128.

6. Goto, H.; Terao, Y.; Akai, S. Synthesis of various kinds of isoflavones, isoflavanes, and biphenyl-ketones and their 1,1-diphenyl-2-picrylhydrazyl radical-scavenging activities. *Chem. Pharm. Bull.* **2009**, *57*, 346–360.

7. Mitra, I.; Saha, A.; Roy, K. Chemometric modeling of free radical scavenging activity of flavone derivatives. *Eur. J. Med. Chem.* **2010**, *45*, 5071–5079.

8.   Umehara, K.; Nemoto, K.; Kimijima, K.; Matsushita, A.; Terada, E.; Monthakantirat, O.; De-Eknamkul, W.; Miyase, T.; Warashina, T.; Degawa, M.; *et al*. Estrogenic constituents of the heartwood of *Dalbergia parviflora*. *Phytochemistry* **2008**,*69*, 546–552.

9.   Umehara, K.; Nemoto, K.; Matsushita, A.; Terada, E.; Monthakantirat, O.; de-Eknamkul, W.; Miyase, T.; Warashina, T.; Degawa, M.; Noguchi, H. Flavonoids from the heartwood of the Thai medicinal plant *Dalbergia parviflora* and their effects on estrogenic-responsive human breast cancer cells. *J. Nat. Prod.* **2009**, *72*, 2163–2168.

10.   Promden, W.; Monthakantirat, O.; Umehara, K.; Noguchi, H.; de-Eknamkul, W. Structure and antioxidant activity relationships of isoflavonoids from *Dalbergia parviflora*. *Molecules* **2014**, *19*, 2226–2237.

11.   McCord, J.M.; Fridovich, I. The reduction of cytochrome c by milk xanthine oxidase. *J. Biol. Chem.* **1968**, *243*, 5753–5760.

12.   Prior, R.L.; Hoang, H.; Gu, L.; Wu, X.; Bacchiocca, M.; Howard, L.; Hampsch-Woodill, M.; Huang, D.; Ou, B.; Jacob, R. Assays for hydrophilic and lipophilic antioxidant capacity (oxygen radical absorbance capacity (ORAC(FL))) of plasma and other biological and food samples. *J. Agric. Food Chem.* **2003**, *51*, 3273–3329.

13.   Blois, M.S. Antioxidant determinations by the use of a stable free radical. *Nature* **1958**, *181*, 1199–1200.

14.   Walker, J.D.; Jaworska, J.; Comber, J.H.I.; Schultz, T.W.; Dearden, J.C. Guidelines for developing and using quantitative structure-activity relationships. *Environ. Toxicol. Chem.* **2003**, *22*, 1653–1665.

15.   Perkins, R.; Fang, H.; Tong, W.; Welsh, W.J. Quantitative structure-activity relationship methods: Perspectives on drug discovery and toxicology. *Environ. Toxicol. Chem.* **2003**, *22*, 1666–1679.

16.   Benigni, R.; Giuliani, A. Putting the predictive toxicology challenge into perspective: Reflections on the results.*Bioinformatics* **2013**, *19*, 1194–1200.

17.   Cherkasov, A.; Muratov, E.N.; Fourches, D.; Varnek, A.; Baskin, I.I.; Cronin, M.; Dearden, J.; Gramatica, P.; Martin, Y.C.; Todeschini, R.; *et al*. QSAR modeling: Where have you been? Where are you going to? *J. Med. Chem.* **2014**, *57*, 4977–5010.

18.   Torrens, F.; Castellano, G. QSPR prediction of retention times of phenylurea herbicides by biological plastic evolution.*Curr. Drug Saf.* **2012**, *7*, 262–268.

19. Torrens, F.; Castellano, G. QSPR prediction of chromatographic retention times of pesticides: Partition and fractal indices. *J. Environ. Sci. Health Part B* **2014**, *49*, 400–407.

20. Wright, J.S.; Johnson, E.R.; DiLabio, G.A. Predicting the activity of phenolic antioxidants: Theoretical methods, analysis of substituent effects, and application to major families of antioxidants. *J. Am. Chem. Soc.* **2001**, *123*, 1173–1183.

21. Fujita, T.; Ban, T. Structure–activity study of phenethylamines as substrates of biosynthetic enzymes of sympathetic transmitters. *J. Med. Chem.* **1971**, *14*, 148–152.

22. Gaulton, A.; Bellis, L.J.; Bento, A.P.; Chambers, J.; Davies, M.; Hersey, A.; Light, Y.; McGlinchey, S.; Michalovich, D.; Al-Lazikani, B.; *et al.* ChEMBL: A large-scale bioactivity database for drug discovery. *Nucleic Acids Res.* **2012**, *40*, D1100–D1107.

23. López, A.; Faus, V.; Díez-Sales, O.; Herráez, M. Skin permeation model of phenyl alcohols: Comparison of experimental conditions. *Int. J. Pharm.* **1998**, *173*, 183–191.

24. Cramer, R.D., III. BC(DEF) parameters. 1. The intrinsic dimensionality of intermolecular interactions in the liquid state. *J. Am. Chem. Soc.* **1980**, *102*, 1837–1849.

25. Leeson, P.D.; Davis, A.M. Time-related differences in the physical property profiles of oral drugs. *J. Med. Chem.* **2004**,*47*, 6338–6348.

26. Ghose, A.K.; Viswanadhan, V.N.; Wendoloski, J.J. Prediction of hydrophobic (lipophilic) properties of small organic molecules using fragmental methods: An analysis of ALog *P* and CLog *P* methods. *J. Phys. Chem. A* **1998**, *102*, 3762–3772.

27. Leo, A.; Jow, P.Y.C.; Silipo, C.; Hansch, C. Calculation of hydrophobic constant (log *P*) from π and f constants. *J. Med. Chem.* **1975**, *18*, 865–868.

28. Fichert, T.; Yazdanian, M.; Proudfoot, J.R. A structure-permeability study of small drug-like molecules. *Bioorg. Med. Chem. Lett.* **2003**, *13*, 719–722.

29. Valentin, J.P.; Hammond, T. Safety and secondary pharmacology: Successes, threats, challenges and opportunities. *J. Pharmacol. Toxicol. Methods* **2008**, *58*, 77–87.

30. Campbell, N.A.; Williamson, B.; Heyden, R.J. *Biology: Exploring Life*; Pearson Prentice Hall: Boston, MA, USA, 2006.

31. Kamlet, M.J.; Abboud, J.L.M.; Abraham, M.H.; Taft, R.W. Linear

solvation energy relationships. 23. A comprehensive collection of the solvatochromic parameters, π*, α, and β, and some methods for simplifying the generalized solvatochromic equation. *J. Org. Chem.* **1983**, *48*, 2877–2887.

32. Ertl, P.; Rohde, B.; Selzer, P. Fast calculation of molecular polar surface area as a sum of fragment-based contributions and its application to the prediction of drug transport properties. *J. Med. Chem.* **2000**, *43*, 3714–3717.

33. Prasanna, S.; Doerksen, R.J. Topological polar surface area: A useful descriptor in 2D-QSAR. *Curr. Med. Chem.* **2009**, *16*, 21–41.

34. Hughes, J.D.; Blagg, J.; Price, D.A.; Bailey, S.; DeCrescenzo, G.A.; Devraj, R.V.; Ellsworth, E.; Fobian, Y.M.; Gibbs, M.E.; Gilles, R.W.; *et al*. Physicochemical drug properties associated with *in vivo* toxicological outcomes. *Bioorg. Med. Chem. Lett.* **2008**, *18*, 4872–4875.

35. Gleeson, M.P. Generation of a set of simple, interpretable ADMET rules of thumb. *J. Med. Chem.* **2008**, *51*, 817–834.

36. Bickerton, G.R.; Paolini, G.V.; Besnard, J.; Muresan, S.; Hopkins, A.L. Quantifying the chemical beauty of drugs. *Nat. Chem.* **2012**, *4*, 90–98.

37. Segall, M. Advances in multiparameter optimization methods for *de novo* drug design. *Expert Opin. Drug Discov.* **2014**, *9*, 803–817.

38. Lipinski, C.A.; Lombardo, F.; Dominy, B.W.; Feeney, P.J. Experimental and computational approaches to estimate solubility and permeability in drug discovery and development settings. *Adv. Drug Deliv. Rev.* **2012**, *64*, 4–17.

39. Veber, D.F.; Johnson, S.R.; Cheng, H.Y.; Smith, B.R.; Ward, K.W.; Kopple, K.D. Molecular properties that influence the oral bioavailability of drug candidates. *J. Med. Chem.* **2002**, *45*, 2615–2623.

40. Norinder, U.; Haeberlein, M. Computational approaches to the prediction of the blood-brain distribution. *Adv. Drug Deliv. Rev.* **2002**, *54*, 291–313.

41. Wenlock, M.C.; Austin, R.P.; Barton, P.; Davis, A.M.; Leeson, P.D. A comparison of physiochemical property profiles of development and marketed oral drugs. *J. Med. Chem.* **2003**, *46*, 1250–1256.

42. Besalú, E. Fast computation of cross-validated properties in full linear leave-many-out procedures. *J. Math. Chem.* **2001**, *29*, 191–203.

CITATION

CHAPTER 1

Liu, X.-T.; Wang, X.-G.; Xu, R.; Meng, F.-H.; Yu, N.-J.; Zhao, Y.-M. Qualitative and Quantitative Analysis of Lignan Constituents in Caulis Trachelospermi by HPLC-QTOF-MS and HPLC-UV. Molecules 2015, 20, 8107-8124.

CHAPTER 2

Sobrero P, Schlüter J-P, Lanner U, Schlosser A, Becker A, Valverde C (2012) Quantitative Proteomic Analysis of the Hfq-Regulon in Sinorhizobium meliloti2011. PLoS ONE 7(10): e48494. doi:10.1371/journal.pone.0048494.

CHAPTER 3

Bernitt E, Koh CG, Gov N, Döbereiner H-G (2015) Dynamics of Actin Waves on Patterned Substrates: A Quantitative Analysis of Circular Dorsal Ruffles. PLoS ONE 10(1): e0115857. doi:10.1371/journal.pone.0115857.

CHAPTER 4

Kuo W-T, Lin W-C, Chang K-C, Huang J-Y, Yen K-C, Young I-C, et al. (2015) Quantitative Analysis of Ligand-EGFR Interactions: A Platform for Screening Targeting Molecules. PLoS ONE 10(2): e0116610. doi:10.1371/journal.pone.0116610.

CHAPTER 5

Zhang B, Shimada Y, Kuroyanagi J, Umemoto N, Nishimura Y, Tanaka T (2014) Quantitative Phenotyping-Based In Vivo Chemical Screening in a Zebrafish Model of Leukemia Stem Cell Xenotransplantation. PLoS ONE 9(1): e85439. doi:10.1371/journal.pone.0085439

CHAPTER 6

Chan CXJ, Joseph IG, Huang A, Jackson DN, Lipke PN (2015) Quantitative Analyses of Force-Induced Amyloid Formation in Candida albicans Als5p: Activation by Standard Laboratory Procedures. PLoS ONE 10(6): e0129152. doi:10.1371/journal.pone.0129152.

CHAPTER 7

Yujie Chen, Liang Xu, Yuancen Zhao, Zhongzhen Zhao, Hubiao Chen, Tao Yi, Minjian Qin and Zhitao Liang, "Tissue-specific metabolite profiling and quantitative analysis of ginsenosides in Panax quinquefolium using laser microdissection and liquid chromatography–quadrupole/time of flight-mass spectrometry," Chemistry Central Journal20159:66, DOI: 10.1186/s13065-015-0141-0.

CHAPTER 8

Chen et al.: Quantitative analysis of the major constituents in Chinese medicinal preparation SuoQuan formulae by ultra fast high performance liquid chromatography/quadrupole tandem mass spectrometry. Chemistry Central Journal 2013 7:131. doi:10.1186/1752-153X-7-131.

CHAPTER 9

Guy Schwartz, Eyal Ben-Dor, and Gil Eshel, "Quantitative Analysis of Total Petroleum Hydrocarbons in Soils: Comparison between Reflectance Spectroscopy and Solvent Extraction by 3 Certified Laboratories," Applied and Environmental Soil Science, vol. 2012, Article ID 751956, 11 pages, 2012. doi:10.1155/2012/751956.

CHAPTER 10

Guo et al.: Quantitative LC-MS/MS analysis of seven ginsenosides and three aconitum alkaloids in Shen-Fu decoction. Chemistry Central Journal 2013 7:165. doi:10.1186/1752-153X-7-165.

CHAPTER 11

Castellano, G.; Torrens, F. Quantitative Structure-Antioxidant Activity Models of Isoflavonoids: A Theoretical Study. Int. J. Mol. Sci. 2015, 16, 12891-12906.

INDEX

A

absorption, distribution, metabolism, excretion (ADME) 261
aconitine 244, 245, 251, 252, 253, 256
Aggregation assays 159, 163
Alpiniae Oxyphyllae 205
antioxidant 259, 260, 261, 264, 265, 266, 267, 269, 270, 272, 273, 274
applicability domain (AD) 260
atomic force microscopy (AFM) 104

B

blood-brain barrier (BBB) 260, 269

C

Candida albicans 149, 150, 162, 170, 171, 172, 173, 175, 278
central nervous system (CNS) 267
chemometrics 225
Chinese medicinal materials (CMMs) 179
Chinese National Knowledge Infrastructure (CNKI) 200
chronic myelogenous leukemia (CML) 122
Circular Dorsal Ruffles (CDRs) 75, 76
collision energy (CE) 253

D

Dictyostelium discoideum 76

E

electrospray ionization (ESI) 15
epidermal growth factor (EGF) 103, 104
Epidermal growth factor receptor (EGFR) 103, 104
extraction 223, 224, 225, 226, 229, 234, 240

F

flow cytometry 149, 155, 160, 161
fluorescence-activated cell sorting (FACS) 123

G

ginsenosides 177, 178, 179, 180, 184, 190, 197, 198, 199, 200, 201, 202, 203, 278
glutamine synthetase (GS) 59

H

homodimers 104
hydrocarbons 223, 224, 225, 226, 229, 239, 240, 241, 242